Springer Series on Polymer and Composite Materials

T0075471

Series Editor

Susheel Kalia, Army Cadet College Wing, Indian Military Academy Army Cadet College Wing, Dehradun, India

The "Springer Series on Polymer and Composite Materials" publishes monographs and edited works in the areas of Polymer Science and Composite Materials. These compound classes form the basis for the development of many new materials for various applications. The series covers biomaterials, nanomaterials, polymeric nanofibers and electrospun materials, polymer hybrids, conducting polymers, composite materials from macro- to nano-scale, and many more; from fundamentals, over the synthesis and development of the new materials, to their applications. The authored or edited books in this series address researchers and professionals, academic and industrial chemists involved in the areas of Polymer Science and the development of new Materials. They cover aspects such as the chemistry, physics, characterization, and material science of Polymers, and Polymer and Composite Materials. The books in this series can serve a growing demand for concise and comprehensive treatments of specific topics in this rapidly growing field.

More information about this series at http://www.springer.com/series/13173

Ololade Olatunji

Aquatic Biopolymers

Understanding their Industrial Significance and Environmental Implications

 Springer

Ololade Olatunji
Chemical Engineering Department
University of Lagos
Lagos, Nigeria

ISSN 2364-1878   ISSN 2364-1886   (electronic)
Springer Series on Polymer and Composite Materials
ISBN 978-3-030-34711-6   ISBN 978-3-030-34709-3   (eBook)
https://doi.org/10.1007/978-3-030-34709-3

This Springer imprint is published by the registered company Springer Nature Switzerland AG
The registered company address is: Gewerbestrasse 11, 6330 Cham, Switzerland

Preface

The field of polymer science is a very broad one, due to the fact that polymers are ubiquitous. Polymers are used in every industry, from food industries where they are used as food additives or make up the bulk of the food itself, the biomedical industry where polymers are used as scaffolds for tissue regeneration to the energy industries where they are used as the starting feedstock for bioethanol production or as substrates in the new generation of solar cells. For this reason, there have been numerous texts which present polymers in various forms.

A review of the literature shows that there are currently no books which focus on polymers from the aquatic environment as a whole. While there are texts which look at different polymers which can be sourced from the aquatic environment, none of these texts have considered the polymers with a focus on the aquatic source. This is of much important in present time as the aquatic environment continues to be faced with different challenges from pollution to increased acidification. It is important that while considering the production, application and processing of these polymers that the environment from which they are obtained and the impact of the process of extraction are also taken into consideration. This ensures that the future polymer engineers, producers, suppliers and consumers have a broader understanding of polymers which includes the global economy and environment.

Lagos, Nigeria   Ololade Olatunji

Contents

Chapter 1
Introduction to Aquatic Biopolymers

Polymers refer to macromolecules made up of repeating units of smaller molecules covalently bound together to form a bigger molecule. For example, proteins are polymers made up of amino acids, cellulose is made up of repeating units of glucose and polyethylene is made up of repeating units of ethylene. The units which join together to form polymers are referred to as monomers, and the process through which this is achieved is termed polymerization. Biopolymers refer to polymers which are produced by living organisms. The term aquatic biopolymers refers to biopolymers which are produced by living organisms that inhabit the aquatic ecosystem.

Since the beginning of the ages, polymers have played a significant role in human life in providing the most basic need, food, in the form of carbohydrates and proteins. As life advances, polymers have come to be of great importance to modern civilization, from the food we eat, the portable plastic bottles that give us convenient access to clean drinking water, the medicines we take, basic hygiene, tools and the fuels that run our engines. Polymers are fundamental to our existence as modern humans. Considering that the polynucleotides which form our DNA that carry the instructions for every life function are also polymers, it can therefore be said that polymers indeed make life. The aquatic environment is a rich source of a wide range of natural resources, and among those resources are polymers. They are found in the cell walls, tissues, exoskeletons, secretions and anatomical regions of aquatic organisms across the five kingdoms.

A large variation of factors exists in the aquatic environment. These variable factors include salinity, depth, light intensity, temperature, pressure, density, pH, prey/predator presence, nutrients, dissolved gases, water flow rate, flora and fauna. These factors determine the types of biochemicals produced by the different organisms to survive their environment, even more so where they are in such direct contact with the continuous medium (Christophe et al. 2015). Such variation in the environmental factors also translates to a diverse range of polymers. Over the course of evolution of life, living organisms both plants and animals have devised means to produce, absorb, store and process different forms of biopolymers. This ranges from polysaccharides used for energy and structure, proteins for metabolism and cell renewal and polyesters for water repelling among others. These are stored in

© Springer International Publishing 2020
O. Olatunji, *Aquatic Biopolymers*, Springer Series on Polymer and Composite Materials,
https://doi.org/10.1007/978-3-030-34709-3_1

different anatomical regions of the plant. Carbohydrates, for example, are stored in chloroplasts, cytoplasm, periplastic compartment, peri-plastid membranes, in the cytoplasm, and vacuole localized in the cytoplasm (Prabhu et al. 2019).

The diverse chemistry and bioactivity that can be derived from naturally sourced polymers add to their significance. This is of particular importance as increasing occurrence of cancer and other degenerative disease and resistance to existing antimicrobial drugs calls for more diverse and complex compounds existing in nature. The rain forests are often described as the medicine chest of the world. This is attributed to the presence of a diverse range of bioactive compounds which can be obtained from the diversity of organisms in the rain forest. The aquatic ecosystem consists of even more diverse range of organisms, and its existence and subsistence are crucial to that of the rain forest and life on earth. This book provides the reader with a rich understanding of the wide range of polymers which can be sourced from the aquatic environment, some of which have bioactivities such as anticancer and antimicrobial activities.

Due to limited access compared to land where humans are better adapted to inhabit, the aquatic world is relatively underexplored. As technology advances, various tools have been developed to better explore the world that exists below sea level. Today, submersibles have been developed which can reach the deepest part of the ocean which have previously been inaccessible. High-resolution specialized cameras which can capture the faintest light underwater have also been developed alongside more sophisticated analytical tools and techniques (Linley et al. 2016). This improved access to the world below sea level has significantly improved knowledge and access to the aquatic world in recent years. Within this book, we also explore some of the polymers which have been discovered in recent years from deep-sea organisms.

When evaluating the commercial implications of production of biopolymers from aquatic sources, it should be noted that apart from the cost of raw materials, the downstream processing constitutes around 60% of the production cost. Downstream processing is where the bulk of the value addition to the final product occurs. This includes solid–liquid separation, purification, solvent recovery and product recovery. Therefore within this book, some of these stages are included when the extraction process for each polymer is described. This is with the aim to understand the processes involved in the production of the biopolymers and how these impact on the environment.

As the aquatic ecosystems of the world are increasingly facing threat from pollution as a result of human activity and poor waste management (Hitchcock and Mitrovic 2019), it is important, therefore, to understand what is at stake as far as the aquatic environment is concerned. This book aims to provide an understanding of the biopolymers existing within the aquatic ecosystem, how they are obtained, their role in ecosystem as well as their existing and potential economic value.

The benefits of aquatic biopolymer production can be summarized as follows:

- Utilize aquatic waste from fisheries and aquaculture especially as aquatic activities have shown an increase in recent times due to increased demand for aquatic food.

- Source biopolymers with more diverse chemistry to address complex health issues, meet the demand for novel materials and address bacteria resistance.
- Replace non-biodegradable plastics which presently are having deteriorating impact on aquatic life and water resource by augmenting the feedstock for bioplastic production with aquatic biopolymers which have faster biomass accumulation rate.
- Alternatives to depleting fossil resources which do not require use of land or freshwater to cultivate.
- Value addition to aquatic resources to serve as a means of improved income for developing countries and emerging economies where a lot of the fishing activities occur.
- Development of biorefinery through increased feedstock from aquatic waste and excess abundant resources.

This book takes a new approach to developing understanding of polymers by focusing on polymers derived from specifically the aquatic ecosystem. In so doing, it presents a much broader view of polymers, the opportunities, impact and key issues in the polymer and aquatic industry. Each chapter introduces the reader to the different chemical structures of polymers and how these could vary from simple linear to quite complex non-uniform structures. In exploring the production processes, the reader learns about different techniques which are used in the polymer industry and where possible some case studies of specific companies are included.

In the process of discussing the sources of polymers, a variety of species of organisms are explored, showing how polymers exist across the five kingdoms and all over the food web. The ubiquity of polymers is further established through reviewing the numerous applications of each aquatic biopolymer covered. The current issues facing the aquatic environment and the environmental impact of the extraction processes are also discussed. This gives the reader a rich understanding of the biopolymer resource of the aquatic ecosystem, their economic and environmental significance.

References

Christophe M, Rachid A, Mario L (2015) Fish as a reference species in different water masses. Aquat Ecotoxicol, 309–331

Hitchcock JN, Mitrovic M (2019) Microplastic pollution in estuaries across a gradient of human impact. Environ Pollut 247:457–466

Linley TD, Gerringer ME, Yancey PH, Drazen JC, Weinstock CL, Jamieson AJ (2016) Deep sea research part I: oceanographic research papers, 114:99–110

Prabhu M, Chemodanov A, Gottlieb R, Kazir M, Goldberg A (2019) Starch from the sea: the green macroalga *Ulva ohnoi* as a potential source for sustainable starch production in the marine biorefinery. Algal Res 37:215–227

Chapter 2
Overview of the Aquatic Ecosystem

Abstract The chapter gives an overview of the aquatic ecosystem as a whole. The aquatic ecosystem is classified based on different criteria: depth, water flow, salinity and features. The organisms within the aquatic environment are also described; these include algae, aquatic plants and animals. In the process, some current issues facing the aquatic environment are also discussed. Some of the terms described here are used in other chapters of the book; therefore, this chapter also serves as a reference point for the rest of the book. The chapter also includes illustrations and images of some aquatic ecosystems from different parts of the world.

Keywords Biopolymers · Marine · Freshwater · Littoral · Pelagic · Estuaries

2.1 Introduction

Evidence suggests that water was brought to earth by water carrying meteorites and asteroids colliding with the planet during its formation. This water formed the oceans of the earth which then spill into other bodies of water. Therefore, every drop of water on earth is billions of years old (Takir et al. 2018). This water is recycled through the water cycle and changes state between gas, liquid and solid through the processes of evaporation, condensation and freezing. The oceans absorb a significant amount of energy from earth's revolution which would otherwise cause extremely high wind speeds which could have destructive effects on life on land. The aquatic ecosystem is made up of the living and non-living components that exist within a body of water. It consists of such diverse life-forms that there exists more diversity of life in the water than on land. The survival of an aquatic ecosystem depends on maintaining a balance between the existence and activities of the producers, consumers, decomposers as well as the abiotic components. This is same in terrestrial ecosystem except that the aquatic ecosystem has an abundance of moisture, continuity of the medium (water), limitation of air and variation in temperature and light.

To say that water plays an important role in sustaining life is almost an understatement. Water is indeed life. The aquatic world is full of diverse life-forms, and it makes up 70% of the earth. Water has served as the main evidence of life on other planets; to ascertain the existence of life, one must first confirm the presence

© Springer International Publishing 2020   5
O. Olatunji, *Aquatic Biopolymers*, Springer Series on Polymer and Composite Materials,
https://doi.org/10.1007/978-3-030-34709-3_2

of water. Humans, plants, animals, prokaryotes and protists all depend on water to exist. More than just hydration, many biochemical reactions and processes depend on water. Dissolving of nutrients for uptake, the maintenance of erect upright form in plants and much more require water, such is the extent of the relevance of water for life on earth.

Aquatic ecosystems refer to the complex web of relationships between living and non-living organisms which exist around a body of water. In order to understand the industrial and environmental significance of aquatic biopolymers, it is also important to understand the ecosystem within which these biopolymers exist. The aquatic environment holds a world of resources which play important roles in both the present and future industries. In addition, it is a vital food source and means of generating income for many communities. Biopolymers form a significant part of the aquatic environment, and various aquatic sourced biopolymers are explored in the different chapters of this book.

The United Nations Convention on climate change sets the global goal of ensuring that the global temperature rise does not go 2 °C above the preindustrial temperature. The global temperature rise is due to many factors, part of which is hypothesized to be the continuous dependence on fossil fuel which results in the release of gaseous compounds which result in global warming. To this end, this book seeks to explore the range of available resources in the aquatic environment, their production process as well as their economic value both existing and potential. It explores how the aquatic environment could indeed offer alternatives to the biopolymers which we require in various aspects of life.

Optimum utilization of the vast resources of the aquatic world requires an understanding of the different resources and the role they play in the environment and the economy. The sustainable millennium goal SDG 14 states: "Conserve and sustainably use the oceans, seas and marine resources for sustainable development" (FAO 2018). Even where one is interested in only a single resource, it is important to have a general overview of the range of biopolymers and how they are connected. This helps to explore alternatives, co-production and the connection of that particular biopolymer to the rest of the aquatic ecosystem.

While still at its infancy, the marine biorefinery could be built around developing a synergy with the already-existing offshore facilities. This will enable a more seamless transition from mostly fossil-based polymer products to more diverse polymer industry that includes the use of aquatic biopolymers sourced from aquatic waste and by-products in a sustainable manner. Furthermore, deep-sea mining, oil and gas drilling and intensive unsustainable fishing methods are reported as the major threats to the aquatic ecosystem. For example, a recent study revealed that the increase in the copper concentration in the water during deep-sea mining results in reduced metabolic activities and immune response in *B. azoricus*, a mussel which inhabits the deep sea (Martins et al. 2017). By utilizing the biopolymers from aquatic wastes and by-products, the demand for these unsustainable sources of polymers is reduced, and through value addition to the aquatic resource, there is less pressure on fishermen to adopt unsustainable fishing methods.

This predicted rise in population is necessitating the expansion of the food industry in general. Aquatic food serves as a key source of protein and essential nutrients in which many developing and advanced economies rely on. This will mean a rise in the amount of aquatic waste and by-products being generated—thus indicating long-term availability of feedstock such as shrimp and lobster shells, fish scales and internal organs to power the marine biorefinery. Going further, we shall explore the availability trends of the respective aquatic feedstocks for biopolymer production.

This chapter gives the reader an overview of the aquatic environment and in doing so provides an insight into the diversity of the aquatic ecosystem which serves as a source of a wide variety of biopolymers which will be discussed in the subsequent chapters.

2.2 Types of Aquatic Ecosystems

Through exploring the various aquatic biosystems, we can then understand the various roles played by these biopolymers within the living organisms and within the aquatic environment as well as how to extract and utilize them. Aquatic ecosystem can refer to either inland or marine waters. Inland waters refer to aquatic ecosystem within the land borders, while the marine waters refer to generally the seas and oceans. These coastal lands/waters refer to the boundary between the aquatic ecosystem and terrestrial ecosystem. These are generally referred to with terms such as beach, shore, coastlines and waterside.

Based on source and salinity, aquatic ecosystem can be classified as marine and freshwater. Brackish water is formed where the saltwater from the marine mixes with freshwater. There are five oceans on earth: the Atlantic, Pacific, Indian, Arctic and Antarctic oceans. These feed into the numerous seas and inland waters (Table 2.1).

Aquatic habitat can further be characterized by other factors, an example is characterization by flow rate of water. Based on flow rate aquatic habitat can be either flowing or standing waters. Organisms living in either of these have features which allow them to cope at the respective flow conditions (Table 2.2).

Table 2.1 Classification of aquatic ecosystem based on water salinity

Aquatic ecosystem	
Marine	Freshwater
Examples: ocean, sea, estuaries, lagoons, salt marshes	Examples: ponds, lakes, rivers, springs streams, wetlands

Table 2.2 Classification of aquatic ecosystem based on water flow

Aquatic ecosystem	
Lotic-flowing waters	Lentic-still waters
Rivers, streams, springs	Ocean, lakes, ponds, swamps

Table 2.3 Different Lake zones

Lake zones	Subdivision	Depth range	Organisms and features
Littoral zone	Epilimnetic	Surface to 6 m	Aquatic plants
	Hypolimnetic	6–10 m	
Limnetic zone		From 10 to depth of effective light penetration	Planktons
Profundal zone		From point beyond light penetration to bottom	Heterotrophs

The aquatic environment can also be classified based on depth. Table 2.3 shows the classification of lake zones (Peters and Lodge 2009), and Table 2.4 shows the classification of the ocean depths (Honjo 2009; Linley et al. 2016; Staby and Salvanes 2019). The lake zones are similar for rivers; however, in ponds, the profundal zones do not exist since they are relatively shallow and light can penetrate to the bottom. Different aquatic organisms have adapted to live in waters at different depths, turbulence, salinity, temperature, light and pressure. Other adaptations include defense mechanisms and competition for food. For example, organisms living in the deepest region of the water where no light penetrates have adapted features which enable them to create light and/or sense prey in the dark.

Other factors by which aquatic ecosystem can be characterized include pH, turbidity, dissolved gases, temperature, light penetration and areal distribution.

Types of aquatic ecosystem include:

- Aquifers
- Springs
- Rivers
- Lagoons
- Streams
- Lakes
- Seas
- Ponds
- Wetlands
- Beaches
- Bays
- Estuaries
- Oceans
- Coral Reefs

All of these aquatic systems play a significant role in the survival of different life-forms and the accumulation of a wide range of biopolymers and other resources on earth. The following sections describe each type of aquatic ecosystem and the different life-forms which exist within the respective ecosystems.

Table 2.4 Different zones of the ocean

Ocean zones	Depth range	Organisms and features
Epipelagic zone (Sunlight zone)	Surface to 200 m	Abundant sunlight, minimal pressure, higher temperatures. Coral reefs and photosynthetic organisms, more ocean life and human activities
Mesopelagic zone (Twilight zone)	200–1000 m	Lower light intensity (mostly blue light). Zooplankton and rarer sea life such as the Barreleyes, ribbonfish and snipe eels. Vertical migration of organisms and gradient of oxygen level, light intensity, temperature and salinity
Bathypelagic zone (Midnight zone)	1000–2000 m	Dark, pressure up to 5.858 lbs/square inch, bioluminescent organisms. Nematodes, bivalve molluscs, polychaetes, crustaceans, sipunculid worms and gastropods
Abyssopelagic zone (Abyssal zone)	2000–6000 m	Near freezing point, high pressure, rare species and invertebrates such as the giant squid. Includes some parts of the ocean floor
Hadalpelagic zone (The trenches)	6000–11,000 m (valleys below the ocean floors)	Deepest known is the Mariana Trench in Japan. temperature and pressure, few rare species recently discovered include the Mariana Snailfish and Ethereal snailfish

2.2.1 Estuaries

Estuaries are very important in the maintenance of the aquatic biodiversity. They are water bodies which are enclosed by structures such as vegetation and the organisms; they have easy access to the sea (Dame 2008). The water in the estuaries is brackish as a result of ocean water mixing with freshwater. There is also variation in turbulence depending on the contact with the sea. These estuaries serve as nurseries for young species. The underwater plant roots protect them from bigger prey simply by serving as a hiding place that only allows entry of organisms below a certain size. This allows them a better chance of survival and growing to a safer size after which they can then easily access the sea and begin their journey to their final home in the deeper waters.

In recent times, some estuaries are under threat from environmental pollution, disease outbreak and overfishing (Dame 2008). Some of these are due to the release of toxins from factories and homes. This in turn distorts the balance in the ecosystem. Losing estuaries means losing several species of aquatic plants and animals.

One of the recent threats studied in the estuaries of the Clyde, Bega and Hunter estuaries in east coast of Australia revealed that increased human activities resulted in higher microplastics in the estuaries (Hitchcock and Mitrovic 2019). In the three estuaries studied, the microplastic pollution was as high as 1032 parts per m^3 in the coast with the most populated Hunter estuary, to 98 part per m^3 in the Clyde estuary which the coastal land is least populated by humans while Bega estuary had a microplastic pollution of 246 parts per m^3. Most of the microparticles had a particle diameter below 200 μm.

2.2.2 Springs and Aquifers

In some parts of the earth, water accumulates under layers of rocks and sediments. These are often artificially obtained by digging wells deep into the earth's core and either drawing out or pumping. Such water is relatively high quality as it is naturally filtered through layers of rock and sand on the way out of the aquifer. They are used for human consumption, agriculture and other activities important for life. In order to conserve water and ensure the availability of aquifer-sourced water for future use, it is important that the rate at which water is returned to the aquifer is balanced with the rate at which it is being removed. Aquifer forms a relatively low energy requiring source of clean water. Recharge zones are points in which rainwater seeps back into the aquifer beneath and this makes up for the water removed either through natural springs or by human activity. Recharge zones can be natural or man-made, and these recharge zones experience dry and wet periods. While some aquifers show rapid response to climate change, others show slow response. The response of an aquifer to change in climate depends on the type of aquifer. An example is the Northern Sudan Platform subbasin where average annual precipitation of 85 mm was recorded between 2002 and 2012, and between 2013 and 2016, average annual precipitation recorded was 107 mm (Abdelmohsen et al. 2019).

The underground water trapped within aquifers does not always need to be pumped or drained by human activities. Due to the topology of the ground, the water escapes from aquifers into the surface due to pressure difference and gravitational force. The path through which the water flows is known as springs. This serves as a source of natural spring water. These could either be slow moving or generate enough pressure to form bubbling springs. The water from springs is usually of high quality due to the fact that the rainwater flowing into the aquifers is filtered through different layers of soil and rocks. Water flowing from springs forms a large part of waters in streams and rivers in addition to annual rainfall.

2.2.3 Rivers and Streams

Rivers serve as a means of transportation, food, recreation and biodiversity. Large bodies of freshwater which flow are referred to as rivers. Generally, a flowing body of water that spans a relatively large area can be classified as a river. The Nile River is the longest river in Africa and possibly in the world; it is measured 6825 km in length and runs through 11 countries including Rwanda, Ethiopia, Kenya, Uganda, Tanzania, Sudan and others (Abd-Elbaky and Jin 2019). Other examples are the Volga River, which is the largest river in Europe; the Danube River, which is the second largest river in Europe and runs through the heart of Europe; the Amazon River, which is crucial water flow for the Amazon rain forest and contends with the Nile River in length as the Amazon River also measures over 6000 km in length. Figure 2.1 shows Mbagathi River in Kajiado County in Kenya. The image also shows the terrestrial plants and trees on the land through which the river flows.

The constant flow of a river is due to the force of gravity as a river flows over a sloped floor. It is important for a river to flow rather than be stagnant as flowing water allows air to mix into the water to support life-forms which require a level of dissolved

Fig. 2.1 Mbagathi River, Kajiado county, Kenya. August 2019. Photograph courtesy of Ross van Horn

oxygen to survive. The constant flow also prevents the settling of pathogenic larvae of organisms such as mosquitoes on the surface of the water. Several aquatic and terrestrial life-forms depend on the water cycle in the river. For example, some life-forms such as the tadpole stage of frogs and the dragonflies begin life in the river until they reach adulthood and transform to live on land or fly. Even in adulthood, some of these organisms still depend on the water as a source of food and refuge. Rivers are linked to streams, pools, water channels and floodplains. The life-forms and structure of the abiotic and biotic components around a river vary at different parts of the river. Figure 2.2 is an example of a stream flowing through the town of Stellenbosch, South Africa. The water from this stream also serves as irrigation water for the farms in the area.

2.2.4 Lakes

A large standing body of water surrounded by land is referred to as a lake. Over a thousand lakes exist in the world. A lake is formed by the deposition of water from rainfall, ice formation, river water and other factors which lead to water flow and formation ending up in the lake. Since the lake is formed from these variable factors, the level of water and conditions in a lake therefore varies at different periods as water evaporates and is deposited. Lakes form a habitat for diverse life-forms such as water plants, snails and fish. Waterbirds such as swans, ducks and geese are also a regular feature in lakes, and these in turn serve as food for land animals as well as aesthetics and recreation for humans. Example of lakes are the Lake Chad which is surrounded by the seven Chad basin countries: Nigeria, Cameroon, Central African Republic, Niger, Chad, Algeria and Sudan (FAO 1997), Lake Victoria in Tanzania and numerous other lakes across the world.

The life in the lake is held in a sensitive balance between the photosynthesizing life-forms and the decomposition of the organic matter. A lake is comprised of the oxygen-dissolved zone, the light-penetrating zone and the dark oxygen-depleted layer. These lake zones are classified as littoral, limnetic and profundal zones, respectively (Table 2.3).

Some of the lakes in the world are drying up due to climate change, pollution and some due to the natural cycle of a lake. Lake Victoria, the second largest freshwater lake in the world, is possibly an example of such, which is located in Tanzania and covering an area of over 69,000 km^2 as of 2018. It was reported to have shrunk by 203 km^2 from 1984 to 2018 with 0.3% shrinkage in area (Awange et al. 2019). Similarly, the Lake Chad has been reported to have shrunk from a surface area of 25,000 to 2000 km^2 (Mahmood and Jia 2019). This shrinkage of the lakes is attributed to a combination of climate change and human activities.

Fig. 2.2 Three different parts of a stream from shallower low turbulence to slightly deeper slow-moving to the steeper higher turbulence. Stellenbosch Town, South Africa, June 2019 (Winter). Photograph by Ololade Olatunji

2.2.5 Ponds

Ponds refer to water-filled depressions on the earth's surface. Ponds, like lakes, are formed from water deposited from rain, snow, river runoffs or other water formations. However, ponds are much smaller than lakes. While lakes could extend for up to hundreds of acres, ponds only cover a few acres of area. Due to the relative smaller

Fig. 2.3 A partly dried pond on the hill in Stellenbosch, South Africa June, 2019. Photograph by Ololade Olatunji

size, ponds can become dry at different times, depending on the environmental conditions. Ponds are usually shallow and therefore have stronger light penetrations and dissolved oxygen along its depths such that ponds support more aquatic plant and algae growth. Ponds are susceptible to drying up, especially during periods of lower rainfalls. Figure 2.3 shows an example of a pond with a decreased water level in June in the Town of Stellenbosch, South Africa.

2.2.6 Lagoons

Lagoons are bodies of water separated from the ocean by barriers such as reefs. They can be natural or man-made and generally saline, low tide and relatively shallow compared to the sea with minimal mixing with freshwater. Unlike estuaries, the ocean water in lagoons is trapped behind coastal dune systems, sandpits or island barriers and therefore does not provide the aquatic organisms therein the same ease of access as the estuaries (Harris 2008). An example is the Lagos Lagoon in Nigeria (Fig. 2.4), the Vermelha Lagoon in Brazil, the Nichupte Lagoon of the Mexican Caribbean, Venice Lagoon in Italy, Manzala Lagoon in Egypt, Santos Andre Lagoon in Portugal, Tunis Lagoon in Tunisia, Vistula Lagoon of the Baltic Sea and the La Pletera salt marsh Lagoon in northeast Catalonia. The lagoon serves as host to some aquatic life such as fish, amphibians, crustaceans, aquatic plants and waterbirds. Species such as the *Aphanius iberus* fish are endemic to lagoons like the La Pletera salt marsh Lagoon (Casamitjana et al. 2019). Lagoon serves as a source of fish, recreation and transportation and forms a major attraction for tourism.

Fig. 2.4 A view of Lagos Lagoon from the University of Lagos. The third mainland bridge is built over the Lagoon; some aquatic plants are also visible in the image. Photograph by Ololade Olatunji

2.2.7   Wetlands

The way in which the land and water meet varies. In some parts, the sea meets the land at rocky shores or sandy beaches, while in some parts the meeting point is demarcated by wetlands connecting aquatic and terrestrial environments. Flood plains and rain forests are watered by the cycle of the river flowing through them. One of much global importance is the Amazon rain forest in Brazil, through which the Amazon River flows and is one of the most diverse wildlife. Wetlands are important to the environment as they support a diverse range of life-forms such as frogs, birds, turtles and aquatic plants. They are wet soil regions close to a water body. Wetlands play a significant role in abating floods by absorbing rainwater. Examples of wetlands are marshes, swamps, floodplains and bogs.

The amount of water retained in wetlands varies with rainfall, riverflow, season and climate change. The water from the river which flows into wetlands is important for agriculture; for example, growing of floating rice species in rice-growing regions of countries such as Vietnam and Japan is carried out in sync with the wet and dry cycle of flow of water from the river into the wetlands on which the rice is grown (Iizumi and Ramankutty 2015). The amount of water retained by the land from the aquatic environment is measured as the terrestrial water storage (TWS). For example, the average TWS measured for the Nile River basins between September and November was 42.66 and -23.34 mm between March and May. These values are based on average between 2003 and 2016 (Abd-Elbaky and Jin 2019). The TWS at different parts of the year therefore varies with the season depending on rainfall.

Currently, the Amazon rain forest is threatened by deforestation. The felling of trees at a rate faster than the forest can grow new trees denies several species their habitat, cover and a source of food, consequently leading to a decline in population. This results in the loss of plant species which depend on the larger trees for cover and temperature control and species such as woodpecker which depends on the trees for shelter and protection from prey. Likewise as the population of the organisms which are a source of food for the larger prey decline, the larger organisms which prey on them also begin to die out.

2.2.8   The Coral Reefs

The coral reef is commonly referred to as an underwater city as it is made up of living structure which houses a very diverse range of aquatic species with complex relationships. The reef itself is made up of organisms referred to as the corals which build a protective housing made up of limestone within which they permanently live. The corals have a symbiotic relationship with algae where they make use of the energy produced by algae from photosynthesis to build their limestone structure. Coral reefs exist within the epipelagic zone (sunlight zone), usually within the 35 m depth of the sea fringe. Although it is commonly referred to as the coral reef, the term biological reef is also used as a way to better highlight its significance and also provides a term that also credits the significant role of another organism which plays a very important role in creating these reefs, the algae. Algae provide the energy which is required by the corals to build these reefs. The energy produced from photosynthesizing organisms such as the sea grass and algae serves as a power source for the coral reef, such that these underwater habitats are essentially solar powered. Collectively, corals can build the underwater structure known as coral reefs extending up to thousands of miles. The largest coral reef on earth is that which exists in the northern coast of Australia, referred to as the Great Barrier Reef. Organisms which can be found within and around the coral reef include a range of finfish species, sea horses, starfish, anemones, sea cucumbers, clown fishes, mantis shrimp, sea turtles and much more. The algae, the shells of crustaceans, the skin and even scales of the fish all comprise a range of biopolymers which will be explored in different sections of this book. The coral reef serves an important role in maintaining the aquatic biodiversity which in turn is significant toward the availability of diverse range of aquatic biopolymers.

Recent studies reveal that the Great Barrier Reef is a source of net release of CO_2 to the environment (Lonborg et al. 2019). The coral reefs therefore potentially contribute to global warming based on the theory that CO_2 results in the overall increase in the earth's temperature. On the other hand, the Great Barrier Reef is valued at AU$ 56 billion (Pendleton et al. 2019). This is value derived from recreational activities, tourism and fisheries. The coral reefs are also quite fragile and are at risk from factors such as ocean acidification (Pendleton et al. 2019) and nutrient runoff from sugarcane plantations (Deane et al. 2018). The coral reefs therefore serve as an example of how the aquatic life and activities impact on terrestrial life and activities and vice versa.

2.3 Aquatic Organisms

Life inhabiting the aquatic environment includes the five kingdoms: plants, animals, prokaryotes, protista and fungi. These could be large organisms such as the whales and giant squids or they could be microorganisms such as bacteria and microalgae.

2.3.1 Aquatic Plants

One of the differentiating features between aquatic and terrestrial plants is that aquatic plants growing in flowing water possess larger and narrower leaves while land plants or plants grown in poorly oxygenated water had smaller and broader leaves. However, the size or distribution (number of stomata per unit cm^2 of leaf) of the stomata does not vary (Penfound and Earle 1948).

Aquatic plants are also referred to as hydrophytes. Some aquatic plants are invasive while others are extractive primary producers which photosynthesize using nutrients such as phosphorus, manganese, zinc, iron and nitrogen and carbon dioxide from the aquatic environment and releasing oxygen in exchange (Oyedeji and Abowei 2012). In some aquatic environment where oxygen supply from the atmosphere is low, aquatic plants serve as the primary oxygen source for the other aquatic organisms within. Aquatic plants also add some aesthetics to the environment if well controlled. The water lily, for example, blooms with colorful flowers and some accounts of it being grown for its ease of growth and blooming beautifully in the United States after the First World War.

Some aquatic plants can be beneficial to the aquaculture or wild aquatic environment serving as nutritional base to shell and finfish, zooplankton and other invertebrates as well as controlling the growth of invasive phytoplankton and moss by removing the metabolites required by these organisms. Others may pose a nuisance by disrupting water transportation and blocking navigational channels or serve as hosts to pathogens and pests. Reported cases of water hyacinth disrupting water transportation exist in the Niger Delta region of Nigeria (Oyedeji and Abowei 2012). Water hyacinth which was once said to be an admirable plant grown for aesthetic appeal has since then gone from an admirable water plant to one of the biggest pests in the world. Water hyacinth pest problems have resulted in millions of dollars worth of loss in terms of destruction of aquatic life and human activity related to aquatic environment. This includes blocking of drainage, destruction of aquatic food source, impediment of runoff leading to flooding, hindering of water transportation and loss of income from water-based recreational activities. Water hyacinth is a self-pollinating, mat-forming vegetative water plant. Its structure comprises a flowering leaves, inflorescences supported by a stolon and rhizomes with fibrous unbranched roots reaching down to the water. They can grow up to 3 feet high.

The fact that aquatic plants are adapted to cope in the aquatic environment indicates some significant differences between their structure and that of terrestrial plants. This abundance of water means they do not require cutin or suberin in their leaves and roots for the prevention of water loss. They instead have other features which address the need for buoyancy to float on water as well as survive the tides. Other features of aquatic plants include lacunae, which are holes in the root cortex which aid in moderating gas exchange (North and Peterson 2005). Aquatic plants can be of three types based on their biological structure; these are:

- Phytoplanktons
- Periphytons and
- Multicellular Macrophytes

Phytoplanktons are photosynthetic microscopic plants (and animals) which drift in water. They require light for synthesis; therefore, they are located in the littoral zones of water. They serve as primary producers as they can convert nutrients from the water such as zinc, nitrogen and magnesium into carbon and oxygen. Most of the oxygen productions from photosynthesis in the oceans, lakes and rivers are carried out by phytoplankton. Planktons which do not photosynthesize are referred to as zooplanktons. Phytoplankton also comprises algae and bacteria as well as plant phytoplanktons. There are about 43 species of phytoplankton. They serve as a nutrient source to other aquatic organisms such as finfish and shellfish and therefore are required for a healthy aquatic ecosystem. Their environmental and economic significance lies in the sustenance of the other aquatic organisms of economic importance.

Periphytons are aquatic organisms which require a substrate to grow on. They could be heterotrophic or autotrophic (Petters and Lodge 2009). Rocks and plants could serve as such substrate as well as other surfaces. These surfaces provide the structural support for them to survive aquatic environment. They obtain nutrients from the water, and they serve as a food source for herbivores.

Multicellular macrophytes are more independently growing and anchored multicellular aquatic plants. They can be **emergent, submerged or free-floating**. Many submerged aquatic plants do not possess cuticle as they are surrounded by abundant water. Likewise, they partially or completely lack xylem and stomata remains open. Roots may be present in submerged plants to anchor them to the surface or it could be absent. While floating aquatic plants are adapted with air spaces trapped within their structure to enable them to remain afloat, submerged aquatic plants are adapted to survive in low light due to their distance from the surface of the water. The leaves of the floating aquatic plants are flat to aid flotation while their roots are specially adapted to aid oxygen uptake. Emergent aquatic plants grow in the shallowest end of the water. Their roots are buried in the soil at the bottom of the water while the rest of the plant grows above water. Figure 2.5 illustrates these three different types of aquatic plants.

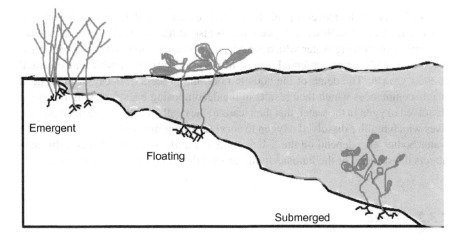

Fig. 2.5 An illustration of floating, submerged and emergent aquatic plant

In some cases, some common features may exist between aquatic plants and terrestrial plants which serve different purposes in aquatic plants. An example of such is the presence of cuticles in some aquatic plants such as duckweed for the purpose of protection against ultraviolet radiation and pathogens rather than for water retention as it is used in terrestrial plants (Borisjuk et al. 2018). There are also plants that are classified as amphibious and plants that are both aquatic and terrestrial such as the marshland and shore weeds (*Littorella uniflora*) (Braendle and Crowford 1999).

Effects aquatic plants can have on aquatic system include reduction in temperature, and decayed plants can cause high biochemical oxygen demand and reduce dissolved oxygen. They can also result in buildup of toxins and bad odor and generally destroy the aesthetics of the environment. Therefore, in order for aquatic plants to serve their beneficial role, there is a need for the aquatic system to be well maintained with controlled plant growth. Some common examples of aquatic plants are wild rice, wasabi, watercress, water caltrop, chinese water chestnut and water mimosa among others. Duckweed and water hyacinth are often used as animal feed. Non-edible aquatic plants include water lilies and water lettuce. The edible water plants include rice and water chestnuts.

While aquatic plants play an important role in the aquatic ecosystem, uncontrolled population growth of any species could cause an imbalance in the ecosystem and eventual destruction of other life-forms in the ecosystem. Water hyacinths are one of such. These invasive plants mostly exist in slow-moving or still waters such as lakes and ponds. It survives in temperature between 12 and 35 °C with optimal at around 25–30 °C. It can also survive in slightly acidic to neutral water of pH between 5 and 7. Water hyacinths do not survive in saltwater; hence, they do not exist on the sea. They can barely survive in brackish water and can be found in some lagoons.

At a low controlled level, aquatic plants add aesthetics to the aquatic environment as some have colorful flowers. Aquatic plants also absorb the water waves and abate the impact of flowing water which may lead to erosion. However, at uncontrolled levels they tend to starve essential aquatic life of light and oxygen which are essential to their growth. The death of the aquatic flora, the process of decay of these dead plants commences which then results in the decomposing bacteria taking up all the dissolved oxygen in the water, this then leads to the death of the fish and other aquatic lives which require dissolved oxygen to survive. The humans who live around these water bodies also depend on the fish as food and trade, once this depletes this also affects the survival of the humans living around the hyacinth-infested waters.

2.3.2 Algae

Algae are polyphyletic; they are a group of photosynthesizing organisms which belong to different kingdoms; some are plants while some are not; some are microscopic in nature while others are meters in length; and some are unicellular and others multicellular. What they have in common is that they possess chlorophyll, they photosynthesize, and however, they grow on water and do not have roots, stems or leaves as terrestrial plants do (Cavalier-Smith 2007).

Although there are thousands of species of algae, they can be classified into three main types: green, red and brown (Chlorophyta, Rhodophyta and Phaeophyta, respectively) (Kadam et al. 2015; Bleakley and Hayes 2017). This classification is mainly based on their domination pigments which give them the visible color. The green algae include species such as *Ulva lactuca*, the red algae include species such as *Kappaphycus alvarezii,* and the brown algae include species such as Laminaria Japonica. Most brown and red algae are found in marine waters while green algae can be found in freshwater. Some freshwater red algae exist albeit rare and brown algae are almost exclusively found in marine water. Fucoxanthin is the pigment responsible for the brown coloration of the phaeophyta, that of red algae are caused by phycobilins while that of green algae is due to the presence of only chlorophyll. The brown algae are all multicellular and could grow to tens of meters in length. The red algae are also mostly multicellular while the green algae comprise unicellular and multicellular species (Bleakley and Hayes 2017).

One of the direct uses of algae, particularly marine macroalgae, is as food for humans and other animals (Fougere and Bernard 2019). Currently, algae have been having both detrimental and potentially revamping effect to food security. Harmful algal blooms threaten aquatic life, so does the potential for algae serving as a more sustainable source of food (FAO, IFAD, UNICEF, WFP and WHO 2018).

Marine algae are usually found in the photic zone of the sea as they require sufficient sunlight for photosynthesis. The importance of light for algae growth is further emphasized in studies where in the absence of sufficient light, with surplus nutrient supply, very little growth occurs even in the presence of abundant nutrients (Prabhu et al. 2019). In the absence of light, the organisms are unable to convert the nutrients for growth and metabolism. Light is therefore as important for aquatic algae life as it is for terrestrial plants.

Algae can either be macroalgae or microalgae. There a thousands of species of both microalgae and macroalgae. Microalgae are photosynthetic, mostly unicellular, microscopic organisms, while macroalgae also called seaweeds are multicellular, photosynthesizing organisms. Algae can either be heterotrophic, autotrophic or mixotrophic. Autotrophic algae can fix inorganic CO_2 from the atmosphere through photosynthesis and produce storage carbohydrates while heterotrophic algae can fix small organic molecules into lipids and proteins while mixotrophic algae can do both. This makes algae potentially suitable for the production of both biodiesel and bioethanol, since the lipids can be taking through transesterification and carbohydrates can be saccharified and fermented. Algae have been explored for several decades for potential commercial production of third-generation biofuel (Nigam and Singh 2010)

Algae are of particular environmental and economic importance for several reasons. They are able to accumulate carbohydrate at a much faster rate than land plants due to much higher photosynthetic efficiency. They can grow in wastewater with high nitrogen content unlike terrestrial plants, and they take up CO_2 at a much faster rate. Algae can also grow in a variety of environments from saline to fresh, from high turbulence ocean water to slow flowing municipal wastewater (John et al. 2010).

Microalgae are more compliant to genetic engineering for the optimization of desired output than higher plants being microscopic in nature. They are more susceptible to control of metabolic pathways and DNA alteration (Rosenberg et al. 2008). Microalgae make for good third-generation biofuel as they contain both lipid and starch, for the production of biodiesel and bioethanol, respectively. Furthermore, microalgae does not contain lignin and hemicellulose (Mahapatra and Ramachandra 2013), which makes the process of extraction of the fermentable carbohydrates less expensive and rigorous as these biopolymers are more complex to saccharify. Like macroalgae. The microalgae also has the ability to remediate wastewater by removing sulfur and nitrates (Lv et al. 2017). Microalgae can be found in fresh and saltwater. Example of freshwater microalgae is the *Chlorococcum* sp.

Green, brown and red algae are distinguished from one another based on their cellular structures and composition. While the cell walls of brown algae contain alginates and fucoidan with laminarin as reserve photosynthetic pigment, red algae contain agar and carrageenan as their cell wall polysaccharide alongside cellulose, those of green contain ulvan in their cell wall and starch within their chloroplast. Diatoms are eukaryotic microalgae that belong to the group of plankton and benthic algae. Their skeleton is made of amorphous silica, and some diatoms are photosynthetic while others are not. They can be found in both marine and freshwater (Chetia et al. 2017).

Algae can either be sourced from natural stocks; some fishermen simply dive into the sea to pick algae by hand. Algae can also be cultivated in open sea or aquaculture. There are three main methods of cultivating algae as illustrated in Fig. 2.6. These are broadcast, off button, and longline methods.

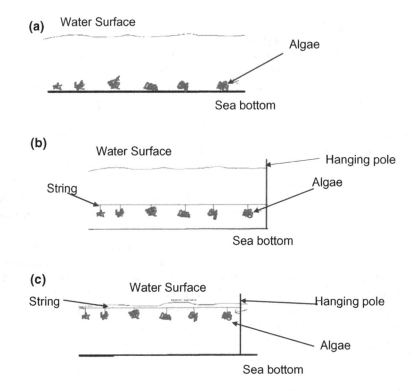

Fig. 2.6 An illustration of the three methods of cultivation of red algae. **a** Broadcast, **b** off-bottom and **c** longline methods

In the broadcast method, algae seedlings are cultivated close to the bottom of the water. This is usually in slow-moving shallow waters in ponds or lakes or artificial water systems. In the off-bottom method, the algae seedlings grow on lines tied to poles such that the algae extend about 20 cm from the bottom of the water, midway in the shallow water depth. In the longline method, the algae seedlings grow closer to the surface hanging from ropes attached to poles. Different algae species have preferences for particular growth method.

2.3.3 Aquatic Animals

These include the vertebrates: amphibians, mammals, birds, fish and reptiles as well as invertebrates such as crustaceans, mollusks, insects, squids, echinoderms. Although some mammals, amphibians and birds spend some, most or all of their time outside of water, they are classified as aquatic if they mostly depend on water for their survival and feeding. Examples are polar bears who mainly feed on seals and the kingfisher birds that feed on small fish from rivers. Some of the animals of particular interests which are further discussed in other chapters are the frogs whose skin serves as a source of some bioactive polymers and other aquatic animals with special features which aid them in surviving in the deepest part of the ocean.

The decline or surge in the population of fish in aquatic environment has long been used as an indication of the presence or absence of contaminants within such aquatic environment. In industrial quality checks, the product safety for human consumption is confirmed by testing on fish in controlled aquatic systems such as aquarium or ponds (Christophe et al. 2015). The aquatic animals serve as a source of food and livelihood for various life-forms including humans. There is need to maintain a balance between the population of different aquatic animals in order to ensure constant supply of aquatic resources.

One of the aquatic invertebrates which recently pose a nuisance to the fisheries industry is the jellyfish. In recent years, giant jellyfish population explosion in the ocean has been reported (ref). The cause of this bloom has been attributed to the decline in the population of the larger aquatic organisms which prey on these jellyfish due to overfishing, eutrophication of coastal waters, global warming, changes to the aquatic habitat and translocation (Dong 2019). The jellyfish bloom has resulted in a decline in the population of smaller finfish as they are poisoned by these jellyfish. This results in devastating loss to fishermen, aqua tourism industry and also destruction of cooling systems of power plants located on the coast. Measures taken to address this include the development of jellyfish bloom detection tools (Azmar et al. 2017), consuming the jellyfish for food and use as feedstock in biopolymer industry as reviewed in the subsequent chapter in this book.

2.3.4 Carnivorous Aquatic Plants

Whether on land or in the water, carnivorous plants are very rare. Globally, there are over 700 species of carnivorous plants from over 5 orders and 10 genera (Lima et al. 2018). Examples of species of aquatic carnivorous plants are *Aldrovanda vesiculosa*, *Utricularia vulgaris*, *U. reflexa* Oliver, *U. stygia* Thor and *U. intermedia* Hayne (Adamec 2010). Two common aquatic carnivorous plants are the water wheels and the pitcher plants. The evolution of plants to feed on insects is thought to originate from the natural defense which plants have against fungi and bacteria (Schulze et al. 2012). Since fungi and insects have the chitin-based structure in common, similar chitin degrading mechanism is involved in breaking down the chitinous shells of insects. Carnivory in plants is thought to be an adaptation to the limitation of nutrients in the growth medium. While to other plants, the roots serve as the main organ for the uptake of nutrients. In carnivorous plants, the leaves have adapted a feature for taking up nutrients from the trapped animal (mostly insects and larvae).

Some carnivorous plants adopt passive techniques for attracting prey, while others adopt active techniques for attracting prey (Forterre et al. 2005). The passive technique involves insect being lured toward a sticky surface where they feed on the plant nectar; however, by the architecture of the plants they are unable to leave due to the adhesive surface and are eventually entrapped within the hairs and tentacles and are eventually digested by the plant enzymes (Schulze et al. 2012). In the active capture system, the plant traps prey via a mechano-electrical sensor and trap system where the movement of the prey in contact with a biological sensor, usually a hairlike sensor, stimulates the closure of a trap which seals and crushes the insect. This is then followed by digestion and the nutrients extracted from the insect. The enzymes produced by the aquatic carnivorous plants are a form of aquatic biopolymers as well as polymeric compounds which these enzymes act on.

2.3.5 Aquatic Insects

Insects form an important part of the aquatic system. Some depend on the aquatic plants for food and in some cases shelter. They also depend on the aquatic environment for nursery where they lay their eggs and larvae. They in turn serve as food and a source of nutrition for other aquatic animals and carnivorous plants. Aquatic insects generally inhabit rocky shores, remain buried in sand or settle on aquatic plants and algae (Merritt and Wallace 2009). They therefore require the aid of other structures or aquatic life-form, in order to inhabit the aquatic environment. Examples of insect which inhabit the aquatic environment include Gerridae, Diptera, Coleoptera and Collembola (Merritt and Wallace 2009).

2.3.6 Aquatic Birds

Some flying and flightless birds inhabit the aquatic environment. These include those which mostly stay afloat on water for most of the time and mainly depend on the aquatic environment for food. They feed mainly on fish and other aquatic organisms. Some are able to fly away from the water and however depend solely on fish for food. Examples are the penguins, ducks, swans, geese and the kingfisher bird.

Some of the challenges faced by these birds include the release of toxins into the aquatic environment. They face risk of toxins present in the water and in the organisms they feed on from the polluted water. One of such is lead toxin, toxins from algal blooms of blue-green cyanobacterium, oil spills, Clostridium botulinum, zinc and lead from industrial waste. These toxins could lead to conditions such as hypochromic anemia, pneumonia, secondary aspergillosis, neuromuscular malfunction and even unexplained death while body is still in good condition (Backues 2015).

2.3.7 Aquatic Microorganisms

Microorganisms exist both on land and in water. They are as important to the aquatic environment as the larger organisms. The microorganisms present in water include bacteria, viruses, protozoa and fungi. They can be of various classification such as filamentous, rod-shaped or helical, aerobic, anaerobic or facilitators, heterotrophs, autotrophs or mixotrophs (Little et al. 2001). Some exist within other organisms where they have a symbiotic relationship with the host organisms. Examples of such are the microbes which produce light within bioluminescent aquatic organisms. These are discussed in later chapters of this book under bioluminescent proteins. Other examples of symbiotic aquatic bacteria include those in the internal organs of fish and the digestive bacteria in the utricle and stolon of the carnivorous aquatic plant *Utricularia breviscapa* (Lima et al. 2018). Such bacteria aid in the digestion of food and organic matter.

Aquatic microorganisms can either drift with the current suspended in the water. These are generally classified as plankton (Meaning vagabond in Greek). Other microorganisms are attached to substrates or within sediments in water. These are known as the benthos (Greek word meaning depth) (Callieri et al. 2019). Planktons are generally stationary; however, some possess some features sufficient to enable them to maintain buoyancy. This allows them to migrate and aggregate as the nutrient content and conditions vary, generally to move toward regions with nutrients and favorable living conditions. Planktons serve as a food source for various species which in turn are also a source of food to other organisms.

Benthos depend on the organic matter settled on the substrate they are attached to. These types of microorganisms form biofilms which are the polymeric extracellular matrix of the aggregates of microorganisms. The biofilms formed by these microbes are crucial to their survival as it serves as a barrier through which nutrients and other compounds are exchanged with the immediate environment (Kostakioti et al. 2013). The biofilm enables the microorganism to adhere to the substrate in water; it also serves a protective purpose to these benthic microorganisms. Outside of the natural aquatic ecosystem, biofilm-producing bacteria also form on wet surfaces in everyday life such as bathrooms, labs and hospitals. Microbes present in the water could be true aquatic microorganisms which only exist in water while others are introduced from land. The introduction of land microorganisms into water could be from seepage from land or deliberate or accidental introduction of fecal waste into the water.

2.3.8 Humans

Since water serves as a source of food, means of transportation, source of energy, recreation and aesthetics, for a very long time in human history, people have settled along aquatic ecosystem. There are also communities in different parts of the world who have settled on water. An example is the Makoko community in Lagos, Nigeria, which is made up of humans who have constructed floating wooden houses on water and use canoe and other floating objects as the primary means of transportation within the water community (Riise and Adeyemi 2015). Similarly, the Satoyama community in the ancient city of Kyoto in Japan have for centuries established a harmonious relationship with the aquatic life which promotes biodiversity and also benefits human health. The humans living in these water-rich community cultivate the aquatic resources such as fish, rice and aquatic plants in a sustainable manner that benefits both humans and aquatic life. This harmonious relationship existing in Satoyama community has recently been reported to be on the decline (Takeuchi et al. 2016). It is therefore important to develop a wider understanding of the importance of the aquatic environment to human life.

Humans use water as a means to move people and goods across continents, countries, cities and communities over the oceans, seas, lakes and rivers. Such means include ships, boats and submarines. Permanent structures such as bridges and underwater tunnels are build to allow land-based transportation systems over water. Water also serves for recreation and major sporting events such as kayaking and water skiing. Deep-sea oil exploration, fishing and mining are some of the major activities which take place in the aquatic environment. An example of human activity is shown in Fig. 2.7, an image of men fishing and a motorized boat moving across the Lagos Lagoon.

Beyond transportation on the surface of water, humans have explored the deepest depth of the ocean and beyond. The location referred to as the Challenger Deep in the Mariana Trench, a depth of 10,994 km below sea level (Kobayashi et al. 2012) has been confirmed as the deepest part of the ocean. Although the conditions are so extreme that no human has physically been at that depth, recently submersibles equipped with high-resolution cameras have discovered life-forms that have adapted to live in what was previously considered to be a barren depth in the ocean. This discovery has been of benefit to humans in ways such as discovering sources of new enzymes these organisms use to digest food and produce light in the dark depths of the sea (Kobayashi et al. 2012; Altun et al. 2008). Such discoveries lead to development

Fig. 2.7 Fishing activity on the Lagoon in Lagos, Nigeria. August 2019. Photograph by Ololade Olatunji

of new ways to produce better environmentally sustainable fuels and new compounds for detecting and treating diseases such as cancers.

Living aquatic resources become available through two main routes: capture from the waters or through fish farming, also referred to as aquaculture. Another form which is a hybrid between these two methods is mariculture, a process where aquatic organisms are cultivated on the sea within enclosed areas. Algae, shellfish and sea cucumber are some of the organisms which are cultivated in mariculture. This is done to obtain either whole biomass or parts of these organisms for use as food, medicine, jewelry or other applications (Engle 2009). The mode of sourcing a particular aquatic resource is important as it affects the logistics involved in sourcing the feedstock for the production process. Some existing manufacturing plants which processing aquatic waste and by-products into other products are located close to the sea or river from where the aquatic resource is sourced or farmed for the reason of nearness to raw material which serves as the feedstock for the production plant. Where there is not such plants situated close by the transport and storage of the raw materials (e.g. shells, scales, viscera and bones) further adds to the cost of production.

The population of aquatic organisms is significantly affected by the level of fishing by humans. According to the FAO, an estimated 94.6 million tonnes of fish were sourced through capture in 2014 (FAO 2016). This includes both marine and inland waters. Aesthetically valued shells and pearls which are used as ornaments for art, decorative, jewelry or other aesthetic application are also sourced by humans from the aquatic environment (FAO 2018).

The human activities have played key roles in the decline of the aquatic ecosystems. This includes overfishing, introduction of species to waters they would not naturally exist in, discharge of polluted water into the waters, agriculture leading to run off of fertilizers into the water, building of hydroelectricity dams, dredging and diversion of the waters for activities such as irrigation, constructions near, on, in water bodies and oil spillage. Pollution of the seas and oceans by dumping of municipal and industrial waste has been shown to lead to diseases and illnesses such as hepatitis, typhoid fever and gastroenteritis. These occur as a result of consuming shellfish such as oysters and mussels which have fed on the contaminants (Kwaasi 2003).

Human activity in and around aquatic environment has also been responsible for the eutrophication of lakes. This refers to the process whereby the nutrient level in the water rises due to run off of nutrients from surrounding land resulting in the increase in the population of aquatic plants which eventually leads to the death of other aquatic life and depletion of the water. In nature, eutrophication occurs as a slow process over long periods of time. At the natural pace, eutrophication does not have such a pronounced negative impact on the environment. When this process is accelerated, it distorts the balance in the ecosystem most often with devastating effects.

The aquatic industry cuts across all three economic sectors: primary, secondary and tertiary. Activities such as fishing, deep-sea mining and trawling belong to the primary sector; the secondary sector comprises fish processing which includes activities such as commercial finfish canning and extraction of biopolymers from aquatic wastes. The tertiary sector of the aquatic industry relates to activities such as boat rental and aqua tourism. Biopolymer resources of the aquatic environment play numerous roles in the life of humans; these range from polymers used in water purification, those with medicinal applications to those used to produce biofuel. These will be discussed in more detail in subsequent chapters of this book.

2.4 Abiotic Components of the Aquatic Ecosystem

The abiotic components of an ecosystem include the physical and chemical parts such as the organic matter such as from the dead organisms, inorganic matter such as the silica of diatoms and the calcium carbonate deposited by the shedding of zooplanktons and exoskeleton of crustaceans (Wilson 2013). The temperature, salinity, pH, water flow rate, size of rocks and soil at the bottom of the water, minerals, light intensity, dissolved oxygen and water level, all contribute to the abiotic factors.

2.5 Conclusion

In this chapter, we have taken an overview of the aquatic ecosystems, the living and non-living components therein, their impact on the environment and economy. The aquatic ecosystems serve as habitat to diverse life-forms and play a significant role even to life on land. The aquatic environment also serves as a source of a variety of natural resources, and of particular interest in this book are the polymer resources of the aquatic environment. In the subsequent chapters, we shall look at the biopolymers which exist within the aquatic environment, their sources, chemistry, industrial processing and applications and the environmental significance of such.

References

Abd-Elbaky M, Jin S (2019) Hydrological mass variations in the Nile River Basin from GRACE and hydrological models. Geodesy and geodynamics (in press)
Abdelmohsen K, Sultan M, Ahmes M, Save H, El Kaliouby B, Emil M, Yan E, Abotalib AZ, Krishnamurthy RV, Abdelmalik K (2019) Response of deep aquifers to climate variability. Sci Total Environ 677:530–544
Adamec L (2010) Mineral cost of carnivory in aquatic carnivorous plants. Flora 205:618–621
Altun T, Celi F, Danabas D (2008) Bioluminescence in aquatic organisms. J Ani Vet Adv. 7(7):841–846
Awange JL, Saleem A, Sukhadiya RM, Ouma YO, Kexiang H (2019) Physical dynamics of Lake Victoria over the past 34 years (1984–2018): is the lake dying? Sci Total Environ 658:199–218
Azmar F, Pujol M, Rizo R (2017) A swarm behaviour for jellyfish bloom detection. Ocean Eng 134:24–34
Backues KA (2015) Anseriformes. Fowler's Zoo and wild animal medicine, 8, pp 116–126

Bleakley S, Hayes M (2017) Algal proteins: extraction, application and challenges concerning production. Foods 6:33 (1–34)

Borisjuk N, Peterson AA, Lv J, Qu G, Luo Q, Shi L, Chen G, Kishchenko O, Zhuo Y, Shi J (2018) Structural and biochemical properties of duckweed surface cuticle. Front Chem. https://doi.org/10.3389/fchem.2018.00317

Braebdle R, Crowford RMM (1999) Plants as amphibians. Perspect Plant Ecol Evol System 2(1):56–78

Callieri C, Eckert EM, Di Cesare A, Bertoni F (2019) Microbial communities. Encycl Ecol 1:126–134

Casamitjana X, Mencio A, Quintana XD, Soler D, Comptee J, Martinoy M, Pascual J (2019) Modeling the salinity fluctuations in salt marsh lagoons. J Hydrol 575:1178–1187

Cavalier-Smith T (2007) Evolution and relationships. In: Brodie J (ed) Unravelling the algae: the past, present, and future of algal systematics. CRC Press, Boca Raton, pp 21

Chetia L, Kalita D, Ahmed GA (2017) Synthesis of Ag nanoparticles using diatom cells for ammonia sensing. Sens Bio-Sens Res 16:55–61

Christophe M, Rachid A, Mario L (2015) Fish as a reference species in different water masses. Aquat Ecotoxicol, 309–331

Dame RF (2008) Estuaries. In: Encyclopedia of ecology, pp 1407–1413

Deane F, Wilson C, Rowlings D, Webb J, Mitchell E, Hammam E, Sheppard E, Grace P (2018) Sugarcane farming and the Great Barrier Reef: the role of a principal approach to change. Land Use Policy 78:691–698

Dong Z (2019) Blooms of the moon jellyfish Aurelia: causes, consequences and control. World seas: an environmental evaluation, 2nd edn., vol III, pp 163–171

Engle CR (2009) Mariculture, economic and social impacts. In: Encyclopedia of ocean sciences, 2nd edn., pp 545–551

FAO (1997) Irrigation potential in Africa: a basin approach. Food land and water bulletin 4. FAO land and water development division. ISBN 02-5-103966-6

FAO (2016). The State of World Fisheries and Aquaculture 2016. Contributing to food security and nutrition for all. Rome. pp 200. ISBN 978-92-5-109185-2

FAO (2018). The State of World Fisheries and Aquaculture (2018). Meeting the sustainable development goals. Rome. Licence: CC BY-NC-SA 3.0 IGO ISBN 978-92-5-130562

FAO, IFAD, UNICEF, WFP, and WHO (2018) The state of food security and nutrition in the world. Building climate resilience for food security and nutrition. Rome, FAO. Licence: CC BY-NC-SA 3.0 IGO. ISBN 978-92-5-130571-3

Forterre Y, Skotheim JM, Dumais J, Mahadevan L (2005) How the venus flytrap snaps. Nature 433:421–425

Fougere H, Bernard L (2019) Effects of diets supplemented with starch and corn oil, marine algae, or hydrogenated palm oil on mammary lipogenic gene expression in cows and goats: A comparative study. J Dairy Sci 102(1):768–779

Harris G (2008) Lagoons. In: Encyclopedia of ecology, 2nd edn., vol. 2, pp 539–545

Hitchcock JN, Mitrovic M (2019) Microplastic pollution in estuaries across a gradient of human impact. Environ Pollut 247:457–466

Honjo S (2009) Biological pump and particle fluxes. Encyclopedia of ocean sciences, 2nd edn., pp 371–375

Iizumi T, Ramankutty N (2015) How do weather and climate influence cropping area and intensity? Global Food Secur 4:46–50

John RP, Anisha GS, Nampoothiri KM, Pandey A (2010) Micro and macroalgal biomass: a renewable source for bioethanol. Biores Technol 102:186–193

Kadam SU, Alvarez C, Tiwari BK, O'Donnell CP (2015) Extraction of Biomolecules from seaweeds. In: Tiwari BK and Troy DJ (eds) Seaweed Sustainability. Academic press. pp 243–269

Kobayashi H, Hatada Y, Tsubouchi T, Nagahama T, Takami H (2012) The hadal amphipod Hirondellea gigas possessing a unique cellulase for digesting wooden debris buried in the deepest seafloor. PLoS One 7(8):e42727, 1–8

Kostakioti M, Hadjifrangiskou M, Hultgren SJ (2013) Bacteria biofilms: development, dispersal and therapeutic strategies in the Dawn of the postantlantibiotic era. Cold Spring Harb Perspect Med 3(4):a01306

Kwaasi AAA (2003) Microbiology: Classification of Microorganisms. In: Caballero B (ed) Encyclopedia of food sciences and nutrition, 2nd edn. Academic Press, USA, pp 3877–3885

Lima FR, Ferreira AJ, Menezes G, Miranda VFO, Dourado MN, Araujo WL (2018) Cultivated bacterial diversity associated with the carnivorous plant *Utricularia breviscapa* (Lentibulariaceae) from floodplains in Brazil. Braz J Microbiol 49:714–722

Linley TD, Gerringer ME, Yancey PH, Drazen JC, Weinstock CL, Jamieson AJ (2016) Deep sea research part I: oceanographic research papers, 114:99–110

Little BJ, Ray RI, Pope RK (2001) Bioactive environments: corrosion. In: Encyclopedia of materials: science and technology, 2nd edn., pp 533–537

Lonborg C, Calleja ML, Fabricius KE, Smith JN, Achterberg EP (2019) The great barrier reef: a source of CO_2 to the atmosphere. Mar Chem 210:24–33

Lv J, Guo J, Feng J, Liu G, Xie S (2017) Effect of sulfate ions on growth and pollutants removal of self-flocculating microalga *Chlorococcum* sp. GD in synthetic municipal wastewater. Bioresour Technol 234:289–296

Mahapatra DM, Ramachandra TV (2013) Algal biofuel: bountiful lipid from *Chlorococcum* sp. proliferating in municipal wastewater. Curr Sci 105:47–55

Mahmood R, Jia S (2019) Observed and simulated hydro-climatic data for the Lake Chad basin, Africa. Data in Brief, 25(1014043):11–15

Martins I, Goulart J, Martins E, Morales-Roman R, Marin S, Riou V, Colaco A, Bettencourt R (2017) Aquat Toxicol, 40–49

Merritt RW, Wallace JB (2009) Aquatic habitats. In: Encyclopedia of insects, 2 edn., pp 38–48

Nigam PS, Singh A (2010) Production of liquid biofuels from renewable resources. Prog Energy Combust Sci https://doi.org/10.1016/j.pecs.2010.01.003

North GB, Peterson CA (2005) Water flow in roots: Structural and regulatory features. Vasc Transp Plants, pp 131–156

Oyedeji AA, Abowei JFN (2012) The classification, distribution, control and economic importance of aquatic plants. Int J Fish Aquat Sci 1(2):118–128

Pendleton L, Guldberg OH, Albright R, Kaup A, Marshall P, Marshall N, Fletcher S, Haraldson G, Hansson L (2019) The great barrier reef: vulnerabilities and solutions in the face of ocean acidification. Reg Stud Mar Sci 31:100729

Penfound WT, Earle IT (1948) The biology of the water hyacinth. Ecol Monogr 18:447–472

Peters JA, Lodge DM (2009) Littoral zone. Encyclopedia of inland waters, pp 79–87

Prabhu M, Chemodanov A, Gottlieb R, Kazir M, Goldberg A (2019) Starch from the sea: the green macroalga Ulva ohnoi as a potential source for sustainable starch production in the marine biorefinery. Algal Res. 37:215–227

Riise J, Adeyemi K (2015) Case study: Makoko floating school. Curr Opin Environ Sustain 13:58–60

Rosenberg JN, Oyler GA, Wilkinson L, Betenbaugh MJ (2008) A green light for engineered algae: redirecting metabolism to fuel a biotechnology revolution. Biotechnology 19:430–436

Scherzer S, Sanggaard KW, Kreuzer I, Knudsen AD, Bemm F, Thogersen IB, Brautigam A, Thomsen LR, Schliesky S, Dyrlund TF (2012) The protein composition of the digestive fluid from the venus flytrap sheds light on prey digestion mechanisms. Mol Cell Proteomics 11:1306–1319

Staby A, Salvanes AGV (2019) Marine life: mesopelagic fish. Encyclopedia of ocean sciences, 3rd edn., pp 283–289

Takeuchi K, Ichikawa K, Elmqvist T (2016) Satoyama landscape as social-ecological system: historical changes and future perspective. Curr Opin Environ Sustain 19:30–39

Takir D, Howard K, Yabuta H, McAdam M, Hibbitts C, Emery J (2018) Linking water-rich asteroids and meteorites: implications for asteroid space missions. In: Primitive meteorites and asteroids, pp 371–408

Wilson B (2013) Benthic shelf and slope habitats. In: The biogeography of the Australian North West shelf, pp 259–265

Chapter 3
Chitin

Abstract Chitin occurs in a variety of organisms in aquatic ecosystems. It is a polymer of acetyl glucosamine monomer. It has wider commercial applications in its deacetylated form as chitosan. Extraction process requires high-temperature treatment under alkaline as well as acidic conditions, depending on the source. Chitin is one of the most explored aquatic biopolymers with existing commercial applications as well as a wide range of applications in research. This chapter covers extraction processes from different sources, the chemistry, conventional and emerging applications, availability of aquatic feedstock for chitin production as well as the environmental and economic impact of chitin.

Keywords Chitin · Polymer · Chitosan · Crustaceans · Shrimps

3.1 Introduction

Chitin is the second most abundant polymer on earth (after cellulose). It is the most abundant aquatic biopolymer and so far has proven itself to be the most ubiquitous of all the biopolymers as it finds application in almost every industry. Chitin is one of the polymers which have a relatively diverse source. It can be found in organisms which are completely restricted to living in water such as shrimps as well as those found in wetlands such as crabs, those in moist environment such as mushrooms and even in insects. Biodegradability is a most desirable feature of polymers, and there is an increasing demand for biodegradable polymers which will degrade into non-toxic compounds once they have completed their service life. Chitin is one of such polymers.

In this chapter, we present the chemistry of chitin, its occurrence in nature, role within the environment and organisms from which they are sourced and the various methods employed in the extraction and processing into other forms, particularly its deacetylation into chitosan, the most commercially relevant form. The chapter also explores the global demand for chitin and its numerous conventional, novel and potential applications as well as the environmental impact of its production. The chapter also identifies some of the companies and countries involved in chitin production from different sources.

© Springer International Publishing 2020   31

O. Olatunji, *Aquatic Biopolymers*, Springer Series on Polymer and Composite Materials,
https://doi.org/10.1007/978-3-030-34709-3_3

3.2 Occurrence in Nature

Chitin is a naturally occurring amino polysaccharide. It is a structural feature in the exoskeletons of arthropods, shells of mollusks and the cell walls of fungi, forming an integral part of the outer shells of crustaceans such as shrimps, crabs and lobsters and mollusks such as snails and slugs. Some research studies have reported the presence of chitin in the scales of certain species of fish (Rumengan et al. 2017). However, the yield of chitin from fish scale is much lower than other sources. Yield of about 20% has been reported for Nile tilapia (Boarin-Alcalde and Graciano-Fonseca 2016). Figure 3.1 illustrates the presence of chitin in the cell wall of mushroom.

Chitin is a non-toxic biopolymer. Crustacean shells generally consist of 20–30% chitin, 30–40% protein and 30–50% calcium carbonate and calcium phosphate (Majekodunmi 2016). Much of the commercial production of chitin is from crabs and shrimp shells as these are more available and more extensive research has been carried out on the extraction of chitin from these sources.

As a lesser explored source, fish scales are a particularly attractive source of chitin as the scales can be collected before the fish is sold or taken for further processing unlike in crustaceans where in some cases such as lobsters and crabs, the shells are served with the food and can only be collected after cooking and serving. However, not all fish scales contain chitin. Fishes which have been reported to contain a detectable amount of chitin include parrotfish (*Chlorurus sordidus*), red snapper (*Lutjanus argentimaculatus*) and Nile tilapia (*Oreochromis nIloticus*) (Rumengan et al. 2017; Boarin-Alcalde and Graciano-Fonseca 2016). However, fish scales have a much lower chitin content than other sources. In contrast, scales of other fish have no chitin present in them. These scales are generally composed of collagen and hydroxyapatite in an orthogonal plywood structure (Gil-Duran et al. 2016). Such fish with no chitin present in their scales include Atlantic tarpon (*Megalops atlanticus*). The composition of fish scales, the presence or absence of chitin, depends on the requirement of the fish in adapting to its environment, and this varies for different

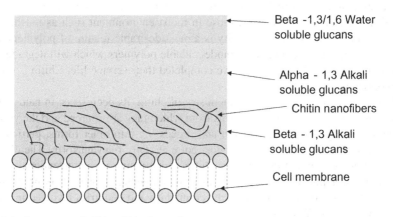

Beta -1,3/1,6 Water soluble glucans

Alpha - 1,3 Alkali soluble glucans

Chitin nanofibers

Beta - 1,3 Alkali soluble glucans

Cell membrane

Fig. 3.1 Occurrence of chitin within the mushroom structure

species of fish. Similarly, the composition of chitin in other aquatic organisms varies for the different species.

The major issue which is presently limiting a larger commercial exploitation of the available chitosan resource is the environmental, technical and economics of scale surrounding its production process. The chitin market is expected to be more than triple in the next 9 years. The major drivers for this growth are expected to be the numerous emerging applications of chitin derivatives, especially chitosan, in the healthcare industry (Future Market Insights 2019). Researchers have presented a plethora of ways to use chitin; however, the major challenge is the cost of chitin production both financially and environmentally. There needs to be further research into more economic yet environmentally friendly means of extracting chitin and converting it into useful products, thus making the chitin industry an exciting one in which new opportunities for researchers and businesses lie in the development of new chitin-based products and discovering new sources of the second most abundant polymer on the planet.

3.3 Chemistry of Chitin

3.3.1 Structure

The repeat unit of chitin is an acetylglucosamine structure linked by a beta 1-4 linkage between the monomer units. Chitin is a linear long-chain polymer with a high molecular weight which varies largely depending on factors such as source, extraction method and extraction conditions and from batch to batch. It is quite similar to cellulose, except that one of the hydroxyl (–OH) groups of glucose as seen in cellulose is replaced by an amine group ($-NH_2$) (Kaya et al. 2017). Chitin is rather unreactive in its native form. However in its derivative forms, it becomes more reactive and much more useful. One of the most important reactions that is often required of chitin is the deacetylation reaction. Chitin in its pure form is about 90% acetylated, and when it is deacetylated to over 60%, it can be classified as chitosan (Kalut 2008). Thus although deacetylation of chitin results in chitosan, chitin also has a level of acetylation. A chitosan polymer chain that has up to 10,000 repeating units will have at least 6000 of these being glucosamine while the other 4000 will be acetyl glucosamine. The level of acetylation affects the properties of the chitosan. At a degree of deacetylation of over 98%, we begin to have a chitosan with good solubility and minimal aggregation. Figure 3.2 shows the structure of a chitin repeating unit.

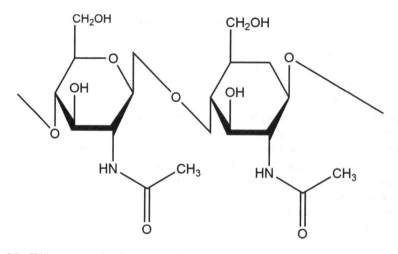

Fig. 3.2 Chitin structure showing the acetylated glucosamine unit

3.3.2 Solubility

For broader applicability, it is important for a polymer to be moldable in order to be reformed into useful products such as films, sheets and other shapes. This is usually achieved in polymers either by melting as in the case of thermoplastics or by dissolving in a suitable solvent. For most natural polymers, the latter is often the case. The polymer needs to readily dissolve in a non-toxic, non-mutagenic and relatively low-cost solvent to be suitable for industrial application. In the solution form, the polymer can then be processed using any of the various polymer processing techniques such as solvent casting, dip coating, spin coating and micromolding (Olatunji and Olsson 2015).

Chitin is insoluble in all the common organic and inorganic solvents; however in its deacetylated state as chitosan, it becomes soluble in acetic acid and in acidic solution below a pH of 6.5 (Roy et al. 2017). Like other polymers, chitosan increases the viscosity of the solvent when dissolved within it. For example when chitosan is added to acetic acid, it forms a viscous sticky solution which gets more viscous as the concentration of chitosan increases. This is a unique property of polymers in general, and it is one of the simple tests used in the laboratory to determine the presence of chitosan following extraction and deacetylation. Chitin has a stronger crystalline form than chitosan and is relatively more hydrophobic. Upon deacetylation, the structure becomes more hydrophilic due to the amino group becoming available to form hydrogen bond with solvents. Chitin will not dissolve in water, acidic, alkaline or other organic solvents. It will however dissolve in rarer solvents such as dimethylacetamide mixed with lithium chloride and hexafluoroisopropyl alcohol and hexafluoroacetone (Agboh and Qin 1997; Rinaudo 2006). However, these solvents are hazardous and not suitable for industrial applications.

While the insolubility of chitin poses a difficulty in processing it into other products, this characteristic of chitin however comes in use during the extraction process. Extraction of chitin from crustacean shells such as crabs involves dissolving the collagenous component of the shell in a strong alkali followed by dissolving the minerals (mostly calcium carbonate) in acid. What's left behind is the only component of these shells that does not dissolve in either acidic or alkali solvents, chitin. Solubility of chitosan increases as the chain length decreases. Chitosan oligomers are chitosan chain with short chain lengths. Degree of polymerization on chitosan could be as low as 5. We could also have chitosan dimers and trimers. This shorter chain allows for improved solubility which aids certain applications (Azuma et al. 2015).

The secondary and tertiary structures of polymers are dependent on the presence of functional groups within the chain and the interactions between the functional groups within the chain and surrounding chains. Chitin has within its chain acetyl, hydroxyl and amino groups. Intermolecular bonding and intramolecular hydrogen bonding between these groups cause chitin to fold in on itself and aggregate, making it unready to form hydrogen bonds with the solvents, hence the insoluble nature of chitin.

3.3.3 Isomers

Chitin exhibits polymorphisms in three different forms, α, β and γ. The α and β forms are the more crystalline forms and have received more attention (Kalut 2008). The chitin tertiary structure comprises of six strands of chitin polymer chain wound together in a protein-like helical structure to form crystalline chitin microfibrils (Roy et al. 2017). The α conformation is the most abundant in nature. In the α conformation, the chitin polymer chains are arranged in alternating manner. In the β chitin, the polymers are arranged in the same direction while in the γ conformation they are randomly arranged.

3.3.4 Variation with Source

The distribution of the acetyl group within the chitin and chitosan structure is completely random. This poses a challenge in achieving a precise standard for solubility and solution properties. Results from one research group may depend highly on the batch being used. There is yet to be a standard correlation between parameters such as degree of acetylation and species such that chitin derived from a particular species would correlate with, for example, molecular weight, solubility and degree of acetylation. The solubility of chitin and chitosan is dependent on the primary, secondary, tertiary and quaternary structure of the polymer; these are affected by the degree of deacetylation which in turn varied from species to species and batch to batch.

For example, the degree of deacetylation of mollusks is different from that extracted from crustaceans. Chitin properties vary even within the same species. For example, different types of mushrooms produce chitin with varying fiber length (Ifuku et al. 2011).

3.3.5 Deacetylation

This is where some or all of the acetyl ($COOCH_3$) functional groups attached to the N-acetylglucosamine units (2-acetamino-2 deoxy-β-D-glucopyranose) are replaced with a hydrogen leaving behind an amine ($-NH_2$) end group of glucosamine (2-amino-2-deoxy-β-D-glucopyranose). This occurs when chitin is reacted with 40–50% w/v of sodium hydroxide (Kalut 2008). The extent to which this occurs within the polymer chain is referred to as the degree of deacetylation (DDA).

3.3.6 Molecular Weight

Chitin from the same source can vary significantly in properties depending on the batch. For example, a study reported molecular weight of chitin extracted from chemical method as 33,400 g/mol while that from lactic acid fermentation was 967,000 g/mol (Castro et al. 2018). Although through methods such as gel permeation chromatography, chitosan can be separated into different molecular weights such that high and low molecular weight chitosans are available in the market. Chitosan of high and low viscosity is also available since molecular weight is related to viscosity.

The molecular weight of chitin plays a very important role in its applicability. Molecular weight affects properties such as solubility, viscosity of the solution, mechanical properties of the film or other products produced and its interaction and reaction with other polymers and materials in either composite forms or reacted forms and other forms. The higher the molecular weight of chitin and chitosan, the lower the solubility. However, this ceases to apply at low molecular weight below 2.43 kDa when the short chain does not aid the formation of hydrogen bonding with the solvent (Roy et al. 2017). Depolymerization of chitin to obtain lower molecular weight oligomers can be achieved by acid or enzyme hydrolysis.

3.3.7 Depolymerization

Another important reaction of chitin is the depolymerization into monomers. Chitin can be completely depolymerized into its monomeric form, acetyl glucosamine, or deacetylated and depolymerized into glucosamine. Glucosamine has become another derivative of chitin which has gained demand over the years. Glucosamine is the monomer sugar unit obtained from depolymerization of chitosan. It has the chemical structure shown below (Hulsey 2018). Glucosamine and acetyl glucosamine have been proven to have some bioactive properties such as anticancer and treatment of osteoarthritis when administered intravenously or orally as a food supplement (Azuma et al. 2015).

3.4 Availability of Raw Materials

The Food and Agriculture Organization of the United Nations reported global production of fish, crustacean, mollusk and other aquatic animals (excluding aquatic mammals and reptiles) of 171 million tonnes in 2016. The highest ever recorded since 1961. This rise in aquatic food production and capture can be attributed to various factors both ecological and economic. There are increased global efforts toward sustainable development of fisheries and aquaculture.

The consumption of fish-based food has also doubled in recent years. This increase in consumption of aquatic sourced food also implies an increase in generation of aquatic waste. These aquatic wastes are a rich source of biopolymers of much environmental and economic significance. With the majority of fish being exported from developing countries as at 2017, further value addition to aquatic resources would significantly boost these economies as non-food products; for example, biopolymers are sourced from the aquatic resources. On the other hand, high-value fish such as lobsters and shrimps are being imported into developing countries as the middle-class demand for higher-quality products as spending capacity increases. The shells for chitin production are usually sourced from post-consumption as the aquatic food such as lobsters is usually served or sold to the consumer with the shell in order to preserve freshness. This implies an increase in the availability of the chitin resource in these regions or at best a wider spread availability in multiple regions.

Up to 100 billion tonnes of chitin can be produced by a species of crustacean annually. The Prince Edward Island located in eastern Canada, for example, is reported to produce around 15 million pounds in weight of shells. This includes around 4 million pounds of lobster shells and 8 million pounds of crab shells. Taken that two-thirds of the mass would be water and based on the reported composition of chitin in crustacean shells, a quarter of this is estimated to be chitin. About 50% would be mostly calcium carbonate, while about 25% would be protein. Additionally, a valuable dye can also be extracted from the shells.

The main sources of chitin are from crabs and shrimp shell. Although research output has shown availability from other sources, these are presently the sources being most explored on a commercial scale. For chitin production for biopolymer to be practical, sustainable and economically viable, there first needs to be a reliable and sustainable source of the aquatic resources which serve as the source of the feedstocks. Here, we review the availability of the main sources of chitin.

Shrimps, lobsters, crabs and gastropods are among the high-value captured seafood species, and the rate of catch has been reported to have shown a record increase from 1998 to 2016. These high-value sea animals are valued between USD 3800 and USD 8800 per tonne. Reduction in shrimp production has been reported in Thailand in recent years due to disease; however, many other countries reported a rise in production of shrimp among other seafood by 2014. A steady catch of 3.5 million tonnes have been recorded for shrimp since 2012 (FAO 2016); this rose to a record high in 2016 with capture of shrimp rising from just over 1 million tonnes in 1970 to 3.4 million tonnes in 2016. Vietnam, for example, exports an estimated USD 7.3 billion worth of aquatic products in 2016 most of which is catfish and shrimp.

The availability of these aquatic resources depends on a number of factors which could be economic or ecological. For instance, increase in per capita income could mean more demand for higher-value seafood, and ecological reasons could be reduction in particular population of sea animal or a particular aquatic disease causing reduction in a particular population. The most abundant species of shrimp, the Argentine red shrimp, has shown the record high in 2016 and an annual increase of 22% since 2011. While a record high of Argentine red shrimp (*Pleoticus muelleri*) of 144,000 tonnes was recorded in 2015 and this species was classified as being fished within biological limit presently, shrimp stock in the southern Indian Ocean is being overexploited. China is presently the highest producer of crustaceans, marine mollusks and aquaculture in general.

The availability status of shrimp varies from region to region; for instance in the Gulf of Mexico, shrimp is maximally fished within sustainable limit while in the Caribbean and Guianas and the Southwest Indian Ocean penaeid shrimps are presently overfished. Deepwater shrimps are maximally sustainably fished, however tending toward $n =$ being overfished. Aquaculture has been on the rise in recent years as a means to meet the demand for fish and make certain species of aquatic food available in regions where they are otherwise scarce and to meet the high cost of deep-sea fishing. The aquatic organisms commonly farmed using aquaculture are salmon, shrimp and bivalves. Aquaculture contributes 7,862,000 tonnes to production of crustaceans globally in 2018. In the same year, 17,139,000 tonnes of mollusks were also produced through aquaculture. Although aquaculture has seen tremendous growth in the past few years, captured species still contribute significantly to availability of crustacean and mollusks. Captured shrimp, crab, lobsters and gastropods recorded for 2016 were 3,400,000, 1,700,000, 315,000 and 170,000 tonnes, respectively (FAO 2016).

Based on the available data, we can generally say that there is sufficient sustainable supply of a variety of crustaceans and mollusks to produce chitin. For example if an estimated two-thirds of the crustacean shells from aquaculture alone is water and 40% of the dry mass comprises of chitin, then an estimate tonne of over 2 million tonnes of chitin annually from crustaceans sourced from aquaculture alone. Other sources report a global production rate of chitin from natural source of up to 100 billion tonnes annually (Elieh-Ali-Komi and Hamblin 2016). However, the present production quantity of 10,000 tonnes reported in 2000 is quite far from the potential production rate. Table 3.1 shows some yields of chitin obtained from different studies.

Table 3.1 Chitin yield reported from different sources

Chitin source	Chitin yield	Extraction method	References
Shrimp shell	13.4%	Ammonium-based ionic liquids	Tolesa et al. (2019)
Sea snail	21.65%	Alkali and acid	Mohan et al. (2019)
Crab shell	6.9% 7.5% 34.4%	Fermentation Alkali and acid Fermentation	Castro et al. (2018) Castro et al. (2018) Flores-Albino et al. (2012)
Nile tilapia	20%	Alkali and acid	Boarin-Alcalde and Graciano-Fonseca (2016)
Beetle (*Holotrichia parallela*)	15%	Alkali and acid	Liu et al. (2012)
Penicillium camemberti	18%	Alkali and acid	Aili et al. (2019)
Mushrooms			
Stipes	7.4%	Alkali and acid	Hassainia et al. (2018)
Pileus	6.4%		
Gills	5.9%		

Other than their use for chitin production, the shells of crustaceans also find use in agriculture for use as soil nutrient and organic fertilizers. While chitosan can act as an adsorbent to remove impurities from water, the shells in crude form just grounded and dried show superior adsorbent property compared to pure chitin. At a concentration of 2.1 mg/mL, the non-treated shrimp shells are capable of removing textile dyes from wastewater (Massimilian and Ludovico 2016). Therefore, the chitin and chitosan manufacturers need to consider the competition from other applications of the aquatic resource.

3.5   Extraction of Chitin

Chitin in nature always exists as a composite or in a complex embedded alongside other compounds such as proteins, glucans, minerals and other compounds within the organism. Therefore, obtaining pure chitin requires treatment stages to isolate it from the other components; some of these could also be useful components for other applications. Extraction processes of chitin from the shells of crustaceans generally involve the same process of demineralization, deproteinization and decolorization followed by further separation and purification. Several concentrations and conditions have been reported from different research groups.

A summary of the processes involved in the extraction of chitin is given in Fig. 3.3. The exact process involved in achieving each stage varies based on biomass used and method being employed.

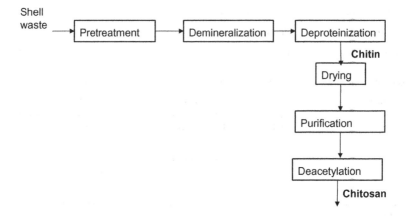

Fig. 3.3 Processes in extraction of chitin

In the following sections, typical example processes are given for extraction of chitin from crustacean shells: mushrooms, insects and fish scales. The difference in extraction processes from one source to another is due to the variation in the structural composition of the shell or other parts of the organisms from which chitin is to be extracted. The yield from a particular source also varies with extraction methods and the efficiency of the process. Chitin could be lost during separation and purification processes.

3.5.1 Extraction from Crustacean Exoskeleton

Crustaceans are characterized by the presence of a hard shell forming an exoskeleton which serves mainly structural roles and some role in calcium regulation in the organism (Ahearn et al. 2004). These hard shells consist of chitin fibers within a protein matrix calcified with calcium carbonate and calcium phosphate (Gadgey and Bahekar 2017). These shells have sites for muscle attachments such that the wastes from their processing may also contain some lipids and other proteins and compounds such as carotenoids.

Extraction of chitin from exoskeletons of crustaceans generally involves three main stages. One stage is removal of the protein in a process referred to as the deproteinization, and this is achieved by immersing the shells in alkali solution. Sodium hydroxide is most commonly used for this. The next stage is then to remove the calcium content which is mainly acid-soluble calcium carbonate. This is referred to as the demineralization step. Since calcium carbonate and most minerals are soluble in acidic solutions, the demineralization process is carried out by immersing the deproteinized residue in acid solution. While hydrochloric acid is most widely used, other acids such as nitric acid (HNO_3), sulfuric acid (H_2SO_4), acetic acid (CH_3COOH) and oxalic acid (HCOOH) can also be used for demineralization. The type of acid used

will determine the extraction time and required condition. The order of demineralization and deproteinization can also be reversed according to the dominant component
in the shells. If the protein content is predominant, then the deproteinization can be
carried out before the demineralization; however if the mineral content is higher,
then the demineralization process can proceed the deproteinization for improved
efficiency of chitin extraction (Kalut 2008). The efficiency of the extraction process
is also affected by the level of contacting between the solid mass and the chemicals
being used for extraction. This can be improved by particle size reduction through
pulverization.

The quality of chitin is significantly affected by the type of acid used, temperature
and the pH. Isolation of the chitin from the dissolved minerals and proteins is achieved
by physical separation by either filtering or centrifugation. It is therefore important
that the chitin does not dissolve in the alkali or acid solution. Chitin is insoluble;
however, it becomes soluble when deacetylated into chitosan such that some of the
chitin may be lost during extraction, if not all. High temperature and high pH may
also result in the depolymerization and hydrolysis of chitin which diminishes the
physical properties, and this is undesirable. Care must therefore be taken to use the
right operating parameters that retain the chemical structure and integrity of the chitin
(Gadgey and Bahekar 2017).

3.5.2 Extraction from Mushrooms

Mushrooms are a species of fungus which form fleshy fruity bodies. They are used in
food and for medicinal applications. They grow in soil, on standing or fallen trees or in
their food source and comprise a stem and a cap with gills. Although not exclusively
aquatic in nature, mushrooms grow naturally near moisture-rich areas such as rain
forests (Ficket et al. 2017). Chitin is present in the cell wall of mushrooms where it
is embedded within the alkali-soluble beta-1,3 glucans. It plays a structural role in
cell wall of fungi.

The morphology of mushrooms varies from that of fish scales and crustacean
shells due to the fact that mushrooms contain glucans alongside proteins and chitin,
thus requiring a slightly different extraction method. The process begins with the
fresh mushrooms being crushed to reduce the particle sizes. This can be done in a
domestic blender or food processor. This is then followed by filtering and washing
in distilled water. The process of washing in water is the first extraction step which
removes the water-soluble parts of the mushroom cell wall, and these are the glucans
and the minerals. This also shows how some of the nutrients in the food are lost
during washing, although in this case it is desirable. The next step is treating with
2% w/v sodium hydroxide for 24 h at a temperature of 100 °C. This deproteinization
stage is similar to that used for crustaceans and fish scales. It is used here not just to
remove proteins but also to remove alkali-soluble glucans. The residue left behind
after separation is then repeatedly washed with distilled water until it is neutral.
What now remains after removal of the glucans and proteins is the residual minerals.

These are removed by treatment with 2M hydrochloric acid for 48 h. The sample which by now consists mainly of chitin is then further treated to remove the pigments followed by further treatment with sodium hydroxide to remove residual proteins and alkali-soluble glucans. Further processing could then follow to obtain chitin fibers or powder.

This method of extraction of chitin from mushrooms has proven effective in five different species of edible mushrooms (Ifuku et al. 2011). These include the common mushroom (*Agaricus bisporus*), king trumpet mushroom (*Pleurotus eryngii*), maitake (*Grifola frondosa*), shiitake (*Lentinula edodes*) and buna-shimeji mushroom (*Hypsizygus marmoreus*). This method yields between 1.3 and 3.5% chitin depending on the variety of mushrooms. Chitin production from edible mushrooms faces the challenge of competing with food, especially in vegan diets where mushrooms are used as a substitute for meat as a source of nutrients and desired meaty texture.

3.5.3   Extraction from Fish Scales

Relatively, fewer studies have gone into extraction of chitin from fish scales than from crustacean shells. This is largely due to the lower yield of chitin in fish scales compared to crustaceans and mollusks. Nonetheless, the extraction of chitin from fish scales is well worth further exploring as the demand for more sources of chitin-based biopolymers rises. Furthermore unlike crustaceans such as crabs and lobsters, fish scales are always removed before the fish is served as food. Therefore, sourcing of fish scales for chitin production requires more sustainable logistics. Increasing worldwide farming of tilapia, a scaly fish would also mean increasing demand for ways to utilize the fish scales generated as waste.

The process of chitin extraction from fish scales begins with separating the scales from the fish. This is usually done at the point of sale at the markets or prior to processing in fish factories or cooking at restaurants. The fish scales are then collected by the processor and washed with water at room temperature to remove fins, skins and other residues.

The main extraction process then commences at demineralization where the minerals present within the structure of the fish scale are dissolved in a 0.5M acid solution, usually hydrochloric acid. The ratio of dried scales to acid solution is generally between 1:10 and 1:20 ratio of grams of dry fish scales to volume of acid solution in ml. Continuous stirring is also applied to ensure even distribution as the demineralization process progresses. A time period of 90 min to 2 h is allowed for this process. This is then followed by washing to remove the acid filter and drying.

The next stage is then to remove the proteins, deproteinization. This is achieved by treating the demineralized residue with 1% solution of sodium hydroxide at 50 °C for 3 h under low stirring (~250 rpm). What remains at this point is mostly chitin with some pigments and impurities (Boarin-Alcalde and Graciano-Fonseca 2016). Further treatment is then carried out to remove the pigments and odor causing impurities. This can be achieved with the addition of sodium hypochlorite or ethanol.

3.5.4 *Extraction of Chitin from Insects*

Although insects are not exclusively aquatic organisms, they can be found in aquatic areas either surrounding stagnant water or releasing their lava on the water. Some aquatic insects such as the dragonfly start life as aquatic larvae form for many years before they move out of the water and fly. These tend to spend most of their life in water, and their life as adults is relatively short (Berg 2009). Making use of the cuticle of insects for chitin production is a potential solution to sustainable pest control or simply making use of an easily cultivated abundant source of chitin.

Chitin extraction from insect is a relatively recent development. A few insect species have been found to contain chitin; these include desert short-horned grasshoppers, green bugs, German cockroach, vespid wasp and yellow jacket wasp (Badawy and Mohamed 2015). To extract chitin from insects, the first step is to kill the insects by freezing or dead insects could be collected. This is then followed by deproteinization with potassium hydroxide at 40 °C for 48 h. The deproteinized mass is then washed repeatedly with distilled water until neutral. The next stage is then to remove the mineral content using acid for this, and 5% acetic acid is used. The use of acetic acid is favorable as it allows use of acid from a sustainable and non-fossil-based source as acetic acid can be produced through fermentation. Dehydration is then carried out with ethanol by series.

Other studies have extracted chitin from bumble bee (Majtan et al. 2007) and beetles (Liu et al. 2012) using the conventional sodium hydroxide deproteinization and hydrochloric acid demineralization. The structure of chitin in terms of demineralization, crystallinity and molecular weight varies for different species of insects. Some studies nonetheless report similarities between the chitins extracted from insects to the commercial chitin (Liu et al. 2012). There is yet to be a defined taxonomic classification of chitin from various species; however, it is concluded in the different studies that the chitin obtained from insect has similar characteristics to those from aquatic sources.

3.5.5 *Microbial Extraction*

Certain microbes are able to metabolize the proteinous parts of the chitin sources, while the acidic condition generated by the acids created by the microbes dissolves the minerals leaving behind a solid crude chitin which can then be further processed and purified to obtain chitin. The obtained chitin can then be further deacetylated to obtain chitosan. The fermentation process achieves partial deproteinization and demineralization. This is then followed by further chemical treatment to obtain purer chitin. Although some acids and alkali are still used in this process, it is at a much lower concentration since most of the demineralization and deproteinization are achieved during the fermentation process. Up to 99.6% and 95.3% of deproteinization and demineralization have been achieved, respectively, from lactic acid fermentation of crab shells to obtain chitin for example (Castro et al. 2018). The fermentation process therefore reduces the environmental impact of chemical chitin extraction.

The process could take up to 80 h as reported by different research studies (Cira et al. 2002; Castro et al. 2018). Another advantage of the fermentation process is the ability to recover the proteins for use as animal feed. Unlike the chemical process where the deproteinization is done with a high concentration of alkali, the protein extracted cannot be recovered and is usually discarded hence generating more waste.

Lactic acid bacteria (*Lactobacillus* sp.) fermentation of crustacean waste, for example, involves the inclusion of a carbon source and in some cases mild acid such as acetic acid to provide the right pH for the growth of the bacteria in the lag phase. The lactic acid bacteria ferment the carbon source which could be whey, sugarcane or other as well as the carbon present on the biowaste. Lactic acid is released as a by-product of the fermentation process. The lactic acid reacts with the calcium carbonate producing calcium lactate which precipitates and can then be separated by washing. Proteolytic enzymes are produced from either the gut bacteria present on the shrimp or strains added to the fermenter or the biowaste. These break down the protein, leaving behind crude chitin. The low pH caused by the presence of lactic acid and other by-products of the fermentation such as acetone also prevent the growth of spoilage bacteria.

Lactococcus lactis, *Terendinobacter turnarae*, *Lactobacillus plantarum*, *Lactobacillus pentosus* and *Lactobacillus salivarius* are some examples of bacteria that have been used in lactic acid fermentation of shell waste to produce chitin. Other than the limited rate of demineralization and deproteinization achieved in chitin produced by fermentation, another concern is the microbial contamination of the chitin produced (Gortari and Hours 2013). Any possible contamination makes this form of chitin not suitable for human or animal consumption. For this reason, despite environmental and economic advantage posed by microbial extraction of chitin, many of the applications of this method of chitin extraction have been limited to research and laboratory experiments.

Other strains of organisms have been explored for production of chitin by fermentation. Filamentous fungi have also been used in the biological production of chitin (Gortari and Hours 2013). The proteolytic enzymes are released by the fungi which results in the deproteinization and demineralization of the shrimp shells. The consequent release of amino acids in the process of deproteinization acts as the nitrogen source which the fungi require for growth and multiplication. This results in the lowering of the pH of the system, thus aiding the demineralization of the shells.

3.5.6 Enzyme Extraction

Enzyme extraction involves the deproteinization of the shell waste through the action of proteolytic enzymes. This has the advantage of not including microbes, thereby minimizing the risk of microbial contamination of the product. However, the proteolytic enzymes only act to hydrolyze the protein; therefore, a pretreatment stage is required to chemically demineralize the shell waste to get rid of the calcium carbonate.

These enzymes are commercially available and can be isolated from a variety of organisms. Examples of such enzymes include Alcalase, Pancreatin, Delvolase, Cytolase, Econase, Maxazime and Cellupulin. Enzymatic deproteinization can achieve between 54 and 97% protein removal from shell waste and minimizes the need for alkali deproteinization (Gortari and Hours 2013).

While enzyme and lactic acid fermentation-based extractions of chitin from shell waste show potential for more environmental and economical alternative to chemical-based extraction using acid and alkali, more research needs to be carried out toward a safe, eco-friendly and reproducible method for commercial extraction of chitin of high quality.

3.6 Environmental Implications

In order to assess the environmental impact of chitin and chitosan production, here we make use of the life cycle assessment report by Munoz et al. (2018). The life cycle assessment was carried out using data from two chitosan production companies: one of which is Mahtani Chitosan and the other an anonymous European company herein referred to as Company X. Table 3.2 summarizes typical consumption in the process of extraction of chitin from shrimp shells.

Mahtani Chitosan produces general-purpose chitosan, while Company X produces chitosan for medical use. They both use the chemical extraction method; however, a number of differences in their operations exist. (1) Mahtani Chitosan is close to the source of the starting material (shrimp shell) while Company X outsources its chitin production to China where it ships the dried crab shells from Canada to china. The chitin produced is then shipped to Europe where the chitosan production takes place.

Table 3.2 Typical consumptions in chitin extraction from shrimp shells

Consumption	Quantity
Water consumption	167 L/kg chitin
Sodium hydroxide	1.3 kg/kg chitin 5.18 kg/kg chitosan
Energy consumption (transportation and electricity)	1.4 L diesel per tonne of shrimp 0.02 L per kg of chitin 1.3 kWh per kg chitin
CO_2 emission	0.7 kg/kg chitin
Shell waste/resource utilization	33 kg shrimp shell/kg chitin
Solid waste generated	1.5 kg calcium salts/kg chitin 4 kg protein/kg chitin

(2) Mahtani Chitosan uses shrimp shell, while Company X uses crab shells as feed material. (3) Mahtani Chitosan uses its protein sludge by-product as fertilizer, while Company X uses theirs as animal feed.

However, it should be considered that the information presented here is based on data from the company's operation and therefore may vary for other companies as such is by no means exhaustive. The goal here is to give an overview of the resource requirements and environmental impact of the chitin production process.

3.6.1 Resource Utilization

The amount of raw material required depends on the type of feedstock being used and the efficiency of the process. Crab and shrimp are most commonly used for commercial chitin production due to their high chitin content. Table 3.1 shows chitin yield/content for different sources of chitin based on reports from the literature. Company X reports using 10 kg dry crab shell per kg of chitin produced, while Mahtani Chitosan reportedly uses 33 kg of shrimp shells per kg of chitin produced.

Use of the waste aquatic materials for chitin and chitosan production diverts the use of these materials as animal feed and as organic fertilizers, the main alternative uses of these materials. The direct environmental impact therefore lies in the pressure; this diversion puts on other resources used for animal feed and fertilizers.

3.6.2 Water Consumption

A lot of water is used in the process of washing the shells, washing after demineralization and deproteinization to get rid of the acids and alkali used to return the solids to neutral. The chemical process generates more wastewater for this reason. The water is either treated and recycled in-house or sent to external water treatment.

An alternative to the use of shell waste in the production of chitosan is to use the shell waste as animal feed. Comparing the quantity of water used in chitosan production to that use in preparation for crab shells as animal feed, less water is actually consumed in the production of chitosan compared to using the crab shells for animal feed. Further consideration must be made to the fact that the diversion of the shell waste from animal feed to chitin production means increased demand for other sources of animal feed such as soybeans and barley.

Mahtani Chitosan reportedly uses 167 L of freshwater per kg of chitin it produces from shrimp shell. A further 250 L of water is then used for each kg of chitosan produced per 1.4 kg of chitin. Company X in Europe reportedly uses 300 L of water per Kg of chitin produced from crab shells. The data for water consumed during chitosan production from chitin by Company X was not provided for confidentiality reasons. At Mahtani Chitosan, the wastewater from the production process is treated on site. This involves neutralization, settling, biological treatment followed by sand filtration before it is then released into the sea.

3.6.3 Solid Waste/By-Products

Protein extracted is recycled and reused as fertilizers or animal feed. The calcium salts produced are used either deposited in landfill sites or spread on the roads as road-filling material. The manner in which the by-products are processed, discarded or reused depends on the regulations in the region of operation and company choice. This will therefore vary from company to company. For every kg of chitin produced, 1.5 kg of calcium salt and 4 kg of protein are produced. The European company is estimated to produce 2.84 kg protein per kg of chitin produced. These solid wastes generated from the extraction of chitin find application as fertilizers and animal feed in the case of the protein and as road fillers in the case of the calcium salt. Shrimp shells are also often used as animal feed, and the diversion of shrimp shells for chitin production which in turn results in additional pressure on other animal feed sources such as barley and soybeans is partly compensated for by the use of the protein from the extraction as animal feed. Similarly, crab shells are often sent to composting sites where they are used as organic fertilizers. The use of the protein sludge as fertilizers also partly compensates for the crab shells diverted from agricultural use to chitin production.

3.6.4 Sodium Hydroxide

From the chemical-based extraction processes discussed, it is obvious that there is a relatively large amount of acids and alkalis used from deproteinization to demineralization and even much more sodium hydroxide (up to 50%) is required for the deacetylation of chitin to its more industrially relevant form, chitosan. This has a lot of environmental impact, and the cost of the chemicals and the cost of disposing them need to also be considered.

A life cycle assessment of sodium hydroxide which evaluated the environmental impact of sodium hydroxide from the production to its use and disposal shows that for every kg of sodium hydroxide used, 3.5 MJ of fossil energy is consumed, 0.6329 kg of carbon dioxide is emitted which contributes to global warming, 1.298 g of 1,4 dichlorobenzene equivalents are released into the aquatic environment, 0.706 of sulfur dioxide equivalents are released into the atmosphere which further contributes to acid rain and 0.4927 g of carcinogens are released to human exposure (Thannimalay et al. 2013).

Mahtani Chitosan reports 1.3 kg sodium hydroxide used per kg of chitin and a further 5.18 kg used per kg of chitosan produced from the chitin. Company X reports

8 kg of 4% vol of sodium hydroxide used per kg of chitin produced; however, no data was provided for the amount used for chitosan production.

Use of proteolytic enzymes in the deproteinization process is an alternative; however, much of this is still in the research stage and is yet to be commercialized. The use of microbial extraction could potentially eliminate the need for sodium hydroxide in the production of chitin. However, the higher concentration is used in the deacetylation process for production of chitosan. An alternative or accompanying process could be enzymatic deacetylation using the enzyme deacetylase. This is presently under research, and the mechanism of action and specificity of this enzyme is still under study. Biotechnology-based deacetylation of chitosan would be a great leap in a truly green process for chitosan production.

3.6.5   Acid Usage and Disposal

In the chemical process, the acid is mainly used in the demineralization stage. According to data produced by Mahtani Chitosan company (Munoz et al. 2018), to produce 1 kg of chitin 8 kg of hydrochloric acid at 32% concentration is used. In the lactic acid fermentation process, lactic acid is produced by the bacteria and poses less environment adverse effect. Since the acid unlike the alkali actually reacts with the calcium carbonate, it is not recovered, and it is converted into calcium chloride, water and carbon dioxide, the products of the reaction between hydrochloric acid and calcium carbonate. The impact of the hydrochloric acid used is therefore the carbon dioxide emission resulting from the process.

3.6.6 CO_2 Emission

In the demineralization stage, carbon dioxide is emitted as the calcium carbonate reacts with hydrochloric acid to form calcium chloride water and carbon dioxide as shown in Eq. (3.1).

$$2HCl + CaCO_3 \rightarrow CaCl_2 + H_2O + CO_2 \qquad (3.1)$$

CO_2 emitted from organic degradation of the shrimp and crab shell waste is rated as zero, while CO_2 emission from clearing of land for building of chitosan factory and the increased pressure on clearing of land for growing of plant-based animal feed such as soybean and barley is rated as GWP-100 (global warming potential) of 1. This is because with organic degradation the carbon being emitted is that which was removed recently during formation to the shells, while that from clearing of long-standing plants at a much larger scale had more impact on CO_2 emission.

Approximately, 0.7 kg of CO_2 is released per 1 kg of chitin produced during the acid mineralization process. This means the production process is resulting in CO_2

emission of 70% of its own weight. Further CO_2 emission occurs as a result of the mineralization of the protein when it is used as fertilizer. Other emissions from this process include ammonia, dinitrogen monoxide and nitrogen oxides.

For chitin production from crab shells, the CO_2 emission during the acid demineralization was estimated at 0.9 kg/kg of chitin produced. For both sources assessed, crabs and shrimp, the chitin production process had more impact on acidification of the environment and climate change than any other process in the supply chain from transportation to chitosan production. However in terms of water use, the shrimp shell supply had more impact on water use than chitin or chitosan production while in crab shell supply chain, chitin production had more impact on water use. Likewise in crab shell production, the chitosan production led to more toxins being released into the environment while in shrimp shell production chitin production and chitosan production had similar level of impact on Fr. Ecotox.

3.6.7 Energy and Electricity

The energy consumption starts from the fuel used in the transportation of raw materials to the factory. For Mahtani Chitosan, a reported 1.4 L of diesel is consumed per tonne of shrimp transported using a tractor. A further 0.02 L of diesel is used up per kg of chitin produced in bulldozer operations. The process is also reported to consume 1.3 KWh of electricity per kg of chitin produced. Conversion of chitin to chitosan through deacetylation consumes a further 1.06 KWh of electricity and 31 MJ of burning wood as fuel per kg of chitosan produced from shrimp shell.

For chitin production from crab waste, Company x reports an electricity consumption of 1.2 KWh and 6 kg of coal fuel for heating per kg of chitin produced. The chitin is then transported to Europe from China, an estimated distance of 22,874 km by sea, adding to the energy consumption.

In developing a greener supply chain for chitin and chitosan production, transportation systems such as bicycles could be considered to transport the aquatic waste serving as the raw materials from the point of generation to the factory. Such is being adopted in some recycle models in countries like Nigeria and India where bicycles, rickshaws and pushcarts are used to transport used plastics from homes and businesses where they are generated to recycle factories.

Therefore although chitin is a biopolymer, the process of extraction may not necessarily be biologically friendly. For this reason, researchers have explored green chemistry for production processes which require the use of less resources and less harmful chemicals in the production of chitin, chitosan and the chemicals used for extraction. Other alternatives for extraction includes the use of enzymes and fermentation by microbes. Although these have been explored by researchers, the chemical process is still preferred in commercial production due to the high cost of the biological extraction methods.

Chitin in its crude form without isolating it from the minerals and collagen has also shown potential in applications such as agriculture and pharmaceuticals. Other than the chemicals used in chitin production, the process of removing chitin from the environment should also be considered. Chitin deposited by molting arthropods, dead diatoms, insects and mushrooms is otherwise used up by microorganisms including bacteria and fungi for metabolism. Removal of these chitinous materials makes them unavailable to these microorganisms within the aquatic environment. This is likely to impact on the biodiversity of microbes and other organisms within the environment, and this would have an effect on the ecosystem. The ecosystem involves every organism playing a role in the carbon and nitrogen cycle as well as the food chain. The reduction or increase in one population will affect that of another and the ecosystem as a whole. We must therefore weigh the benefits of manufacturing chitin and chitin-based products against the environmental cost of removing them from their originating environment.

From the environmental impacts outlined thus far, it becomes apparent that the energy and resources consumed and emissions from chitin and chitosan production vary significantly from company to company. Other than the source of raw material, other factors such as transportation mode, distance, choice of fuel and method of extraction have an effect on the impact the production of chitin will have on the environment. Here, we have compared two processes from two sources (shrimp and crab), from two different countries.

3.7 Applications

The demand for biodegradable polymers lies chiefly in their demand for using alternatives to non-biodegradable, petroleum/fossil fuel-based plastics. Despite the rise in the production of natural derived biodegradable plastics, presently of the 300 million tonnes of plastic being produced annually, only about 1% are from natural source (Brostow and Datashvili 2016).

Presently at 80.413 billion USD and predicted to rise to 105,549 billion USD by 2027, the global demand for chitin-based product is on the rise. The largest demand for chitin derivatives is for chitosan. Glucosamine gained the majority of the market share for chitin derivatives, while chitosan came second. Up to 64.9% of chitin was converted to glucosamine (Future Market Insights 2019).

The global chitin market is as of 2017 is worth 900 million USD with Asia Pacific being the largest producing nations (excluding Japan), North America the second largest producers and Japan being the third largest. There is also chitin production in Europe, Latin America, Middle East and Africa, albeit to a lesser extent. The chitin market is projected to be worth 2.9 billion USD by the year 2027 (Future Market Insights 2019). Most of the chitin produced is converted to chitosan and glucosamine and other derivatives. The demand for chitin and chitin derivatives lies in health care, wastewater treatment and agrochemicals with less although significant value in the food and beverages, cosmetics and toiletries and other applications (Future Market Insights 2019).

Chitin is mainly useful for conversion into other products, namely chitosan, glucosamine and oligosaccharides. Some research studies have been reported in its use as nanofillers; however, its application is largely based on the deacetylated and/or depolymerized forms. Figure 3.3 summarizes the process from chitin to its different derivatives.

With a world population projection of 9.7 billion by the year 2050 and presently over 7 billion (FAO 2016) and rising, every one of these billions of people demanding clean water, processed foods, pharmaceutical products and other daily requirement of modern-day living standards, there is a huge demand waiting for the chitin industry to meet as an alternative to the non-biodegradable, fossil fuel-based polymers which are presently being employed to meet such demands. This is in line with the present challenges of rising cost of fossil fuels from which these plastics are based, the issue of pollution disaster as a result of dumping of plastics in aquatic and terrestrial habitats and the financial distress being experienced globally. There is an urgent need for an alternative versatile polymer which would replace the present ones.

In the present state of technology, the chitin-based products of, for example, food packaging do not meet the same requirements in terms of applicability compared to the conventional counterparts. This has limited the growth of such products. It is therefore important to educate the consumers on the true value in biodegradable products compared to the non-biodegradable and fossil-based options.

Presently, the highest demand for chitin lies in the wastewater treatment. Generally due to the batch-to-batch variation in the quality and properties of chitin, large-scale production is more suitable to low-end applications such as wastewater treatment. For higher-end applications such as pharmaceutical or biomedical applications, each batch needs to be tested; therefore, these are more difficult to scale up.

Chitin is the leading material toward a sustainable industrialized world. As the world's attention is drawn to the environmental concerns raised by petroleum-derived polymers, there is rising demand for non-toxic, biodegradable, biocompatible, affordable, accessible and multifunctional alternatives. Chitin is the first polymer of choice with potential to meet that demand. Figure 3.4 summarizes the value chain of chitin from the raw materials to the final product.

3.7.1 Packaging Films

So far, chitosan has proven to be the natural biopolymer with the best film-forming property. On its own, chitosan forms a uniform film when cast. The figure below shows a film of chitosan. The chitosan used here is chitosan derived from chitin extracted from crab shells and dissolved in 1% acetic acid solution. Pure chitosan dissolved in acetic acid or hydrochloric acid can be formed into films for applications such as edible food packaging. Aside from forming a barrier layer, it serves as an antimicrobial and antioxidant and helps preserve the quality of the food product. Chitosan is applied in the packaging of meat, fish, vegetables and fish and has been shown to preserve the quality of these foods (Kanatt et al. 2013).

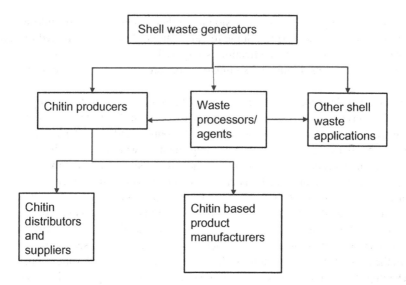

Fig. 3.4 Illustration of the value chain of chitin production

Combining chitosan with other polymers significantly improves its applicability as a packaging material. A recent example of this is the clear plastic film of chitosan combined with other biodegradable polymers, cellulose and polylactic acid (PLA). A synergistic effect can be achieved between two polymers, in this case chitosan (a main chitin derivative) and cellulose (the most abundant natural polymer on earth). A multilayer film made of chitosan nanofibers and cellulose nanofibers coated on polylactic acid achieved a film with up to 73% lower oxygen permeability than when the film is made with PLA only. This is also a significant improvement to films made up of chitosan or cellulose alone (Satam et al. 2018).

While PLA is a naturally derived polymer sourced from lactic acid obtained from milk or bacteria, it is non-toxic, biodegradable and biocompatible; however on its own, it has poor film-forming properties. On the other hand, chitosan forms excellent films; however, it has much higher oxygen and water vapor permeability, and these are not desirable for applications such as food packaging where the food or drug or electrical component needs to be dry and airtight at least over the duration of storage. Cellulose, although hydrophobic, has poor film-forming properties but forms anionic nanofibers with good mechanical strength. By combining the cationic chitosan nanofibers and anionic cellulose nanofibers and spraying layer by layer unto polylactic acid films, a composite film with low oxygen permeability and moderate mechanical property is formed. Alternating three layers of chitosan–cellulose–chitosan nanofiber layers on a PLA layer produced films with the least oxygen permeability and haze resulting in transparent packaging films with excellent physicochemical properties that rival those of petroleum-sourced transparent films. Other methods of creating chitosan composite films include solvent casting and spin coating.

Transparent films are in high demand in the packaging industry. From edible food packaging to transparent films used as screen protectors, there is huge demand for such films. Presently, petroleum-based polymers such as polyethylene offer low cost, easily mass produced and conveniently sourced option. For chitin to compete with this, there needs to be a well-established low-cost production process since chitin already meets the other application demands. Off all the biopolymers available, chitosan and other chitin derivatives demonstrate the most desirable properties suitable for production of transparent packaging films.

3.7.2 Water Treatment

It can be said that one of the best explored applications of chitosan so far is in water treatment. Industrially produced chitosan is used in wastewater treatment for removal of metals and dyes. It also acts as a flocculant and coagulant and in immobilization of microorganisms. This multiple role makes it very useful in wastewater treatment. Chitosan has the ability to remove heavy metals, lipids and overall decrease in turbidity of water. Applicability of chitosan in water treatment can be attributed to its antimicrobial properties, metal chelation and adsorption properties. The antimicrobial property is further discussed in another subsection. The absorption process is widely used in water treatment and is well explored. Chitosan is a commonly used adsorbent alongside others such as cellulose and guarana.

The amine group on the deacetylated chitin plays the role of binding to metals, and this results in the formation of a chitosan–metal complex with multiple metal molecules binding to one chitosan molecule (Shahidi et al. 1999). The amine group also has high capability to bind with minerals, lipids and proteins which gives chitosan its versatility compared to chitin. Since this metal-chelating property is dependent on the availability of the amine group on the polymer for bonding, the higher the degree of deacetylation, the better the chitosan polymer's metal-chelating property.

It is important that a water treatment agent is low cost, non-toxic and biodegradable and can be used at low quantities. Chitosan at a concentration of 0.2 g in 100 ml of water can reduce total dissolved solid content by up to 12.87%. Chitosan also achieved significant reduction in turbidity and conductivity and overall makes water suitable for drinking at relatively low concentration required (Al-Manhel et al. 2018).

3.7.3 Antimicrobial

A full understanding of the mechanism by which chitosan inhibits bacterial cell is still pending. It is believed that it does this by means of blocking the surface of the bacterial cell by chemically binding to the lipids present on the surface. This prevents the bacteria from taking up nutrients necessary for growth and survival and eventually leads to cell death or inhibition. This is made possible by the reactive binding of the deacetylated part of the chitosan polymer chain where the amino ($-NH_2$) group which has been freed of the acetyl group can react with the negatively charged surface of the bacteria cell. Another mechanism proposed is that the chelating property of chitosan actually results in its antimicrobial properties. It is thought that the chitosan binds with the trace metals on the bacteria cell and results in the production of toxins which leads to cell death. The antimicrobial activity of chitosan could be further improved by nanosizing. Nanoparticles of chitosan result in higher surface area of the polymer available to bind with the microbes, hence increased effectiveness (Divya et al. 2017). The antimicrobial activity of chitosan is dependent on the molecular weight of the chitosan. At a sufficiently low molecular weight, chitosan is able to penetrate into the bacterial cell wall and disrupt DNA and RNA replication which results in their inhibition (Varum et al. 2017).

Resistance of microorganism is of huge clinical and industrial significance as microorganisms have shown resistance to almost every antimicrobial agent that exists. Chitosan has recently been of particular interest as there is yet to be reported bacterial resistance to it. Great potential therefore lies in the use of chitosan in various antimicrobial products. Antimicrobial activity of chitosan extends to a broad range of microneedles which includes bacteria, filamentous fungi, yeast and even virus: bacteria such as *Escherichia coli* (*E. coli*), salmonella and staphylococcus, fungi such as *Aspergillus niger*, *Fusarium solani* and *Candida albicans* and viruses such as H1N1 Influenza A and the human cytomegalovirus (9HMCV) strain AD169 (Divya et al. 2017). Chitosan acts against both gram-positive and gram-negative bacteria.

For these antimicrobial properties, chitosan has found application in, for example, antimicrobial food packaging, anti-acne cosmetic formulations, water treatment and antimicrobial film in wound healing. Because chitosan has such diverse characteristics, in a single application more than two or more of these properties can be implemented; for example, in its use as antimicrobial film, the excellent film-forming property is combined with its antimicrobial property.

3.7.4 Biomedical Application

Owing to its biocompatible, non-toxic, film-forming and hydrophilic nature, there are numerous biomedical applications of chitosan currently at different stages of development from basic research to clinical trials. Biomedical applications of chitin and chitosan include tissue repair, wound healing, scaffold production and biomedical implants.

One of the novel applications of chitosan in the photochemical sutureless tissue bonding. Post-surgery, it is required to close the incisions which have been made in tissues such as skin, cornea or peripheral nerves. Manual suturing requires good dexterity and can be quite time consuming with side effects. To address this, photochemical tissue bonding has been introduced. These involve the joining of tissue through activation of chemical cross-linking of collagen fibers using rose bengal as the cross-linker. An improvement to this process is to combine rose bengal, chitosan adhesive and laser at a wavelength of 532 nm (Lauto et al. 2010; Frost et al. 2016).

In the 1988 publication by Lantos, the various applications of plastics in the medical field were reviewed. At a time when there was decreasing price of plastics, it was appealing to evaluate the use of plastics in the medical field and ways of improving the performance in already existing applications as well as expanding the areas of applications. Today 30 years later, plastics have much wider applications in the medical field: from disposable syringe to pacemaker coatings, to waterproof aprons, to sutures to blood bags and protective eyewear. The key issues identified at the time were improved blood compatibility, radiation resistance and improved degradability.

3.7.5 Gene Therapy

In gene therapy, delivery of nucleic acid (DNA/RNA) into cells is often done via viral infection, where the gene to be delivered is planted into a virus and the cell is infected with the virus. Another alternative is to use transpection using polyplexes. Chitosan has been widely investigated as a polyplex in DNA and RNA transfection. The polycationic nature of chitosan, its non-toxic and biocompatibility as well as being of renewable natural source make it a suitable alternative to other synthetic more toxic polymers used as polyplexes such as polyethylenimine and polyamidoamine dendrimers. Chitosan can form non-toxic complexes with DNA and RNA for use in gene transfection. Particularly, chitosan oligomers have been shown to have better physical properties in terms of solution viscosity and stability at physiological pH than high-purity high molecular weight chitosan. Chitosan oligomers with molecular weight between 18 and 20 demonstrate good non-viral gene delivery properties (Koping-Hoggard et al. 2004).

More recently, chitosan/hyaluronic acid nanoparticles have been shown to be effective in the delivery of mRNA to cell. This is promising in the area of cancer treatment through expanding understanding of nucleic acid uptake and metabolism of cancer cells with the potential to send mRNA to target cancer cells to stop their replication and inhibit growth of the tumor. The chitosan/hyaluronic acid nanoparticles were more effective under acidic conditions (Lallana et al. 2017).

3.7.6 Anticancer Application

Cancer treatment is a highly significant area of interest, and chitin finds application even in this field. Different studies have reported antitumor effect of chitin and chitosan as well as their derivatives on a broad range of cancers including melanoma, carcinoma, colon cancer, lung cancer, sarcoma and prostate cancer (Gibot et al. 2015; You-Jin and Kim 2002). The mechanism by which chitin and its derivatives act against cancer has been widely studied and is still presently undergoing studies. Mechanisms reported include increasing the activities of enzymes which inhibit cancer cells or kill them, boosting the body's own immune response to cancerous cells, inhibiting of certain enzymes required for cancer cell growth or altering the pathways for cancer cell proliferation such as interfering with DNA replication. Chitin and chitosan could also act against cancer by accumulating within the cancer cells and partaking in reactions which result in production of toxins which eventually lead to the cancer cell death. The activity of chitin, chitosan and their derivatives vary from preventing cancer cells from forming, inhibiting the growth of existing cancer cells or preventing the metastasis (spreading) of the cancer to other parts of the body thereby making treatment more manageable.

The effectiveness of chitin and chitosan in their various applications has been shown to be significantly affected by factors such as molecular weight and the degree of deacetylation. In fact, one study showed that oligomers of chitosan, which are short-chain chitosan of molecular weight between 1.5 and 5.5 kDa, had more effective antitumor activity against sarcoma than higher molecular weight chitosan and low molecular weight chitosan indicating a specific range of molecular weight required (You-Jin and Kim 2002). Therefore, processing techniques such as gel permeation chromatography to separate into different molecular weights, achieving a higher degree of deacetylation during production and higher purity using methods such as ultrafiltration, becomes important for optimum effectiveness particularly when considering the high-end application such as antitumor and tissue repair application of these biopolymers. Such processing is what adds additional processing costs as the value of the chitin increases from crude chitin to chitosan with low degree of deacetylation and polydisperse to more monodisperse chitosan with high degree of polymerization. The quality criteria for chitin and chitosan are mainly molecular weight, degree of deacetylation, purity and polydispersity. The values for each of these will depend on the intended application.

3.7.7 Anti-inflammatory

Chitosan has been shown to have some anti-inflammatory properties. Inflammations are generally caused by inflammatory cytokines released by the immune cells such as macrophages. In certain inflammatory diseases, things get out of hand, there is excessive production of inflammatory cytokines and these result in pain and discomfort.

Other than the antimicrobial property which also induces anti-inflammatory effect by destroying the inflammatory disease-causing bacteria, chitosan has been shown to have direct anti-inflammatory effect. The mechanism by which this happens is by inhibiting inflammatory cytokine production in the presence of inflammation-inducing microbes. For example in the presence of chitosan-alginate nanoparticles, the inflammatory activity of the acne causing bacteria *P. acnes* which induces the production of inflammatory cytokines by the immune cells was inhibited (Friedman et al.

2013). Acne therapy takes the approach of eradicating the bacteria which results in the inflammation seen as acne spots and also preventing the formation of inflammatory cytokines by the immune cells in the skin which results in the formation of acne on the skin surface. Chitosan with its antimicrobial and anti-inflammatory property therefore is an attractive drug candidate in such therapy. Similar chitosan has demonstrated anti-inflammatory effect on other inflammatory diseases such as inflammatory bowel disease (Friedman et al. 2013).

3.7.8 Antioxidant

The antioxidant activity of the deacetylated form of chitin, chitosan, is attributed to the presence of hydroxyl groups and amino groups which are available to react with free radicals. This antioxidant property of chitosan has been demonstrated. Chitosan extracted from crab shells achieves up to 70% antioxidant activity at a concentration of 1 mg/ml (Yen et al. 2009).

For applicability, compounds used in pharmaceutical or food need to be water soluble. Chitosan solubility can be improved by reducing the chain size to oligosaccharides or dimers, converting to derivatives such as carboxymethyl cellulose or chitosan salts. The antioxidant properties of chitosan have been shown to improve when formed into salts such as chitosan acetate (Charernsriwilaiwat et al. 2012). Even high molecular weight-based chitosan films show antioxidant properties, particularly quaternized high molecular weight chitosan such as N-(2-hydroxyl)propyl-3-trimethyl ammonium chitosan chloride with molecular weights of 400 and 1240 (Wan et al. 2013). Retaining the high molecular weight as well as the antioxidant property makes it applicable to chitosan-based antioxidant material in food and pharmaceutics with good mechanical properties.

3.7.9 Antimalaria

Recently, deacetylated chitin (chitosan) from crab shells combined with silver nanoparticles has demonstrated antimalarial properties. When chitosan combined with silver nanoparticles were spread on a water reservoir, this treatment was effective in eradicating mosquito larvae and pupa (Murugan et al. 2017). The formulation tested non-toxic to fish and also had antibacterial effect. The antimalaria effect therefore comes from the ability to hinder the growth of mosquitos. Furthermore, the silver nanoparticles could be produced using plant extract and this has been explored by several researchers. It is therefore also worth exploring the extent of chitin application in its crude form since this requires less use of chemicals and has less environmental impact.

3.7.10 Papermaking

While their wastewater treatment application already makes chitin useful to the paper-making industry in treatment of wastewater from papermaking, chitin/chitosan has further applications in this industry beyond water treatment, and this includes surface coating, dye fixation and paper strengthening. Antibacterial and greaseproof paper are also in the horizon of future applications of chitosan in the paper industry.

In the wet end application, chitosan and other chitosan derivatives find use as retention and drainage agents. They act as flocculants and coagulants causing the aggregation of fine particles and liquids to aid the separation process. The mechanism by which they do this is through electrostatic attraction facilitated by the attraction between the amino and hydroxyl group and the fine particles thanks to the polycationic nature of chitosan.

Between cellulose fibers exist voids due to the repulsive force between like charges. Due to its excellent film-forming property, chitosan covers this void between the fibers, making for a stronger paper. Chitosan and chitosan derivatives can be applied as coatings on paper. This is aided by its affinity for cellulose. Chitosan coatings give antimicrobial properties and improve water vapor and oxygen barrier properties of the paper. Such papers find use in food packaging and medical uses.

Further application of chitosan in paper production includes its use in dye fixation for improving the dyeing of paper and chitosan nanoparticle derivatives used in producing transparent paper (Song et al. 2018).

3.7.11 Electronics

Ubiquity of chitin is further demonstrated by its applicability in electronics. Flexible electronics, photovoltaic cells and biomedical sensors are some of the applications which employ chitin. Most of these are using chitin in the deacetylated form (chitosan). The electronics industry is one that makes use of a large amount of polymers, from cable coatings to casings to flexible substrates for photovoltaic cells. While many of the applications of chitin involve the deacetylated form, chitosan, some recent applications have explored chitin nanoparticles, for example, in use as biodegradable, eco-friendly flexible substrates for light-emitting diodes (Jin et al. 2016).

Flexible electronics have become desirable for lightweight, low-cost electronics such as screens and photovoltaic cells. Being able to process these electronics on flexible substrates allows for mass reproducible roll to roll processing techniques at low temperatures such as solvent casting and more portable designs. Recently, plastics such as polyethylene terephthalate and polyethylene have been used for such. Even better would be biodegradable polymers for such applications. Chitin and chitosan and their derivatives have demonstrated suitability for such applications (Triyana et al. 2018).

3.7.12 Cosmetics and Toiletries

Chitin is mainly applied in its deacetylated form, chitosan in cosmetics. The antimicrobial property of chitosan makes it useful as a cosmetic anti-acne product ingredient. Its barrier properties, water vapor permeability, make it a good protective barrier to retain skin moisture. It is also used in oral care products as thickener and emulsifier. Due to its antimicrobial, antioxidant property combined with its viscosity-enhancing property, chitosan can play multiple roles in cosmetics and toiletries.

Glucosamine, the monomer of chitin, also finds application in cosmetics. Derivatives of glucosamine have been shown to reduce wrinkles, boost natural production of hyaluronic acid, improve skin suppleness, moisture and rejuvenate overall skin (Jacobs 2007).

3.7.13 Agrochemicals

In agriculture, chitin and its derivatives have found use in various aspects which include: controlled fertilizer release excipients, protective seed coating (frost protection), stimulation of growth and aiding plant defense mechanism. This has both economic and environmental impacts by improving crop yield and optimization of land usage. It also adds nutrients to the soil in a sustainable way replacing the use of energy-consuming agricultural practices.

3.7.14 Biodegradation of Chitin and Its Derivatives

In curbing, the pollution crisis caused by the inadequate dumping of synthetic plastics which pose a serious hazard to aquatic life, research and industry has turned focus on biodegradable plastics. The physical properties which are comparable to those of non-biodegradable synthetic plastics are what make chitin attractive for use as a bioplastic.

Some compounds such as minerals accumulate in nature because their production rate is much higher than their degradation rate. Like many other biopolymers, chitin does degrade because 10–100 billion tonnes of it is being produced annually (Beier and Bertilsson 2014) yet there is no known accumulation of chitin anywhere in nature. This means that biodegradation of chitin is happening at a considerable rate. The microbial degradation of chitin by various microorganisms has been reported; however, it is widely accepted that the predominant degradation of chitin is by bacteria (Beier and Bertilsson 2014; Abd-Aziz et al. 2008).

The process of chitin degradation can occur via different pathways. Chitin degradation enzyme has been detected in various microorganisms including fungi, bacteria, rotifers and some carnivorous plants and digestive tract of higher animals (Gooday

1990). Chitin degradation at physiological condition is predominantly carried out by the enzymes within the physiologic environment. Chitin degradation under standard room conditions is considered for products that are used under such conditions. The degradation mechanism and rate depend on the condition of storage. Its degradation in soil has been observed to be predominant by bacteria; however, some fungi also take part in chitin degradation. In the aquatic environment where degradation of shells and exoskeleton of dead or worn-out aquatic chitinous aquatic organisms occurs, bacteria have been shown to be the main degraders.

Diatoms, microalgae which have quite a sizeable presence in the oceans and other aquatic environment, are also known to have the ability to hydrolyze chitin oligomers (Vrba et al. 1997). Chitinous aquatic organisms and other non-aquatic arthropods occasionally shed their outer shell in order to allow for growth such that the old shell is removed and a new larger one is formed to better accommodate and allow for increase in size of the organism. This process is referred to as molting, and it is thought that the organism releases chitin degradation enzymes during this process (Vrba and Mackacek 1994). It is however yet to be determined, the exact role of these enzymes as they could be either involved in reactive breakdown of chitin or used to hydrolyze dissolved chitin oligomers.

Although here we have considered the degradation of chitin in nature, degradation of chitin when it comes to products which have been made from chitin such as scaffolds, packaging films and water treatment membranes, the environment and conditions within which the degradation is occurring varies significantly. Studies which look at the development of such products have also considered the degradation mechanisms of each of these products.

The process of degradation of chitin is referred to as chitinoclastic. The degradation process could be chitinolytic, and this refers to the breaking of the $(1 \rightarrow 4)$-β-glycosidic bonds. The deacetylation of chitin to chitosan is also a form of chitin degradation as it involves the breaking of the acetyl group. The main difference between deacetylated chitin and glucose is the presence of an amine group in the place of one of the hydroxyl groups on the glucose ring. Deaminization is also another breakdown process where the deacetylated chitin is then converted to cellulose by the removal of the amine functional group. The breakdown of the products of the initial degradation of chitin (chitosan and glucose) then follows their respective degradation pathways to smaller units of glucose, glucosamine and N-acetylglucosamine which can then be returned to the carbon and nitrogen cycle. Enzymes involved in these processes are chitinases, chitosanase and to a lesser extent, cellulases. Lysozyme, an enzyme involved in bacteria cell death as an immune response in animals, has also been shown to degrade chitin (Beier and Bertilsson 2014).

Chitin degradation by microorganisms occurs for the purpose of breaking down the chitin for their metabolism or for the growth of the organism as seen during the molting of arthropods. In the case of lysozyme, it could also occur as an immune response or defense mechanism. The degradation pathway of chitin is illustrated in the chart in Fig. 3.5.

The pathway of degradation of chitin depends on the mix of microbes within the habitat. These have been shown to have more effect than temperature. End product of

Chitin	→	Chitosan	→	Cellulose
↓	Deacetylases	↓	Deaminization	↓
Hydrolysis by Chitinases		Hydrolysis by chitosanase		Hydrolysis by cellulases
↓		↓		↓
Glucosamine		Acetyl glucosamine		Glucose

Fig. 3.5 Degradation pathway of chitin

hydrolysis could either be organic material, or it could be mineralized into inorganic materials depending on the activity of the microbes present. The rate of degradation of chitin is relatively simpler than other polymers such as lignin and cellulose.

The rate of chitin hydrolysis in different environments has been presented in different studies over the years. In soil, the rate of degradation is reported to be between 0.6 and 1.1% per day, in freshwater it could be as high as 30% per day, in brackish water it could be as low as below 1% and up to 8.1%, while marine water records the lowest degradation rate at 0.00043–0.0005 per day (Beier and Bertilsson 2014). In all environments, chitin has a much faster degradation rate compared to fossil-based polymers.

3.8 Commercial Production

The conventional sources of chitin for commercial production are lobster, crab and shrimp shells. However more recently, other sources such as insects and mushrooms are emerging. Companies involved in commercial chitin production from these unconventional sources include KitoZyme which produces chitin from fungi, a non-animal source. Their chitin is used in food supplements and medical products.

Another example of company producing chitin on a commercial scale from shrimp shell is Mahtani Chitosan, a biotechnology company located on the coast of Gujarat, India (Munoz et al. 2018). The company reports producing 300 mt of chitin, 150 mt of glucosamine and 50 mt chitosan per annum. The company primarily produces chitin and the derivatives from waste shells of shrimp species of *Penaeus* spp., *Metapenaeus* spp. and *Parapenaeus* spp. captured from the Arabian Sea (Munoz et al. 2018). They employ the chemical extraction method for demineralization, deproteinization and deacetylation.

A number of companies exist globally which are actively involved in commercial production of chitin and its derivatives, either for conversion into other products or chitin being the end product. In the environmental impact evaluation section of the chapter, we have discussed Mahtani Chitosan and an anonymous European company who produce chitin from shrimp and crab shells, respectively. Another company based in Belgium called KitoZyme S. A produces chitosan from fungi for food and pharmaceutical applications. Another company Primex produces chitosan from arctic shrimp for applications in food. Other companies involved in the production of chitin and/or its derivatives include Asiamerica Group, Inc., Spectrum Chemical Manufacturing Corp, Orison Chemicals Ltd., Parachem, Shandong Laizhou Highly Bio-Products Co. Ltd., Pure Earth Biotechnology Co. Ltd., Qingdao BZ Oligo Biotech Co., Ltd. and Yaizu Suisankagaku Industry Co., Ltd. (Future Market Insights 2019).

Shell-Ex, a biorefinery company located in Canada, is dedicated to converting shrimp waste into products such as liquid fish to serve as soil NPK nutrient and shrimp and crab shell which is mainly targeted at chitosan production company. The specific type of shrimp shell it supplies (*Pandalus borealis*) is regarded as high-quality shell preferred in chitosan production, particularly sourced from the Atlantic Ocean for optimal purity away from industrially polluted water.

Heppe Medical Chitosan GmBH located in Germany develops, produces and sells chitosan alongside other biopolymers targeted at supplying the research and medical sector. For such purpose where high-quality chitosan is required, production occurs in smaller batches. CHitOcean is another company in North America involved in the chitosan manufacturing.

Great opportunities exist in the present and future market for manufacturers of chitin and its derivatives. Opportunities also exist for the raw material collectors and agents as well as the end use industries. As chitin industry expands, this will trigger new product development in all the various areas of applications implementing the diverse chemistry of chitosan into numerous products. Generally, fish consumption in any country is higher in the coastal regions than inland areas. This is an important fact to consider in location of chitin plants near to the raw materials.

3.9 Conclusion

Chitin is almost exclusively sourced from the aquatic environment with the exception of insects, mushrooms and land crabs which are not fully aquatic. It is a versatile biopolymer of much industrial significance. In nature, it plays a significant role in structure within the organisms such as shrimp shells and mushroom. It also plays a role in metal chelation and formation of coral reefs. The amount of chitin being used and made commercially available is much less than what is being generated as waste and how much exists in the aquatic world. Therefore, there is much room for growth in the chitin industry with regard to available raw material. Presently, crustacean shells are the most widely utilized sources used for commercial chitin production although researchers have extracted from various sources such as mushrooms, insects and fish scales. The chemical production method is the method being mainly used in industry for chitin extraction although chitin can also be extracted using microbial fermentation and enzyme extraction. Molecular weight, degree of deacetylation and polydispersity are important parameters for determining quality of chitin. These factors are therefore important in their extraction, processing and applications.

The process of sourcing and extraction of chitin has significant environmental impact from the carbon emission from transportation and preservation of raw materials to the chemical and energy consumed during the production process. Nonetheless, chitin, being the second most abundant polymer on earth, offers great potential as a replacement or at least an alternative to fossil-derived polymers used for similar applications such as packaging films, water treatment, flexible electronics and biomedical products.

References

Abd-Aziz S, Sin T, Slitheen N, Shahab N, Kamaruddin K (2008) Microbial degradation of chitin materials by Trichoderma virens UKM1. J Biol Sci 8(1):52–59

Agboh OC, Qin Y (1997) Chitin and chitosan fibers. Polym Adv Technol 8:355–365

Ahearn GA, Mandal PK, Mandal A (2004) Calcium regulation in crustaceans during the molt cycle: a review and update. Comp Biochem Physiol A Mol IntergrPhysiol 137(2):247–257

Aili D, Adour L, Houali K, Amrane A (2019) Efect of temperature in chitin and chitosan production by solid culture of penicillium camembertii on YPG medium. Int J Biol Macromol 133:998–1007

Al- Manhel AJ, Al-Hilphy ARS, Niamah AK (2018) Extraction of chitosan, characterisation and its use for water purification. J Saudi Soc Agric Sci 17:186–190

Azuma K, Osaki T, Minami S, Okamoto Y (2015) Anticancer and anti-inflammatory properties of chitin and *Chitosan oligosaccharides*. J Funct Biomater 6:33–49

Badawy RM, Mohamed HI (2015) Chitin extraction, composition of different six insect species and their comparable characteristics with that of the shrimp. J Am Sci 11(6):127–134

Beier S, Bertilsson S (2014) Bacterial chitin degradation- mechanisms and ecophysiological strategies. Front Microbiol 4(149):1–13

Berg A (2009) Aquatic insects classification. Encyclopedia of inland waters, pp 128–131

Boarin-Alcalde L, Graciano-Fonseca G (2016) Alkali process for chitin extraction and chitosan production from Nile tilapia (*Oreochromis niloticus*) scales. Am J Aquat Res 44(4):683–688

Brostow W, Datashvili T (2016) Environmental impacts of Natural Polymers. In: Olatunji O (Ed). Natural Polymers, Industry Techniques and Applications. Springer, New York. pp 315–338

Castro R, Guerrero-Legarreta I, Borquez R (2018) Chitin extraction from *Allopetrolisthes punctatus* crab using lactic fermentation. Biotechnol Rep 20:e00287

Charernsriwilaiwat N, Opanasopit P, Rojanarata T, Ngawhirunpat T (2012) In vitro antioxidant activity of chitosan aqueous solution: effect of salt form. Trop J Pharm Res 11(2):235–242

Cira L, Huerta S, Hall GM, Shirai K (2002) Pilot scale lactic acid fermentation of shrimp wastes for chitin recovery. Process Biochem 37:1359–1366

Divya K, Vijayan S, George TK, Jish MS (2017) Antimicrobial properties of chitosan nanoparticle: mode of action and factors affecting activity. Fibers Polym 18(2):221–230

Elieh-Ali-Komi D, Hamblin MR (2016) Chitin and chitosan: production and application of versatile biomedical nanomaterials. J Adv Res (Indore) 4(3):411–427

FAO (2016) The state of world fisheries and aquaculture 2016. Contributing to food security and nutrition for all. Rome, 200 pp. ISBN 978-92-5-109185-2

Ficket Z, Medunic G, Turk FM, Ivanic M, Kniewald G (2017) Influence of soil characteristics on rare earth fingerprints in mosses and mushrooms: example of a pristine temperature rainforest (Slavonia, Croatia). Chemosphere 179:92–100

Flores-Albino B, Arias L, Gomez J, Castillo A, Gimeno M, Shirai K (2012) CHitin and L (+) -lactic acid production from crab (*Callinectes bellicosus*) wastes by fermentation of *Lactobacillus* sp. B2 using sugar cane molasses as carbon source. Bioprocess Biosyst Eng 35:1193–1200

Friedman AJ, Phan J, Schaire D, Champer J, Qin M, Pirouz A, Blecher K, Oren A, Liu P, Modlin RL, Kim J (2013) J Invest Dermatol 133(5):1231–1239

Frost SJ, Mawad D, Higgins MJ, Ruprai H, Kuchel R, Tilley RD, Myers S, Hook JM, Lauto A (2016) NPG Asia Mater 8(73):1–9

Future Market Insights (2019) Chitin market: agrochemical end use industry segment inclined towards high growth—moderate value during the forecast period: global industry analysis (2012–2016) and opportunity assessment (2017–2027). REF-GB-313, pp 1–236

Gadgey KK, Bahekar A (2017) Studies on extraction methods of chitin from crab shell and investigation of its mechanical properties. Int J Mech Eng Technol 8(2):220–231

Gibot L, Chabaud S, Bouhout S, Bolduc S, Augner FA, Moulin VJ (2015) Anticancer properties of chitosan on human melanoma are cell line dependent. Int J Biol Macromol 72:370–379

Gil-Duran S, Arola D, Ossa EA (2016) Effect of chemical composition and microstructure on the mechanical behaviour of fish scales from *Megalops atlanticus*. J Mech Behav Biomed Mater 56(2016):134–145

Gooday GW (1990) The ecology of chitin degradation. Adv Microb Ecol 11:387–430

Gortari MC, Hours RA (2013) Biotechnological processes for chitin recovery out of crustacean waste: a mini-review. Electron J Biotechnol 16(3):1–19

Hassainia A, Satha H, Boufi S (2018) Chitin from *Agaricus bisporus*: extraction and characterization. Int J Biol Macromol 117:1334–1342

Hulsey MJ (2018) Shell biorefinery: a comprehensive introduction. Green Energy Environ 3:318–327

Ifuku S, Nomura R, Morimoto M, Saimoto H (2011) Preparation of chitin nanofibers from mushrooms. Materials 4:1417–1425

Jacobs E (2007) Topically applied glucosamine sulfate and all its related, precursor, and derivative compounds significantly increases the skin's natural production of hyaluronic acid for the rejuvenation of healthier younger-looking skin: while phosphatidylcholine is required to replace its deficiency caused by topical dimethylaminoethanol (DMAE). US Pastent US2007/0092469 A1. Vienna (US)

Jin J, Lee G, Im H, Han YC, Jeong EG, Rolandi M, Choi KC, Bae RS (2016) Chitin nanofiber flexible transparent paper for flexible green electronics. Adv Mater 28(26):5169–5175

Kalut SA (2008) Enhancement of degree of deacetylation of chitin in chitosan production, B. Chemical Engineering, Universiti Malaysia Pahang, pp 14–15

Kanatt SR, Rao MS, Chawla SP, Sharma A (2013) Effects of chitosan coating on shelf-life of ready-to-cook meat products during chilled storage. LWT-Food Sci Technol 53:321–326

Kaya M, Mujtaba M, Ehrlich H, Salaberria AM, Baran T, Amemiya CT, Galli R, Akyuz L, Sargin I, Labidi J (2017) On chemistry of gama- chitin. Carbohydr Polym 176:177–186

Koping-Hoggard M, Varum KM, Issa M, Danielsen S, Christensen BE, Stokke BT, Artursson P (2004) Improved chitosan-mediated gene delivery based on easily dissociated chitosan polyplexes of highly defined chitosan oligomers. Gene Ther 11:1441–1452

Lallana E, Rois de la Rosa JM Tirella A, Pelliccia M, Gennari A, Stratford IJ, Puri S, Ashford m, Tirelli N (2017) Chitosan/hyaluronic acid nanoparticles: rational design revisited for RNA delivery. Mol Pharm 14(7):2422–2436

Lantos PR (1988) Plastics in medical applications. J Biomater Appl 2(3):358–371

Lauto A, Mawad D, Barton M, Gupta A, Piller SC, Hook J (2010) Photochemical tissue bonding with chitosan adhesive films. BioMed Eng Online 9(47):1–11

Liu S, Sun J, Yu L, Zhang C, Bi J, Zhu F, Jiang MQC, Yang Q (2012) Extraction and characterization of chitin from the beetle Holotrichiaparallela Mots chulsky. Molecules 17:4604–4611

Majekodunmi SO (2016) Current development of extraction, characterization and evaluation of properties of chitosan and its use in medicine and pharmaceutical industry. Am J Polym Sci 6(3):86–91

Majtan J, Bilikova K, Marovic O, Grof J, Kogan G, Simuth J (2007) Isolation and characterization of chitin from bumble bee (*Bombus terrestris*). Int J Biol Macromol 40:237–241

Massimilian F, Ludovico P (2016) Use of non-treated shrimp shells for textile dye removal from wastewater. J Environ Chem Eng 4(4):4100–4106

Munoz I, Rodriguez C, Gillet D, Moerschbacher BM (2018) Life cycle assessment of chitosan production in India and Europe 23:1151–1160

Murugan K, Anitha J, Suresh U, Rajaganesh R, Panneerselvam C, Aziz AT, Tseng L-C, Kalimuthu K, Saleh Alsalhi M, Devanesan S, Nicoletti M, Sarkar SK, Benelli G, Hwang J-S (2017) Chitosan-fabricated Ag nanoparticles and larvivorous fishes: a novel route to control the coastal malaria vector *Anopheles sundaicus*? Hydrobiologia 797(1):335–350 doi: 10.1007/s10750-017-3196-1

Olatunji O, Olsson RT (2015) Microneedles from fishscale-nanocellulose blends using low temperature mechanical press method. Pharmaceutics 7:363–378

Rinaudo M (2006) Chitin and CHitosan: properties and applications. Prog Polym Sci (Oxford) 31(7):603–632

Roy JC, Salaun F, Giraud S, Ferri A, Chen G, Guan J (2017) Solubility of chitin: solvents, solution behaviours and their related mechanisms. In: Solubility of polysaccharide (xxx Editor). Intechopen

Rumengan IFM, Suptijah P, Wullur S, Talumepa A (2017) Characterization of chitin extracted from fish scales of marine fish species purchased from local markets in North Sulawesi, Indonesia. IOP Conf Ser: Earth Environ Sci 89:012028

Satam CC, Irvin CW, Lang AW, Jallorina JC, Shofner ML, Reynolds JR, Meredith JC (2018) Spray-coated multilayer cellulose nanocrystal—chitin nanofiber films for barrier applications. Chemistry. https://doi.org/10.1021/acssuschemeng.8b01536

Shahidi F, Arachchi JKV, Jeon Y (1999) Food application of chitin and chitosans. Trends Food Sci Technol 10(2):37–51

Song Z, Li G, Guan F, Liu W (2018) Application of chitin/chitosan and their derivatives in the papermaking industry. Polymers 10(4):E389

Thannimalay L, Yusoff S, Zawawi ZN (2013) Life cycle assessment of sodium hydroxide. Aust J Basic Appl Sci 7(2):421–431

Tolesa LD, Gupta BS, Lee MJ (2019) Chitin and chitosan production from shrimp shells using ammonium based ionic liquids. Int J Biol Macromol 130:818–826

Triyana K, Sembiring A, Rianjanu A, Hidayat SN, Riowirawan R, Julian T, Kusumaatmaja A, Santoso I, Roto R (2018) Chitosan-based quartz crystal microbalance for alcohol sensing. Electronics 7(9):181–190

Varum TK, Senani S, Jayapal N, Chikkerur J, Roy S, Tekulapally VB, Gautam M, Kumar N (2017) Extraction of chitosan and its oligomers from shrimp shell waste, their characterization and antimicrobial effect. Vet World 10:170–175

Vrba J, Machacek J (1994) Release of dissolved extracellular Beta-N-Acetylglucosaminidase during crustacean molting. Limnol Oceanogr 39:712–716

Vrba J, Kofro˘nová-Bobková J, Pernthaler J, Simek K, Macek M, Psenner R (1997) Extracellular, low-affinity beta- N-acetylglucosaminidases linked to the dynamics of diatoms and crustaceans in freshwater systems of different trophic degree. Int Rev Gesamten Hydrobiol 82:277–286

Wan A, Xu Q, Sun Y, Li H (2013) Antioxidant activity of high molecular weight chitosan and N, O-Quaternized chitosans. J Agric Food Chem 61(28):6921–6928

Yen MT, Yang JH, Mau JL (2009) Antioxidant properties of chitosan from crab shells. Carbohyd Polym 74(4):840–844

You-Jin J, Kim S (2002) Antitumor activity of chitosan oligosaccharides produced in ultrafiltration membrane reactor system. J Microbiol Biotechnol 12(3):503–507

Chapter 4
Alginates

Abstract Alginates are obtained from brown algae, which are mainly found in marine aquatic environment. Alginate chemical structure is characterized by the presence of mannuronic or guluronic acid repeating unit in either alternating or block forms within the polymer chain. They form a part of the cell wall where they provide strength and flexibility to the cell wall. The chapter reviews a number of extraction methods used and then discusses the environmental impact of the extraction process. Alginates find application in a variety of industries which includes food, biomedical, textiles and others. Alginates can be said to be one of the well-explored aquatic biopolymers. The chapter discusses the current state and some of the limitations to its commercial production and presents future perspectives on alginates.

Keywords Alginates · Galactose · Polymer · Brown algae · Aquatic · Phycocolloid

4.1 Introduction

Alginates are natural polysaccharides which are present in the cell walls of brown algae (brown seaweed). They are extracted mainly using chemical methods with additional processing such as gel pressing and filtration. Alginates are a general term which refers to salts of alginic acid as well as derivatives of alginic acids and its salts. Within this chapter, alginates refer to salts of alginic acid: sodium alginate, calcium alginate and magnesium alginate. Another term, alginocytes, refers to sources of alginates. Brown marine algae are the main known alginocytes.

Various forms of alginates exist; however, sodium alginates are most common due to its solubility in cold water which makes it easier to extract and apply. As alginates gain more commercial value, the cultivation of alginocytes (alginate producing algae) for alginate production is on the rise. FAO reports an annual increase of over 6% in the cultivation of brown algae, the main source of alginates. A large proportion of the seaweeds grown commercially is grown from aquaculture where they are grown in controlled environment for optimal yield and profitability. The increase in demand for seaweed can in part be attributed to increased health consciousness and demand for more non-animal-based food and pharmaceuticals. Alginate competes with the

O. Olatunji, *Aquatic Biopolymers*, Springer Series on Polymer and Composite Materials,
https://doi.org/10.1007/978-3-030-34709-3_4

human and animal edible algae as a lot of the algae being cultivated globally are also used for human consumption and animal feed. By 2024, alginate market is predicted to increase in value up to about 87 billion USD.

Biopolymers with diverse applications offer the producer and supplier an equally diverse distribution channel, whereby the product is in demand in several industries. Alginate is one of such polymers, and its applications include in food, textiles, pharmaceutics and more. Some of these applications are already being made into commercial products while others are still at an early research stage. Alginates are applied in textiles, cosmetics, paper, food and pharmaceutics.

Alginates are one of three phycocolloids which can be obtained from algae. The two others are agar and carrageenan. Alginates are made from brown algae while agar and carrageenan are obtained from red algae. Alginates are one of the commercially well-explored biopolymers of aquatic source. Since they are produced primarily from algae, they are one of the polymers which are exclusively sourced from the aquatic environment. The continued availability of alginates is therefore significantly affected by the dynamics of the aquatic environment. The production of these phycocolloids competes with the production of other products such as biofuel, fertilizers and sewage treatment materials from algae. It is therefore important to understand the production and impacts of alginate production from these algae in order to assess the adequate utilization of this particular aquatic resource.

4.2 Occurrence in Nature

Brown seaweeds are the only known source of alginates. Although alginates from seaweed were not discovered until the later part of the 1800s, for centuries, coastal communities have grown around gathering of seaweed. Seaweeds are aquatic plants which grow naturally in different aquatic conditions, from slow-moving to more turbulent waters. There are three classifications of seaweeds: brown, red and green. Brown seaweeds are largest in size, up to 20 m long; however, smaller ones exist ranging in 2–4 m long or even smaller 30–60 cm length (McHugh 2003). Red seaweeds are smallest with length ranging under a meter while green seaweeds are the smallest. Here, we focus our interest on the brown seaweed species. These grow best in cold waters around 20 °C in countries such as South Africa, Argentina, Canada, Australia, Chile, Norway, Ireland, Mexico, Scotland, Northern Ireland, USA and other countries in the Southern and Northern Hemisphere. Factors which affect growth of seaweeds include temperature, nutrient content, salinity, light penetration and turbulence of water.

Despite aquaculture dominating the global seaweed production, most of the alginophytes are from species which grow wild such as laminarin and Ascophyllum. Most of the brown seaweeds used for alginate production have a complex reproductive cycle which makes their cultivation less cost-effective than when harvested from wild stock (McHugh 2003). However, as further understanding of the diverse reproductive cycle of these algae is developed, more effective methods for cultivation of alginophytes in aquaculture systems have been introduced in recent years, particularly in Asia (Charrier et al. 2017). Cultivation of some species of wild seaweed has been restricted as a result of overexploitation. Other than being a source of alginate and other phycocolloids, seaweed is a rich source of nutrients such as proteins, minerals and vitamins, it is also the only non-fish source of omega-3 fatty acid. It therefore has a high value as a food product for direct consumption by humans and other animals.

Alginate provides seaweeds with the flexibility and strength needed to survive at sea. Within the seaweed cell walls, alginates provide flexibility and mechanical strength, needed by the algae to survive at sea, particularly where the water is particularly turbulent. They exist in the cell wall attached to metals such as magnesium, calcium and sodium. The seaweed also contains cellulose which provides rigidity. Alginate contents tend to be higher in seaweeds cultivated in more turbulent water while seaweeds grown in calmer waters have lower alginate content. The properties of the alginates extracted from seaweed such as viscosity and gel strength depend on the structure of the alginate, and this in turn is affected by factors such as species of seaweed, growth environment and the method of extraction. Alginate concentration has also been shown to be affected by the period of harvest (Schiener et al. 2014; Taelman et al. 2015).

While seaweeds naturally grow in aquatic environment, much of the world's seaweeds are grown on slow stagnant water with the right nutrients and environmental conditions, to meet the growing demands for alginates and other products of brown seaweed. Depending on the region and its demand, majority of the brown seaweed cultivated is used for alginate while the rest is used for food. However, other statistics show that in China, for example, most of the seaweed cultivated in this region is used as food.

There are a variety of brown seaweed species which act as a source of alginates. These include Laminaria spp, *Macrocystis pyrifera*, *Durvillaea antarctica*, *Lessonia flavicans*, *Ecklonia maxima* and *Ascophyllum nodosum*. The most commercially exploited ones are Laminaria, Macrocystis and Ascophyllum. Alginate is also present as a protective extracellular structure of some bacteria such as Azotobacter and pseudomonas (Aleksandra and Sanja 2017).

4.3 Chemistry of Alginates

4.3.1 Polymer Chain Structure

Alginates are linear ionic heteropolysaccharides copolymer of (1-4) linked beta-D-mannuronic acid and alpha-L-guluronic acid arranged in block or alternating manner along the polymer chain. Such that along the alginate polymer chain, some segments of mannuronic acid, segments of guluronic acid and segments of mannuronic acids and guluronic acid alternating exist. This is illustrated in Fig. 4.1.

-M-M-M-M-M-M-M-G-G-G-G-G-G-G-M-G-M-G-M-G-M-M-M-M-M--G-M-G

Fig. 4.1 A section of one possible block and alternating patterns of mannuronic acid (M) and guluronic acid (G) within an alginate chain

The manner in which the mannuronic acid and guluronic acids are arranged along the polymer chain varies in different seaweeds. So does the ratio of mannuronic: guluronic acid of the alginate. The GG blocks confer more stiffening effect on the polymer chain and are also more soluble at lower pH than the alternating GM blocks (Aravamudhan et al. 2014). An example composition of alginate extracted from the brown algae Laminaria digitata M and G blocks is 0.47, 0.41, 0.06 and 0.06 MM, GG, MG and GM (Fertah et al. 2017). Where MM and GG refer to the fraction of the chain which is made of block arrangements of mannuronic acids and guluronic acids within the polymer chain and MG and GM refers to the fraction of the chain with alternating arrangements. Figure 4.2 shows the structure of the alginate repeating units.

Some bacteria containing alginates have also been reported. However, the structure of alginate obtained from bacteria is quite different from that obtained from seaweed (algae). Bacteria alginate is generally high degree of polymerization hence high molecular weight while that from seaweed could be a variety of molecular weight. The G sequence also varies in bacteria (Windhues and Borchard 2003). Bacteria derived alginate has a molecular weight in the range of 154,600–730,000 g/mol much higher than that of seaweed sourced alginate in the range of 48,000–186,000 g/mol (Nagarajan et al. 2016; Clementi et al. 1999; Donnan and Rose 1950). Nonetheless, molecular weight of alginate varies widely for different sources.

Fig. 4.2 Structure of **a** mannuronic acid and **b** guluronic acid

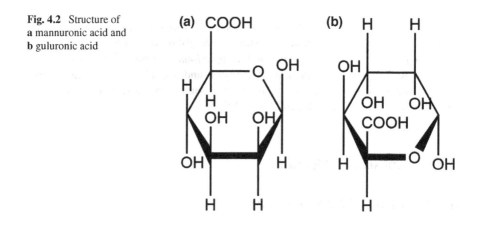

4.3.2   Rheological Properties

Polymers have the unique property of increasing the viscosity of a solvent when they are dissolved in the solvent. Alginates increase the viscosity of the solution and form irreversible heat-stable gels. This forms the basis of most of their applications in food and pharmaceuticals. The viscosity-enhancing properties and the gelling properties of alginates are dependent on the amount of mannuronic acid and guluronic acid, respectively, within the polymer chain. Alginates with higher mannuronic acid content have higher viscosity while those with more guluronic acid have higher gel-forming properties.

Alginates produce non-thermoreversible ionic gels, once formed into gels, and unlike for example gelatine, they retain these gel forms at different temperatures unlike for example gelatine which losses its gel form when the temperature is increased. The ability to form non-thermoreversible ionic gels serves as the basis of much of the applications of alginates. Alginate forms a gel in the presence of bivalent cations through ion exchange. It does not form a gel with monovalent cations and magnesium ion. Calcium, strontium and barium are the bivalent ions used due to their non-toxicity. Sodium, potassium and ammonium alginates are water soluble while the others are insoluble.

4.3.3   Characterization of Alginate

Molecular weight of extracted sodium alginate reported varies in the ten thousands (40,680 g/mol using the chemical method with a yield of 44.32% (Helmiyati and Aprilliza 2017). For chemical characterization of polymers, methods commonly used include FTIR, XRD, DSC and SEM (Olatunji and Olsson 2016). Typical FTIR spectra of sodium alginate show peaks at 939 and 884 cm^{-1} for uronic acid and mannuronic acid functional group, respectively. The alginate chains also include OH functional group and the CH stretching which also appear in the FTIR scan at around 3200–3400 cm^{-1} for OH and 2928 cm^{-1} for the CH stretching. The absorbance and transmittance depend on the composition of the functional groups within the alginate structure. Pure sodium alginate has a decomposition temperature of 251.12 °C (Helmiyati and Aprilliza 2017). This is obtained from differential scanning calorimetry. Extracted sodium alginate should therefore show values close to this depending on the level of purity achieved during extraction process.

Crystallinity of alginate affects properties such as solubility and thermal properties. Pure sodium alginate has a crystallinity of about 35.62%. Crystallinity of raw extract of alginate is usually less than this, and the purer it is, the closer it gets to the crystallinity value of pure form, for example, a crystallinity 29.292% is reported in one study for sodium alginate extracted from brown algae (Helmiyati and Aprilliza 2017). Furthermore, when the SEM of alginate containing brown seaweed is compared to that of extracted seaweed, a distinguishing factor is the visible alginate fibrils which can be seen in the extract but not visible in the intact seaweed due to the presence of the other components such as lipids, minerals and other non-alginate carbohydrates. The appearance of distinct fibrils can also therefore be used as an additional characterization of alginate extract (Fertah et al. 2017).

The color of alginate can vary from whitish to brown. The brown pigment indicates the presence of fucoxanthin pigment and varies for different types of alginophytes (Mushollaeni 2011). Another main carbohydrate present in seaweed is laminarin, a storage polysaccharide which is also discussed in a separate chapter in this book.

4.3.4 Decomposition of Alginate

The decomposition of alginate is important for two reasons: Firstly, if the biodegradable polymer is going to degrade within the body or environment, the safety of the products of degradation is as important as the safety of the product itself. For example, when used as a thickener in food what are the products likely to be formed in the body after consumption. Secondly, the degradation product could be a useful way to extend the usage of the product after the first use, for instance, using an alginate packaging film for the production of organic acids after their use as food packaging.

According to its structure, alginate can be broken down into smaller chains of either purely mannuronic or guluronic acid oligomers or alternating chains of mannuronic and guluronic acid units using enzymes which are specific to the respective sites. For example, an alginate lyase enzyme, poly-1 → 4-alpha-L-guluronate lyase enzyme extracted from the mollusk Lamis sp., catalyzes the cleavage of the 1 → 4 glycosidic bond of alpha-L-guluronic acids (Sil'chenko et al. 2013).

The degradation of alginate salts and gels is particularly important as these are the forms in which alginate exists in many products. Dense alginate hydrogels have a half-life of 4–6 days in model tissues (Shkand et al. 2016). Under hydrothermal conditions, alginate degrades into its monosaccharides, mannuronic and guluronic acids through the hydrolysis of glycosidic bonds. Lactic acids and glycolic acids as well as sodium carbonate are also produced as products of alginate decomposition (Aida et al. 2010). Controlled depolymerization of alginates can be carried out to obtain other organic acids (Aida et al. 2012) and alginate lyase enzymes (Zhu and Yin 2015). The products of the depolymerized alginates are safe and in some cases are of commercial importance such as the alginate oligosaccharides which are discussed in the section on applications of alginate.

Alginate is also produced by some species of bacteria such as pseudomonas and Azotobacter (Chen and Long 2018). These bacteria also produce the enzyme alginate lyases to break down the alginate for use. The alginate oligosaccharides derived from the action of alginate lyases on alginate has potential applications such as protection against disease-causing organisms (An et al. 2009), growth-promoting agents in plants (Iwasaki and Matsubara 2000) and antioxidants (Falkeborg et al. 2014; Nagarajan et al. 2016; Chen and Long 2018).

4.4 Availability of Raw Material

Recent years have seen significant increase in seaweed cultivation. The global sea-weed market is projected to exceed 17.59 Billion USD by 2021 (Monagail et al. 2017) and 87 Billion USD by 2024. About 96.5% of the aquatic plant production is dominated by seaweed. In 2016, about 30.1 million tonnes of seaweed was produced, both cultivated and caught wild, and this has shown an average increase of over 7% between 2010 and 2016 (FAO 2018), such that there exists a very active seaweed industry to serve the demand for raw material for alginate production. As of 2015, 99% of the global farmed seaweed output was from eight Asian nations while 150 other countries with coastal waters which could take part in seaweed farming exist, thus indicating the under exploitation of seaweed farming in coastal waters (Radulovich et al. 2015; Monagail et al. 2017).

Seaweed harvests go up to tens of thousands as of 2009 for example in 2009 France, UK, Norway and Ireland produced 30,500 tonnes of *Laminaria* spp. In the same year, Chile and Peru produced 27,000 tonnes of *Lessonia* spp. A drastically lower harvest from 35,000 tonnes in 1991 to 5000 tonnes in 2009 was recorded for *Macrocystis pyrifera* in 2009. This can be attributed to the closure of the International Specialty Products facility in San Diego. Ascophyllum also showed a sharp reduction in harvest between the specified period going from 13,500 tonnes in 1999 to only 2000 tonnes in 2009. This is due to the increase in demand for alginates in higher gel strength which diverted interest to seaweeds which produced higher guluronic acid content (Bixler and Porse 2011; Hernandez-Carmona et al. 2013).

China and Indonesia are the leading producers of aquatic plants from aquaculture. The seaweed cultivation has been on an increase from 13.1 million tonnes produced in 1995 to 30.1 million tonnes produced in 2016. Indonesia, one of the world's leading seaweed producers, reportedly increased production from 4 million tonnes in 2010 to 11 million tonnes in 2016 (FAO 2018). This increase in production is credited to the increase in demand for seaweed for production of its polysaccharides, agar, alginate and carrageenan.

Despite the millions of tonnes of farmed seaweed available, the alginophyte species such as Laminaria and Ascophyllum are obtained from natural stock due to the expensive cultivation process required for these species which cannot be grown through the vegetative method but rather must go through a more complex sexual reproductive process which requires strict control of the environment to facilitate the alternation of generations at different phases of the production process (McHugh 2003). Such expensive cultivation process can only be compensated by selling the cultivated seaweed as food rather than as raw material for alginate production. This leaves the wild seaweed annual production rate at around 1.3 million tonnes, and this fluctuates annually (Monagail et al. 2017).

10% of the alginophytes are supplied by Chile, and these are from the wild. Although rarely used for alginate production, when in surplus, cultivated seaweed can be sold at a subsidized rate for alginate production. In Europe, 99% of the seaweed produced are from wild harvest, and this makes up 1% of the global seaweed production with Norway leading the production in Europe at 154,230 tonnes of brown seaweed harvested from natural stock annually. France harvested 33,919 tonnes wild seaweed in the same year (Monagail et al. 2017). Therefore, although aquaculture has become more dominant, wild seaweed still plays a significant role, and however, these are more economical sold as food for direct consumption rather than as sources for alginate. 800,000 tonnes of wild seaweed has been recorded in 2014 from a total of 32 countries. Compared to, as of 2003, 35–42 countries, out of the 195 countries in the world were reported to be involved in seaweed production (McHugh 2003). As of 2016, FAO reports 30.1 million tonnes of farmed aquatic plants globally, a large proportion of this being seaweeds and smaller proportion of microalgae.

It can therefore be concluded that a smaller proportion of the seaweed available is diverted to alginate production while a larger proportion is used for food as has been for centuries. This is mainly due to the need for the alginophytes to be sourced from natural stock. Nonetheless, the alginate market is thriving on brown seaweed production in the hundred thousand tonnes which presently meets the global demand estimated at an annual rate of 26,500 MT (Konda et al. 2015).

4.5   Extraction of Alginates

First Isolated by Dr E. C. C Stanford in 1881 in Britain (Draget 2009; Parreidt et al. 2018), alginate makes up about 10–45% of the dry mass of seaweed (Chen and Long 2018). Extraction yield of alginate reported ranges from 16.93% from Padina to 30.3% from *S. crassifolium* (Mushollaeni 2011). Commercially chemical extraction methods are more widely used in industries. Novel methods are being explored by researchers which are more cost-effective and also more environmentally friendly. This includes the use of enzymes such as alginate lyase (Chen and Long 2018). Alginate is often extracted as sodium alginate or in other forms such as calcium alginate. These are explored separately in the following subsections.

4.5.1   Extraction of Sodium Alginate

Within the cell walls of brown algae, alginate occurs as salts of alginic acid as calcium, magnesium and sodium alginic acid salts. Sodium alginate is mostly extracted since it is the only water-soluble salt of alginic acid. It is therefore of more economic significance. This solubility in water makes it more practical to extract and also apply. The extraction process involves isolating the sodium alginate by dissolving it out in water. The other alginates (calcium and magnesium) must also be converted to sodium alginates in order to have a good yield of alginate from the source. Three methods of extractions are outlined as follows. In all three methods, the seaweed is first prepared by washing, cutting or milling to reduce size and the pigments removed. Removal of pigment can be achieved with formaldehyde, and the cell walls may be softened by soaking in 0.2 M HCL overnight. An alkali, sodium carbonate is then added under high temperature (~80 °C) to extract sodium alginate in an aqueous mixture with the seaweed residue. The solid residue is then separated by filtering (Fertah et al. 2017). The process then proceeds using either of the two methods below to recover the sodium alginate from the aqueous solution.

4.5.2 Method 1

The first method involves the addition of calcium salt to the liquid extract which results in the formation of calcium alginate which is insoluble in water. This can therefore be separated from the mixture by filtration or pressing. This is followed by adding acid to form alginic acid. Alcohol is also added to render the sodium alginate to be formed, insoluble. Sodium alginate is then formed by the addition of sodium carbonate which precipitates and is separated and dried followed by further processing depending on the end use.

4.5.3 Method 2

In this method, acid is first added to the liquid extract. This results in the formation of alginic acid. This alginic acid is insoluble in water forming a soft gel which can be separated from the aqueous mass. This is followed by addition of alcohol and then an alkali, sodium carbonate to form sodium alginate from the alginic acid. The alcohol added to the water allows the sodium alginate to precipitate out of the water once formed, thus making it easier to separate. The product, sodium alginate can then be dried and used for further processing and application. This method makes use of mainly acid and alcohol.

4.5.4 Method 3

In this method following the sodium carbonate extraction, the sodium alginate is precipitated using ethanol (Fertah et al. 2017). This method is more commonly used for laboratory-scale extraction where dewatering processes might be impractical at the small scale and alginate of high purity is required. Furthermore, this method can be used where cost saving is not of concern, and the alginate is being extracted, mainly for research purposes.

All these methods generally involve the addition of acid to dissolve the sodium alginate from the seaweed mass. This alkaline extraction takes about two hours and requires a hot solution of sodium carbonate. The seaweed residue is then separated from the liquid. The main task that follows is therefore to separate the sodium alginate from the water. Water has a rather high latent heat of evaporation and therefore requires lots of energy to evaporate such that less energy-consuming and economic methods are employed to recover the sodium alginate. This is either method 1 where acid is added followed by dewatering and reforming of sodium alginate precipitate or method 2 where calcium salt such as calcium chloride is added followed by reforming and precipitation of sodium alginate. Figure 4.3 summarizes the extraction process in a flowchart.

In between, the methods as outlined above exist some major challenges mainly in separation. The alkali treatment process yields a very thick slurry made up of the sodium alginate solution and the solid residue of the seaweed, comprising mostly cellulose. This cellulose can also be extracted for other uses (discussed in chapter dedicated to cellulose) (Lakshmi et al. 2017). To ease filtration of the thick slurry, large volume of water is added to reduce the viscosity and improve flow in the filter. Filtration aids are also often used to improve the filtration process and prevent clogging of the filter as the slurry also contains small size particles which can clog the filter pore.

Depending on whether it is method 1 or 2 that is used to recover the sodium alginate the process of separation varies. If the calcium chloride method is used, the fibrous calcium alginate formed can be separated by filtration with a sieve and washed. Additional step involves conversion to a fibrous form of alginic acid which is then further screw pressed to reduce water content. Conversion to sodium alginate

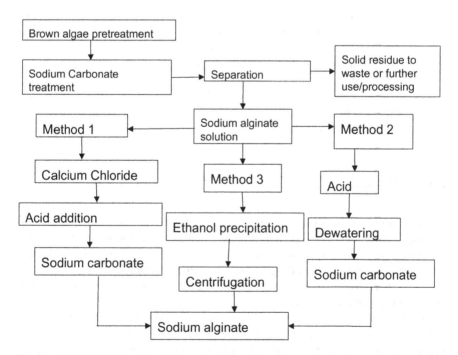

Fig. 4.3   Flow chart illustrating alginic acid extraction from brown algae using method 1 and 2

then follows as outlined in method 2. If the acid method is used, alginic acid formed is gelatinous and cannot be separated by filtration, and flotation is used in this case. The gelatinous alginic acid is then dewatered followed by reconversion to sodium alginate as outlined in method 1.

The color of the sodium alginate extract is very important in enhancing or diminishing its market value. Lighter colored alginate attracts better price. This necessitates a decoloring process either before or after the alkali treatment. Decoloring can be achieved by either bleaching the final product with sodium hypochlorite to remove color or soaking the seaweed in formalin prior to extraction in order to adhere the pigment to the cellulose such that the pigment is retained in the cellulose while the sodium alginate is extracted.

4.5.5 Extraction of Calcium Alginate and Other Salts of Alginic Acid

The process of extracting sodium alginate also involves formation of calcium alginate, and therefore, if calcium alginate is the desired product, following the initial treatment of the seaweed with hot sodium carbonate, the calcium salt route of method 1 can be taken (McHugh 1987). Addition of calcium salt such as calcium chloride will result in the formation of calcium alginate which is an insoluble fibrous alginate that can be separated by filtration. Although sodium alginate is more widely produced, calcium alginate finds increasing applications in food and pharmaceuticals. The other salts of alginate are obtained following neutralization of alginic acid using alkali like potassium carbonate and ammonium hydroxide. The alkali used depends on the desired salt.

However, propylene glycol alginate is made in a different way. It is an ester of alginic acid, and its application varies from that of the alginate salts. The process of producing propylene glycol alginate begins from the moist alginic acid which has partially reacted with sodium carbonate. This partially reacted alginic acid is then treated with propylene oxide at a temperature of 80 °C for two hours under pressure.

4.5.6 Enzyme Extraction of Alginate

Although the chemical extraction method is almost the only method used in commercial production of alginate, other methods of extraction are possible greener alternatives to obtain alginates. For such process, alginate lyases can be employed, and however, these alginate degrading enzymes are primarily used in breaking down the alginate chains into oligosaccharides of alginate, mannuronic and guluronic acids (Rhein-Knudsen et al. 2015). These have been shown to have some bioactive properties and show potential for applications in biomedicine and biofuel production. The enzyme extraction method is based on the alginate lyase breaking down the alginate within the cell walls into shorter chains and monomers which are more water soluble. These are then released in the aqueous form and can be separated from the rest of the solid. Enzyme extraction is rarely used for alginate.

4.6   Environmental Implications

Having established an understanding of the source, chemistry and production process for alginate, here we evaluate the environmental impact of the stages involved in the extraction process. Some estimates on material and energy consumption during extraction are given based on results from extraction of alginate in reported studies. Table 4.1 summarizes the consumption for a typical bench-scale extraction of sodium alginate from *L. digitata* (Fertah et al. 2017).

4.6.1   Brown Algae Cultivation and the Environment

Since alginates are obtained from brown seaweeds which are mainly sourced from wild, this section focuses on the environmental impact of harvesting of wild seaweed for alginate production. Seaweeds have a major environmental benefit being extractive species, and they serve the benefit of extracting waste material from the environment. They produce their food using the process of photosynthesis like other

Table 4.1 Typical consumption for bench-scale extraction of alginate from brown algae

Consumption	Quantity
Brown algae	1.93 g per gram alginate produced
Water	60 ml[a] per gram algae
Formaldehyde	15 ml[a] of a 2% solution/gram algae
HCl	15 ml[a] of a 0.2 M solution/gram algae
Sodium carbonate	15 ml[a] of a 2% solution/gram algae
Ethanol	15 ml[a] of 95% solution per gram algae
Energy	
Centrifugation	3500 rpm for 15 min[b]
Drying	60 °C for 8 h
Heat for extraction	40 °C for 3 h[b]
Agitation	200 rpm for 5 h[b]

[a] Assumes a 1:15 g/ml, mass-to-volume ratio of algae dry mass-to-volume of liquid; [b] typical values of time and rpm are used where not specified in the reported study

plants, and they synthesize their food from nutrients extracted from the aquatic system they grow in. These could be the waste produced by other species, thereby cleaning up the environment which helps to promote and sustain the growth of other species. Such that farming of seaweeds alongside other fed species of aquatic animals within the same culture is encouraged in order to benefit from the synergy between extractive and fed species for optimal use of resources. This has become common place as the farming of extractive species is estimated at 49.5% as of 2016 (FAO 2018).

Another advantage of wild seaweed as a source of alginate is that it requires even much less intensive care such as watering, fertilizer application and weeding as required by land crops. On the other hand, the cost of sending vessels out to sea and into rocky coasts, the machinery and manpower required for harvesting contribute to the cost of the raw material which varies around 50–100 USD per MT depending on the quality and source (Konda et al. 2015).

Although seaweed grows naturally in the wild, meeting the demand for future commercial alginate production cannot be sustained by the natural stocks alone. This means new technologies need to be developed for more profitable cultivation of alginocytes on commercial scale. There are also issues concerning the overexploitation of natural seaweed resource as the demand for seaweed continues to rise. Recent concerns of overexploitation, environmentally harvesting methods and diminishing seaweed beds have lead to restriction of harvesting of some species of seaweeds in countries such as Canada and Portugal (Monagail et al. 2017).

The seaweeds in nature play a significant role in sustaining the aquatic flora and fauna. They help remove waste products of aquatic animals preventing toxic levels which could be harmful to the animals and other organisms. They also provide shade from light to aquatic organisms which thrive better under low light, and they serve as habitat and refuge to some aquatic organisms, offering protection from predators. Seaweeds also have the impact of dampening tidal waves by absorbing the wave energy, thereby preventing or reducing coastline erosion (Monagail et al. 2017). Seaweed harvesting has also been shown to significantly alter biodiversity and population of some species in areas such that commercial-scale harvesting of wild seaweed for alginate production pose some significant impact on aquatic ecosystem.

Global climate change has been shown to have affected seaweed species distribution. The reported 0.6 °C increase in temperature has had notable impact on species distributions in areas such as Spain and Portugal, while extreme weather conditions like earthquakes, tsunamis and el nino have resulted in total eradication of seaweed beds in areas such as Chile (Castilla et al. 2010).

Brown algae are typically harvested by cutting the upper part where much of the useful carbohydrates are found, leaving behind the lower part which allows the regrowth of the plant within a year or two. The rate of regeneration after harvest depends on factors such as the efficiency of the harvest method, species, environment and others. The harvest could be done by hand, using rakes with boats, diving, use of cutters or mechanical means. Mechanical harvesting has been discontinued in some areas due to adverse effect on the seaweed population and environment (Monagail et al. 2017).

Seaweed cultivation, although requires less land space, does not always involve less resource usage when compared to terrestrial crops such as corn, sugar beet or potatoes. The resource consumption varies for different methods of cultivation, and brown algae cultivation is particularly more resource intensive as the reproductive process of brown algae is a more complex alternating reproductive cycle compared to those which reproduce vegetatively. Aquaculture of cush algae generally requires more resource use than wild-sourced seaweed. Because some seaweed cultivation involves permanent or partial occupation of aquatic environment, this could have further impacts such as blockage of sunlight for aquatic life below and mechanical displacement. Cultivation of seaweed, however, reduces problems such as loss of seaweed species through overharvesting of wild stocks.

Contrary to algae grown for direct food consumption which are mostly grown on a commercial scale in aquaculture, seaweed for algae production is best grown wild and harvested from natural source. Seaweed harvest for algae production, therefore, requires less resource consumption outside of transport and the much less temporary impact of the process of harvesting.

4.6.2 Chemicals

Based on the sample case study (Konda et al. 2015) which is representative of a typical bench-scale extraction process, producing 26 mega tones of sodium alginate pellets requires 720 MT formalin at a concentration of 0.1%, 835 MT of HCl at 0.38% concentration, 73 MT of sodium carbonate at 10% concentration for extraction, 500 MT of calcium chloride at 10% concentration for the formation of insoluble calcium alginate salt and a further 4.2 MT of sodium hypochlorite at 5% for bleaching. More HCl is then used for neutralization after bleaching requiring a further 150 MT of 3.6% concentration. Sodium carbonate is then added again to reform the sodium alginate salt and this requires about 80 MT at 10%. Finally, the extracted product is purified and dried using around 140 MT of air and 281 MT of a mixture of ethanol, methanol and acetone at around 48% or as required. Although alginate is a biopolymer from a natural renewable source, the use of mineral-based acids and alkali and other chemicals may have contradictory impact on the objective of alginate as a greener alternative to synthetic fossil-based polymers.

4.6.3 Land Use

A general advantage of aquatic sourced raw materials for biopolymers is that the raw materials do not require large land areas for production; rather, they make use of aquatic space. This is significant as land spaces for human habitation are becoming more limited due to the increasing population of humans and the need for more land space to grow food crops and rear land animals for survival of the population.

Table 4.2 Compositions of Brown seaweed (*S. latisimma*) and corn stover

	Component	Composition (%)
Brown algae (*S. latissima*)[a]	Alginate	32
	Protein	~0
	Laminarin	15
	Mannitol	18
	Ash/salt	35
Corn Stover	Hemicellulose	19
	Lignin	16
	Protein	3
	Cellulose	35
	Other	27

[a]For the same species, values vary significantly with harvest period, growth conditions

In addition to this, the biomass harvested for alginate production does not remove nutrients from the soil. On the contrary, seaweeds remove waste from the water and carbon dioxide from the air. Therefore, their growth, whether wild or cultivated, contributes positively to the environment.

Seaweeds also grow more rapidly than other terrestrial plants such as sugarcane or maize as they carry out photosynthesis 3–4 times more efficiently. Seaweed could grow between 30 and 80 dry MT per hectares annually while other terrestrial plants grow at a rate of 3–30 dry negatonned per hectares per year. By virtue of its size only, brown seaweed among the seaweed serves as an abundant source for carbohydrate biomass and has been shown to have little to no lignin content. This makes extraction of the carbohydrate content relatively less demanding than biomass with higher lignin content. Table 4.2 compares the carbohydrate content of brown seaweed to that of corn stover on dry basis. Note that this is composition for a specific species (*S. latissima*) and the compositions also vary with season. The compositions here are based on three harvests between December 2010 and August 2011 (Konda et al. 2015).

4.6.4 Energy Consumption

The seaweed is often processed with a starting moisture content of around 25% drying which is energy intensive, and therefore, some of the cost of drying can be reduced by drying in the sun. The cultivation and harvesting have been determined to be the most energy-consuming stage. This requires sending out vessels to source the seaweeds which could be several miles offshore.

The initial alkali extraction stage for both methods is carried out at 80 °C (Fertah et al. 2017). Advancement in enzyme assisted alginate extraction could offer lower energy-consuming process in the future. This is the stage in the extraction where the most heat is consumed. Further energy is consumed in running the equipment for dewatering processes which could be done by filtration or gel pressing. These consume less energy than evaporation of the sodium alginate from aqueous solution.

Assessment of algae seaweed production in Ireland revealed that 75.1% of seaweed sourcing resource consumption is attributed to fossil fuel consumption required for transportation and machinery to operate aquaculture (where the algae are not from wild stock) (Taelman et al. 2015).

4.6.5 Water Consumption and Wastewater Generation

An estimated 1500 MT of water used in the extraction process and further 3900 MT of water for dilution in production of 26 MT of sodium alginate dry pellets (10% moisture content) in the case study presented by Konda et al. (2015). Harvesting brown algae from wild stocks makes little direct use of water, and alginate production makes use of a lot of water to the extent that the survival of an alginate production factory is largely affected by reliable availability of abundant water. The wastewater generated has high pH from the initial extraction stage while the later extraction stage will generate wastewater containing calcium or acid, depending on the method used. It also contains the compounds used to decolorize. The generated wastewater can either be recycled in-house or sent to wastewater treatment plant before releasing it to the sea in accordance with the local laws.

4.6.6 Carbon Dioxide Emission

Commercial alginate extraction process results in CO_2 emission from the running of the alginate production facilities. On the other hand, seaweeds are photosynthetic aquatic plants, whose rate of photosynthesis is 3–4 times more efficient than that of terrestrial plants. The CO_2 emission in the extraction process is therefore partly compensated for by the CO_2 removal during photosynthesis. This is even more advantageous for aquatic plants as the plants do not remove nutrients from the soil.

4.6.7 Solid Waste Generated

The main solid waste generated is the seaweed residue after the alkali sodium alginate extraction. The solid residue also contains the formalin and filtration aid if used. Although details of the process used by commercial producers in the treatment and deposition of the solid waste generated from alginate production from seaweed are not provided here, some research publications have presented innovative ways of processing and reusing such solid wastes. These include reuse of adsorption of heavy metals in waste treatment (Romero-Gonzalez et al. 2001) and the use of the laminarin and mannitol content for bioethanol production (Horn et al. 2000). The utilization of the solid wastes generated following the extraction of alginate further adds value to the feedstock which the alginate producer could either use in a coproduction process or sell to biorefineries.

4.6.8 Diversion of Resource for Alginate Production

Seaweeds for centuries have been harvested for food, especially in Asia where they have constituted an essential part of the cuisine for centuries such as seaweed wraps, salads and soups. Wild seaweed has high market value as direct consumption as food. The process of converting the harvested seaweed into food requires much less capital investment, and the cost of production is covered by the payment in service rendered, compared to the extraction of alginate from seaweed which requires additional facilities, materials, energy and skilled labor. The justification of these additional inputs could lie in the role alginate which plays as a food additive in improving food quality and shelf life which contributes toward the sustainable development goal of making more food available for more people, in cases where alginate does play such a role.

Brown seaweeds are also used as fertilizers, and therefore, the alginate industry competes with these other products for the brown algae resource. For example, in Europe brown seaweed species such as Laminaria and Ascophyllum are used for fertilizers, and these are the same species widely used for alginate production. Other than alginate seaweeds also have potential in biorefinery as bioethanol for fuel production and other uses. Although requiring much more running cost and capital investment than for example ethanol, alginate production cost is compensated for by the relatively higher selling price estimated at around 14–30 USD/kg (Konda et al. 2015) compared to ethanol which sells at around 2.5 USD/gal.

4.7 Applications

Global demand for alginate is estimated at 26,500 MT per year (Konda et al. 2015). Application of alginates ranges from the use as a thickener in baked food, stabilizer of beer foam, bandages, textiles and biomaterials for tissue repair. Today the demand from more products to be made from naturally sourced polymer encourages the production of polymers such as alginate.

4.7.1 Alginates in Food

One of the widest applications of alginate is in food. Its excellent gelling and viscosity-enhancing properties make it suitable as a thickening, gelling agent, emulsion stabilizer and texture modifier in foods such as dressings, ice creams, noodles, beer, jelly and others. Alginate is FAO, WHO and FDA approved for food applications. The forms of alginate used in the food industry are alginic acid, sodium alginate, potassium alginate, ammonium alginate, calcium alginate and propylene glycol alginate. These are labeled E400–E405, respectively Featherstone (2015).

Alginates in food industry allow storage of food in cans, jars, hard gels, etc. for longer periods while retaining or even improving the texture and consistency. In modifying the texture, retaining moisture and improving the texture of foods, alginate serves the role of extending food shelf life and preserving the organoleptic properties over the food shelf life period. This has the economic and environmental impact of reducing food waste and serving as an alternative to more energy-intensive preservation methods such as freeze drying or freezing. Foods such as vegetables, fruits, cheese, seafood and meat can be stored at room temperature when preserved with alginate in methods such as canning and sealing.

The use of alginate can also be used to reduce the amount of other components needed due to its superior viscosity enhancing and gelling property. For example, where 5% starch is used as a thickener, about 4% of the starch can be replaced with 1% low viscosity sodium alginate with calcium salt. This allows processing such as improved heat transfer during heat sterilization and mixing, unlike starch which is more sensitive to heat. The sodium alginate then reacts with the calcium salt over time after processing to achieve the desired viscosity in the sealed food. The viscosity difference can be up 10x Featherstone (2015). So with sodium alginate, you can process the food at lower viscosity for easy mixing and heat transfer and then package and seal while the sodium alginate in the presence of calcium salt continues to increase in viscosity over a period of several hours or days.

Sodium alginate is the alginic salt which is soluble in water and is used in the food industry for its viscosity-enhancing property. It also forms a non-thermoreversible stable gel at room temperature on addition of calcium salt such as calcium chloride or other polyvalent metal ions. This property makes it useful as a thickener, gelling agent and stabilizer in the food industry. Polypropylene glycol alginate is also used in the food industry as an emulsion stabilizer and stabilizer.

Even better than biodegradable packaging is edible packaging in the form of films or coatings. Being indigestible edible polysaccharide alginate packaging can contribute to the fiber intake. Alginate can serve as a sacrificial moisture agent, whereby the moisture which would otherwise evaporate from the food evaporates from the alginate coating instead such that the food retains its moisture content. Alginate is used to produce edible films with thickness no more than 0.3 mm which can be eaten with the food or easily removed. This packaging is to be as safe as the food itself and should not alter the organoleptic properties of the food. Methods for producing alginate films include solvent casting and extrusion (Olatunji and Olsson

2016) while methods of applying coatings could be dipping, spraying and vacuum impregnation (Parreidt et al. 2018).

Alginate coatings can also improve the aesthetic appeal of the food by improving gloss, reducing the chances of bruising by giving a more rigid surface and even hiding surface imperfections. Aesthetics are important in reducing food wastage as some foods are rejected simply due to their appearance whereas the food still has all its nutrients, taste and texture intact. Therefore, improving the aesthetics will reduce the chance of the food being rejected and left to spoil.

To enhance other properties such as antibacterial and antioxidant properties, alginate-based edible films and coatings usually incorporate other compounds with these desired properties. The incorporation of these compounds in the coating or film has the advantage of not being included in the main food in case where for example the direct contact with the food could initiate some undesirable interactions which alters the food property or quality. Mechanical properties such as elasticity and film-forming ability of alginate could be improved with the use of plasticizers or forming composites with other compounds (Parriedt et al. 2018).

4.7.2 Textiles

Alginate serves as a more environmentally friendly printing substrate for laying down color patterns on cotton, rayon and jute in the textiles industry, as it is a more biodegradable option. Its viscous nature allows formation of printing pastes of good consistency and being hydrophilic as well as its ability to form a paste at relatively moderate water temperature. Sodium alginate is commonly used in printing pastes as it prevents the otherwise hardening of the fabric once reacted with the dye. When sodium alginate is used as a printing paste, a better fabric feel is achieved. Sodium alginate also achieves good color yield and maintain stability. Application of sodium alginate in textiles has been practiced for many decades (Rompp et al. 1983; Wang et al. 2014), and as demand grows for novel fabric technologies and more biodegradable fabrics, alginate is likely to find increasing application in the modern textile industry. Low to medium molecular weight sodium alginate is more suitable for textile applications. Alginate printing pastes can be applied for screen printing, roll printing and inkjet printing. In terms of printing paste, sodium alginate has superior properties to other options such as starch.

4.7.3 Lowering Blood Sugar Level

Diabetes has become a medical condition of increasing significance with increasing incidence annually across the globe. Use of alginate-based dietary supplements as either a preventive or therapeutic measure further indicates high demand for alginate-based products in the pharmaceutical and nutraceutical industry.

Alginates in forms such as calcium alginate, sodium alginate and alginate gels have been shown to have the effect of reducing blood sugar level when consumed orally (Hisni et al. 2016). The mechanism by which this is achieved has been studied to be due to the suppression of starch digestion while calcium alginate did not affect the permeability of glucose across membranes nor bind to the glucose to prevent absorption into the bloodstream as is the mechanism for some blood sugar lowering substances, calcium alginate acts by inhibiting the action of the enzyme glucosidase which breaks down starch to sugars. It is thought that at a dosage of 5% body weight with a particle size of 53 μm, calcium alginate can aid in reducing blood sugar levels (Idota et al. 2018). Some studies in human subject also indicate potential for alginate gels to aid in reducing the uptake of cholesterol as well as glucose as a means of controlling levels in the blood at a dosage of 1.5 g, reduced cholesterol and glucose levels were measured in the test subjects. The mechanism by which this is achieved is thought to be through the delay of the uptake by the strong gel (Paxman et al. 2008). Retention of the glucose and cholesterol within the gel or acting as a barrier between the intestine wall and the glucose or cholesterol could prevent or delay their uptake into the bloodstream resulting in more of them being passed along the alimentary canal along with the alginate fiber.

4.7.4   Biomedical Application

In tissue engineering, different forms of alginate are applied for the production of biopolymer-based extracellular matrix (ECM) to promote tissue regeneration. Their ability to provide stability in hydrogel formulations and maintain an aqueous environment by absorbing and retaining biological fluids makes alginates attractive for such purpose. Advanced studies up to clinical trials on animals and humans have been carried out on a number of alginate-based hydrogel implants for cardiac regeneration (Liberski et al. 2016). Presently, there is no cure for cardiac failure, and treatments exist to manage cardiac illnesses. Among the potential approaches to treat heart failure is the use of alginate-based biomaterials to achieve self-repair of the cardiac tissue. The properties of alginate which allows them to form viscous fluids when dissolved in water and to form hydrogel when reacted with calcium salts such as calcium chloride make them suitable for application as injectable implants for cardiac repair. Here, the unique hydrogel-forming property of alginate comes as an advantage as it can form hydrogels in physiological fluid under mild temperatures ~40 °C). These hydrogels can mimic the biomechanical properties of the cardiac tissue. The biodegradability of alginate is also important for such purpose as it remains stable enough to allow formation of the ECM, and once the tissue is regenerated, it can biodegrade. Companies such as Life Technology Inc., LoneStar Heart. Inc and Bellerophon BCM LLC have developed alginate-based products for cardiac tissue regeneration, some at various stages of clinical trials such as AlgiMatrix, Algisyl-LVR and PRESERVATION. Different salts of alginates with varying structures and molecular weight are used and tailored for specific mechanism of action. In the area of cardiac tissue regeneration, alginates have also been used as a pharmaceutical excipient for the release of angiogenesis drugs, cell transfer, and promote interaction between implant and cardiac cell.

Alginate-based hydrogels are also used as injectable fillers in cosmetic dermatological procedures to alter facial features such as lip plumpness and appearance. One such product is commercially available since 2010 under the brand name Novabel (Moulonguet et al. 2011).

4.7.5 Alginate Oligomers and Monomers

Oligosaccharides of alginates are of commercial significance as they offer a range of high-value bioactive properties. These bioactivities include antioxidant, antiinflammatory and antiproliferative effects (Wan et al. 2018). For this reason, they have found applications in food (animals and humans), agriculture, pharmaceutics and cosmetics industry.

In agriculture, alginate oligosaccharides are used for their growth-promoting abilities to promote endothelial cells and multiple cytokines. When used as animal feed supplement, alginate oligosaccharides boost serum hormone levels and antioxidant activity and also improve intestinal functions and integrity in pigs (Wan et al. 2017).

The method of degradation of alginate by alginate lyase has been shown to be a factor in influencing the bioactivity of the resulting oligomer. Alginate lyase-derived oligomers had superior bioactive properties such as antitumor effect while acid hydrolyzed alginate oligomers only had the reduced chain length without the bioactive properties (Takeshita and Oda 2016).

4.7.6 Other Applications

Other uses of alginates include use in animal feed to improve texture and stability and in welding rods where sodium silicate is coated unto the surface of the metal rods. Alginate is used to bind the coating and hold it in place hence aiding a uniform coating upon drying. This is achieved by the gelling property of alginate. In the paper, industry alginate is used as a partial replacement for rosin pulp. It is used to achieve a smoother surface and better absorption of ink during printing. It also aids production of crumple resistant paper by promoting absorption of wax and grease. Alginates are attractive in micro- and nanoparticle formulation due to their ability to form stable structure (Paques 2015). Alginate is also used in the encapsulation of live bacteria and active enzymes (Qin et al. 2018). Another well-known application of alginate in the cosmetic industry is for use as a color retainer in lipstick.

4.8   Commercial Production

Alginate was discovered in 1882, and its commercial production was first established in San Diego in the USA in 1927. Production of alginate in the UK followed between 1934 and 1939 years later began in Norway (McHugh 1987). Today alginate is being produced commercially in countries in Europe, Americas, Asia-Pacific and Asia. From its first use as a binder for briquette production to now being used in various industries from food to textiles to pharmaceuticals alginate application has advanced through the years across different industries.

Compared to some lesser-explored biopolymers, alginate is moderately well explored commercially in various industries. Annual production of alginate from seaweed is estimated at 38,000 tonnes (Venkatesan et al. 2014). Alginate production from bacteria is yet to be commercialized. The world alginate market is estimated to be worth 624 million USD as of 2016. The highest demand for alginate to date is in the food industry where they find various applications (note this is different from the direct consumption of seaweed as food). As the world population increases and the level of civilization increases, demand for processed foods is expected to consequently increase and alongside this will be increased in demand for food processing agents such as ice cream, beer and packaged meat, all of which require the use of alginate of one form or the other in their processing. Propylene glycol alginate has shown steady rise in market revenue between 2014 and 2018. It is expected to show further increase in market size for the next 7 years. Sodium alginate which is also in high demand in the food industry shows similar steady reported and potential increase. Although calcium and potassium have lower rate of production in the industry, they are also expected to show increase in revenue over the coming years, especially as more novel applications for these forms of alginate emerge.

Cost of production of algae is still relatively higher compared to the commercial production of well-established mechanized crop farming of crops such as corn and soya bean. Each dry tonne of *L. japonica* costs between 650 and 700 USD to produce in China and producing the carrageenan-producing species of seaweeds in Mexico costs around 689 USD. This is higher than the estimated cost of producing corn and soybeans in the USA which are estimated at 195 USD and 408 USD, respectively (Forster and Radulovich 2015).

Alginates with higher guluronic acid content are in higher demand in the food industry since this offers better gelling properties. While alginates with high M offer better viscosity-enhancing property, other polymers exist which offer similar, and however, the superior gelling properties are more unique to alginate. The demand for alginate with higher guluronic acid is estimated to grow at a rate of 4.8^ CAGR between 2016 and 2025.

Manauronic rich alginate finds more use in the textiles and paper industry where their viscosity-enhancing properties are more valued. The superior binding properties of alginate also makes it useful as welding rod as more developing countries begin to focus on the development of manufacturing and construction, easily sourced raw materials from renewable sources such as alginate begins to increase in demand.

Calcium alginate, more commonly used in the pharmaceutical industries for applications such as wound dressing, is estimated to reach a market value of 154.3 million USD by 2025. The use of calcium alginate as a wound healing aid also makes it applicable in food as a meat binder. This increased product use in newer applications in the industry is suspected to result in increased demand and hence commercial production.

Propylene glycol alginate, commonly used in the food industry as an emulsifier, thickener and stabilizer, is also used in the beer industry for foam stabilization. The beer industry is one of the largest industries from large-scale beer manufacturing across the globe in both developed and developing countries and rising such application further points to expected growth in the alginate industry. Propylene glycol alginate is also used in fruit juices and foamy dairy products. 26% of the demand for alginate is allocated to propylene glycol alginate (Grandview Research 2017).

The commercial production of alginates can be said to be well explored across different countries. Key players in the alginate market includes as follows: ISP Alginates Ltd. (UK), FMC Biopolymer USA, Degussa Texturant Systems in Germany, Kimica Corporation in Japan, Danisco Cultor in Denmark, Figu Chemical Industry Co. Ltd. in Japan, Algisa Compania Industrial de Alginatos S.S. in Chile and China Seaweed Industrial Association In China (McHugh 2003), Seasol International, Acadian Seaplants Ltd., Yan Cheng, Chase Organics GB Ltd., Indigrow Ltd., Mara Seaweed, CP Kelco, Aquatic Chemicals and Pacific Harvest. Many others exist operating at small and larger scales across the globe.

Commercial alginate is priced around 12USD per Kg (Hermandez-Carmona et al. 2013), making it the second most expensive seaweed sourced polysaccharide, second to Agar. By region, Europe took the lead in alginate production as of 2003 producing 16,000 tonnes of alginate (McHugh 2003), followed closely by Asia-Pacific which produced 15,600 tonnes and then the Americas which recorded a production of 4500 tonnes. There was no alginate production recorded in Africa. As of 2013, the combined production of alginates, agar and carrageenan from seaweed was worth 1.018 Billion USD as of 2013 alginate was valued at 12 USD per Kg (Hernandez-Carmona et al. 2013).

The main issue which surrounds future market of alginate is the future availability of seaweed for alginate production as the natural stocks available are a limited resource. If well managed, with harvest rate balanced by the rate of regeneration and recovery, long-term availability of raw material for alginate production should be of little concern. The sustainable strategies being adopted by the alginate market stakeholders which includes the manufacturer, government and regulatory bodies and environmentalists include sustainable harvesting of seaweed for alginate production and research and development into alternative and sustainable alginate sources. Other factors which threaten the natural stock of seaweed are environmental disaster which could destroy large stocks of seaweed at an instance such as tsunamis. These factors are all important when considering the prospects of commercial alginate production.

Cost analysis of alginate production is presented in Table 4.3. The estimate used below is obtained from the scenarios presented in literature (Konda et al. 2015) where cost analysis is carried out for simulated scenario of the coproduction of

Table 4.3 Cost estimates for production of 130,000–220,000 MT/year of alginate

Cost	Estimates in Million USD/annum
Facilities cost	155
Raw materials	300
Waste treatment	30
Utility	40
Annual operating cost	520

ethanol and alginate from brown algae. The estimated cost for alginate is obtained by subtracting the cost estimates for ethanol production from brown algae from the cost of coproducing of ethanol and alginate. An assumption is made that the additional cost is from the additional process for alginate production and is independent of ethanol production. The figures given here could overestimate or underestimate the real cost of production, and however, it is used here merely for comparison of the costs. The estimated output of 130,000–220,000 MT annually assumes a yield of 50–88% (extracted alginate/theoretical content).

The potential for coproduction of alginate alongside ethanol production from seaweed biomass is also a possibility. However, assessments show that for such biorefineries, the alginate production rate required to make the process economically viable would exceed the current global demand for alginate which could result in a crash of the alginate market resulting from surplus supply (Konda et al. 2015).

4.9 Conclusion

Alginates have relatively diverse applicability which cuts across several industries and therefore has a large commercial significance. Extraction of alginate requires use of acids, alkali and other chemicals which pose some environmental impacts. The process of harvesting and cultivating brown algae for use in alginate production offers some benefits to the environment such as CO_2 removal from the atmosphere and removal of nutrients from water thereby abating eutrophication, when well managed. Future advancement could possibly see the use of alginates in bioethanol production, and this is dependent on further advancement in developing enzymes to catalyze the breakdown of alginate into fermentable sugars. Alginate is one of the aquatic biopolymers with a well-established market, extraction and applications. Nonetheless, there still exist several aspects where advancements are required, such as the development of enzymes for controlled hydrolysis and more environmentally and economically sustainable extraction processes.

References

Aida TM, Yamagata T, Abe C, Richard L, Smith J (2012) Production of organic acids from alginate in high temperature water. J Supercrit Fluids 65:39–44

Aida TM, Yamagata T, Watanabe M, Smith RL Jr (2010) Depolymerization of sodium alginate under hydrothermal conditions. Carbohydr Polym 80(1):296–302

Aleksandra R, Sanja SI (2017) The Influence of nanofillers on physical- chemical properties of polysaccharide- based film intended for food packaging. In: Food packaging, pp 637–697

An QD, Zhang GL, Wu HT, Zhang ZC, Zheng GS, Luan L, Murata Y, Li X (2009) Alginate-deriving oligosaccharide production by alginase from newly isolated Flavobacterium sp. LXA and its potential application in protection against pathogens. J Appl Microbiol 106:161–170

Aravamudhan A, Ramos DM, Nada AA, Kumbar GS (2014) Natural polymers: polysaccharides and their derivatives for biomedical applications. In: Kumbar SG, Laurencin CT, Deng M (eds) Natural and synthetic biomedical polymers. Elsevier Press, pp 67–89

Bixler HJ, Porse H (2011) A decade of change in the seaweed hydrocolloids industry. J Appl Phycol. https://doi.org/10.1007/s10811-010-9529-3

Castilla CJ, Manriquez PH, Camano A (2010) Effects of rocky shore coseismic uplift and the 2010 Chilean mega-earthquake on intertidal biomarker species. 418:17–23

Charrier B, Abreu HM, Araujo R, Bruhn A, Coates JC, De Clerck O, Katsaros C, Robaina RR, Wichard T (2017) Furthering knowledge of seaweed growth and development to facilitate sustainable aquaculture. New Phytol 216:967–975

Chen F, Long J (2018) Influences of process parameters on the apparent diffusion of an acid dye in sodium alginate paste for textile printing. J Clean Prod 205:1139–1147

Clementi F, Crudele MA, Parente E, Mancini M, Moresi M (1999) Production and characterization of alginate from *Azotobacter vinelandii*. J Food Sci Agric 79(8):602–610

Donnan FG, Rose RC (1950) Osmotic pressure, molecular weight, and viscosity of sodium alginate. Can J Res 28b(3):105–113

Draget KI (2009) Alginates. Handbook of hydrocolloids, 2nd edn. Woodhead Publishing, Sarston, pp 807–828

Falkeborg M, Cheong LZ, Gianfico C, Sztukiel KM, Kristensen K, Glasius M, Xu X, Guo Z (2014) Alginate oligosaccharides: enzymatic preparation and antioxidant property evaluation. Food Chem 164:185–194

FAO (2018) The state of world fisheries and aquaculture 2018—meeting the sustainable development goals. Rome. Licence: CC BY-NC-SA 3.0 IGO. ISBN 978-92-5-130562-1

Featherstone S (2015) Ingredients used in the preparation of canned food. In: A complete course in canning and related processes, 4th edn. 2(Microbiology, packaging, HACCP and ingredients) pp. 147–211

Fertah M, Belfkira A, Dahmane A, Taourirt M, Brouillette F (2017) Extraction and characterization of sodium alginate from Morooccan *Laminaria digitata* brown seaweed. Arab J Chem 10:53707–53714

Forster J, Radulovich R (2015) Seaweed and food security. In: Tiwari BK, Troy DJ (eds) Seaweed sustainability. Elsevier Academic Press, pp 289–313

Grandview Research (2017) Alginate market analysis by type (High G, High M), by product (Sodium alginate, calcium alginate, potassium alginate, propylene glycol alginate). By application, and segment forecasts, 2018–2025. Report ID: GVR-2-68038-244-0, pp 1–127

Helmiyati, Aprilliza M (2017) Characterization and properties of sodium alginate from brown algae used as an ecofriendly superabsorbent. In: IOP conference series: materials science engineering, 188 012019 https://doi.org/10.1088/1757-899x/188/1/012019

Hernandez-Carmona G, Freile-Pelegrin Y, Hernandez-Garibay E (2013) Conventional and alternative technologies for the extraction of algal polysaccharides

Hisni A, Purwanti D, Ustadi A (2016) Blood glucose level and lipid profile of streptozotocin-induced diabetes rats treated with sodium alginate from sargassum crassifolium. J Biol Sci 16(3):58–64

Horn SJ, Aasen IM, Ostgaard K (2000) Ethanol production from seaweed extract. J Ind Microbiol Biotechnol 25(5):249–254

Idota Y, Kato T, Shiragami K, Koike M, Yokoyama A, Takahashi H, Yano K, Ogihara T (2018) Mechanism of suppression of blood glucose level by calcium alginate in rats. Biol Pharm Bull 41(9):1362–1366

Iwasaki KI, Matsubara Y (2000) Purification of alginate oligosaccharides with root growth-promoting activity toward lettuce. Biosci Biotechnol Biochem 64:1067–1070

Konda M, Singh S, Simmons BA, Klein-Marcuschamer D (2015) An investigation on the economic feasibility of Macroalgae as a potential feedstock for biorefineries. Bioenergy Res 8:1046–1056

Lakshmi SD, Trivedi N, Reddy CRK (2017) Synthesis and characterization of seaweed cellulose derived carboxymethyl cellulose. Carbohyd Polym 157:1604–1610

Liberski A, Latif N, Raynaud C, Bollensdorff C, Yacoub M (2016) Alginate for cardiac regeneration: from seaweed to clinical trials. Global Cardiol Sci Pract 4:1–25

McHugh DJ (1987) Production, properties and uses of alginates In "Production and utilization of products from commercial seaweeds. FAO Fish Tech Paper 288:58–115

McHugh DJ (2003) A guide to the seaweed industry. FAO fisheries technical paper 441. FAO, Rome. ISBN 92-5-104958-0, Chapter 5, pp 1–12

Monagail MM, Cornish L, Morrison L, Araujo R, Critchley AT (2017) Sustainable harvesting of wild seaweed resources. Eur J Phycol 52(4):371–390

Moulonguet I, de Goursac C, Plantier F (2011) Am J Dermatopathol 33(7):710–711

Mushollaeni W (2011) The physicochemical characterization of sodium alginate from indonesian brown seaweeds. Afr J Food Sci 5(6):349–352

Nagarajan A, Shanmugam A, Zackaria A (2016) Mini review on alginate: scope and future perspectives. J Algal Biomass Util 7(1):45–55

Olatunji O, Olsson RT (2016) Processing and characterization of natural polymers. In: Natural polymers, industry techniques and applications. Springer, Switzerland. ISBN 978-3-319-26412-7

Paques JP (2015) Alginate nanospheres prepared by internal or external gelation with nanoparticles. In: Macroencapsulation and microspheres for food application, pp 39–55

Parreidt TS, Muller K, Schmid M (2018) Alginate-based edible films and coatings for food packaging applications. Foods 7(170):1–38

Paxman JR, Richardson JC, Pw Dettmar, Corfe BM (2008) Alginate reduces the increased uptake of cholesterol and glucose in overweight male subjects: a pilot study. Nutr Res 28(8):501–505

Qin Y, Jiang J, Zhao L, Zhang J, Wang F (2018) Applications of alginate as a functional food ingredient. Biopolymers for food Design. Handb Food Bioeng, pp 409–429. https://doi.org/10.1016/b978-0-12-811449-0.00013-x

Radulovich R, Neori A, Valderrama D, Reddy CRK, Cronin H, Forster J (2015) Farming of seaweeds. In: Tiwari BK, Troy DJ (eds) Seaweed sustainability: food and non-food applications, pp 27–59

Rhein-Knudsen N, Ale MT, Meyer AS (2015) Seaweed hydrocolloid production: an update on enzyme assisted extraction and modification technologies. Mar Drugs 13(6):3340–3359

Romero-Gonzalez ME, Williams CJ, Gardiner PHE (2001) Study of the mechanisms of cadmium biosorption by dealginated seaweed waste. Environ Sci Technol 35(14):3025–3030

Rompp W, Axon G, Thompson T (1983) Sodium alginate: a textile printing thickener. Am Dyestuff Rep 72(2):1–16

Schiener P, Black KD, Stanley MS, Green DH (2014) The seasonal variation in the chemical composition of the kelp species *Laminaria digitata*, *Laminaria hyperborea*, *Saccharina latissima* and *Alaria esculenta*. J Appl Phycol 27(1):363–373

Shkand T, Chizh MO, Sleta IV, Sandomirsky BP, Tatarets AL, Patsenker LD (2016) Assessment of alginate hydrogel degradation in biological tissue using viscosity sensitive fluorescent dyes. Methods Appl Fluoresc 4(4):044002

Taelman SE, Champenois J, Edwards MD, De Meester S, Dewulf J (2015) Comparative environmental life cycle assessment of two seaweed cultivation systems in north west Europe with a focus on quantifying sea surface occupation. Algal Res 11:173–183

Takeshita S, Oda T (2016) Usefulness of alginate lyases derived from marine organisms for the preparation of alginate oligomers with various bioactivities. Adv Food Nutr Res 79:137–160

Venkatesan J, Nithya R, Sudha PN, Kim S (2014) Role of alginate in bone tissue engineering. Adv Food Nutr Res 73:45–57

Wang L, Liu B, Yang Q, Lu D (2014) Rheological studies of mixed printing pastes from sodium alginate and modified xanthan and their application in the reactive printing of cotton. Color Technol 130(4):320–335

Wan J, Zhang J, Chen D, Yu B, He J (2017) Effects of alginate oligosaccharide on the growth performance, antioxidant capacity and intestinal digestion-absorption function in weaned pigs. Anim Feed Sci Technol 234:118–127

Wan J, Zhang J, Chen D, Yu B, Huang Z, Mao X, Zheng P, Yu J, He J (2018) Alginate oligosaccharide enhances intestinal integrity of weaned pigs through altering intestinal inflammatory responses and antioxidant status. RSC Adv 8:13482–13492

Windhues T, Borchard W (2003) Effect of acetylation on physicochemical properties of bacterial and algal alginates in physiological sodium chloride solutions investigated with light scattering techniques. Carbohyd Polym 52(1):47–52

Zhu B, Yin H (2015) Alginate lyase: review of major sources and classification, properties, structure-function analysis and applications. Bioengineered 6(3):125–131

Chapter 5
Fucoidan

Abstract Fucoidan is extracted from brown algae and echinoderms, most commonly, sea cucumber. It is a heteropolymer with fucose as its main repeating unit. However, a variety of monomers and functional groups are also present within its polymer chain. The presence of sulfate groups is attributed to many of its bioactive properties. The extracts are highly polydisperse; therefore, additional processes are often required to obtain fucoidan with uniform molecular weight. The molecular weight, degree of sulfation and monomeric unit vary significantly with species, extraction method and growth parameters of the organisms. This diversity of fucoidan also limits its pharmaceutical and biomedical applications. This chapter discusses the processes involved in fucoidan production, its chemical structure, environmental issues associated with fucoidan production, applications and industrial significance.

Keywords Fucoidan · Sulfated · Polysaccharide · Fucus · Echinoderms

5.1 Introduction

Fucoidan refers to a group of sulfated polysaccharides made up of mainly fucose alongside deoxy sugars and other monosaccharides with varying levels of branching and acetylation. They are commonly found in brown algae and echinoderms such as sea cucumber and sea urchins (Berteau and Mulloy 2003). Although there is yet much unknown about the exact structure of these fucose-rich sulfated polysaccharides, the broad range of potential applications which have been demonstrated by fucoidans provide very promising prospects, particularly in the biomedical and pharmaceutical industries. While fucoidans have gained approval as functional foods commercially available as supplements with health benefits marketed including cancer prevention, the limitation in their advancement for clinical use, despite findings that fucoidans possess such bioactivities, lies in the variation in the chemical structures which result in a variation in the bioactivity of fucoidans from different sources, seasons, fractions and extraction forms as well as expensive purification requirements. Current research efforts are therefore exploring different approaches to address these limitations. Such attempts toward developing standard fucoidans include purification techniques, development of axenic cell culture for production

© Springer International Publishing 2020   95
O. Olatunji, *Aquatic Biopolymers*, Springer Series on Polymer and Composite Materials,
https://doi.org/10.1007/978-3-030-34709-3_5

of fucoidans with controlled chemical structure and development of enzymes for hydrolysis of fucoidans toward controlling of molecular weight and attaining better understanding of fucoidan structure.

Brown algae which serve as the main known source of fucoidan production are relatively available but are also used in direct consumption as food and production of other polysaccharides of commercial value such that fucoidan is not a main priority extract presently. Strategies to promote commercial-scale fucoidan production from algae could be developing special strains with increased fucoidan content for commercial fucoidan production.

The fact that the structural variation in fucoidans results in variation of bioactivity provides opportunity for tuning fucoidan bioactivity for improved specificity of fucoidan-based drugs. This is only achievable after gaining a full understanding of the structure of fucoidan and mapping out the correlation between factors such as source and extraction methods on the chemical properties such as molecular weight, branching, monomeric structure, acetylation and sulfation. In the present state of technology on fucoidans, these parameters vary with no particular identified correlation or trend.

Sea cucumbers which are increasing in demand particularly in China are now declining in population in different seas. In China, they are no longer available in the sea, while in the coastal areas in countries like Madagascar, they are fast declining in population. They usually inhabit the epipelagic (sunlight) zone of the sea which makes them a relatively easy catch for fishers who can simply dive into catch them. They are very valuable for their bioactive properties which can be attributed to the fucoidan content among other compounds such as lactones and triterpenes (Qin et al. 2018).

Understanding the active compounds in these sea cucumbers such as fucoidan as being present in other more available resources such as brown algae could decrease the pressure on sea cucumber by providing more sources for the same active compound which yields the same health benefits. To this end, this chapter explores fucoidan, another valuable aquatic biopolymer.

5.2 Occurrence in Nature

Fucoidan can be regarded as a relatively recently discovered and naturally occurring biopolymer. Unlike, for example, cellulose, which has been in use for centuries, fucoidan appeared in the literature for the first time in 1913 (Kylin 1913). Fucoidan occurs in the extracellular matrix of brown algae and marine invertebrates, where it plays important biological functions such as maintaining cell wall integrity, tissue hydration, aiding in communications between cells, regulating osmotic pressure and serving as a defense mechanism for the organism (Deniaud-Bouët et al. 2017). Fucoidan distinguishes from the other biopolymers in the ECM of brown algae—alginates and cellulose as the sulfated fucose-rich polysaccharide. Like other polymers present in algae, the fucoidan content in brown algae shows seasonal variation. Fucoidans further show more complex seasonal variation in their structure and bioactivities (Fletcher et al. 2017).

Although not conventional sources of fucoidan, one study discovered the presence of fucoidan-like compounds in sea grasses (Kannan et al. 2013). The sulfated polysaccharide extracted from the sea grass *Halodule pinifolia* contained fucoidan-related monomers such as mannuronic acid, fucose and high level of uronic acid and showed antioxidant activity, thus suggesting the possibility that fucoidans might not be limited to brown algae and echinoderms alone.

Sea cucumbers are marine invertebrates which produce sulfated fucoidan as one of their main polysaccharide components, the other being fucosylated glycosaminoglycan (Qin et al. 2018). Species of sea cucumber include *Phylloporus proteus* (Qin et al. 2018) and *Holothuria tubulosa* (Chang et al. 2015).

5.3 Chemistry of Fucoidans

Fucoidans are sulfated polysaccharide with a sulfate group attached to some of the monosaccharide units. The polysaccharide backbone is made up of mainly $1 \rightarrow 3$ linked or alternating $1 \rightarrow 3$ and $1 \rightarrow 4$ linked fucose units with some sulfate groups attached to the oxygens. Branching occurs linking the mainly fucose backbone with other monosaccharides such as rhamnose, xylose, arabinose, galactose and uronic acid (Weelden et al. 2019). The level of branching, sulfation and polymer chain configuration and monosaccharide units present within a given fucoidan depends on the species, growth condition of the species and extraction method. These structural variations also result in significant variation in the activities of the fucoidans.

Peng et al. (2018) reported fucoidan from *Kjellmaniella crassifolia* which contains 71.68% carbohydrate and 20.04% sulfate with 31.89% of the monosaccharides being fucose and 23.54% galactose. Wei et al. (2019) reported fucose and galactose contents of 77.4% and 13.9%, respectively, in fucose extracted from brown algae. Figure 5.1 shows different backbone structures of fucoidans from different sources showing $1 \rightarrow 3$ and $1 \rightarrow 4$ linkages and example of a sulfate attachment (Weelden et al. 2019).

Fucoidans are polydisperse such that within any given sample, the molecular weight and degree of polymerization of each fucoidan polymer chain varies. This is not uncommon characteristics of polymers. Methods such as mass spectrometry, gas chromatography, MALDI-TOF and NMR are available for characterization of fucoidan, and these have been employed in understanding the chemical structure of fucoidan from different sources. In addition to the polydispersity, the average molecular weight, configuration and degree of branching vary for fucoidans from different sources (Fitton et al. 2015a, b). The fucoidans from sea cucumber tend to have less complex structure and species relationship. For example, fucoidans from

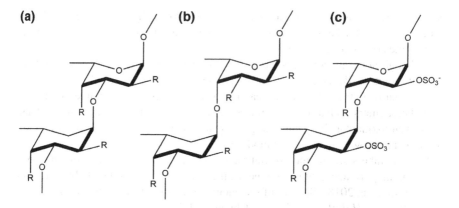

Fig. 5.1 Different backbone structures of fucoidans showing **a** 1 → 3 inked fucose units, **b** alternating 1 → 3 and 1 → 4 linked fucose units and **c** showing 1 → 4 fucose linkages and sulfate group attachments. Here R could be any monosaccharide units found in fucoidans or a sulfate group

sea cucumber of species *Isostichopus badionotus*, *Thelenota ananas* and *Acaudina molpadioides* are all linear, and those from sea cucumber species of *Apostichopus japonicus* and *Stichopus japonica* are all branched. Fucoidans from seaweeds have more complex and branched structures which could vary within the same species (Chang et al. 2015). The monomer units of the fucoidan polymer structure also vary for different brown algae sources. For instance, fucoidan from the fucaceae family consists of high fucose content relative to other monosaccharides present within the polymer structure, while in another fucoidan source, *Undaria pinnatifida*, the fucoidan chains contain more of galactose. The sulfite and acetyl contents as well as uronic acid component also vary in different fucoidan sources (Fitton et al. 2015a, b).

This wide variation presents a diverse range of properties of fucoidans from various sources. The properties of different fractions of fucoidans from the same source also vary. Fucoidans of different properties can be isolated and separated from the same batch of extract. For example, batch of fucoidan from *Sargassum muticum* from 5 to 100 kDa showed a range of properties. While fractions with molecular weight above 100 kDa contained the highest level of sulfates and phenolics, the fractions with molecular weight between 50 and 100 kDa contained 25% of the solubles and highest oligosaccharide contents. The fraction with the highest sulfates and phenolics also showed the highest antiradical activity, while the fraction with molecular weight in the range 10–30 kDa had the highest cytotoxic effect on cervical cancer cells (Alvarez-Vinas et al. 2019). Although several studies have reported relationships between bioactive properties such as anticoagulant activity and antitumoral activity of fucoidan and their structure and molecular weight, there is yet to be direct correlations between the structural properties of fucoidans and their bioactivities.

Fucoidans can be classified according to their molecular weights such as low molecular weight, middle molecular weight and high molecular weight fucoidans (LMWF, MMWF and HMWF). Low molecular weight fucoidans have molecular weight less than 10 kDa, medium molecular weight fucoidans have molecular weights in the range 10–10,000 kDa, while high molecular weight fucoidans have molecular weights above 10,000 kDa. The bioactivities are strongly affected by molecular weight; for instance, high molecular weight fucoidans have been associated with better anticancer effects (Miyazaki et al. 2018) and low molecular weight fucoidans offer better therapeutic enhancement and abate side effects when combined with chemotherapy (Chen et al. 2015). Fucoidans obtained by hydrothermal treatment of *Sargassum muticum* are separated into different molecular weights through fractionation which results in fractions with different sulfate and phenolics contents and with different radical scavenging properties (Alvarez-Vinas et al. 2019). The medium molecular weight fraction (50–100 kDa) had the highest oligosaccharide content, the >100 kDa fraction had the highest amount of sulfates and phenolics, and this fraction also showed highest antiradical properties, while the low molecular weight fractions with molecular weights in the range 10–30 kDa were more toxic against cervical cancer. Furthermore, it is possible to obtain low molecular weight fucoidan from high molecular weight fucoidan through enzyme hydrolysis. This method of obtaining LMWF leads to better bioactivity than acid-extracted LMWF from crude (Hwang et al. 2017; Sanjeewa et al. 2017).

The presence of sulfate groups attached to the sugars within the fucoidan polymer chain also affects the bioactivity of fucoidan. Generally, highly sulfated low molecular weight fucoidans have better anticancer activity than unsulfated low molecular weight fucoidan or sulfated high molecular weight fucoidan (Cho et al. 2011).

5.3.1 Degradation of Fucoidan

The biodegradation of fucoidan is important for several reasons such as production of fractions of fucoidans with improved or varied bioactivity from the crude fucoidans, understanding the structure of fucoidan and determining the biodegradation products and timeframe of such degradation. There are no known enzymes in the human body which degrade fucoidans. They are also not degraded by the enzymes present in the intestine. However, there are different means by which fucoidan can be degraded and then further utilized by humans for its bioactive properties.

Fucoidan is degraded by acid hydrolysis into smaller molecular weights and then into its sugar units. Orally ingested fucoidan can be detected in the urine (Tokita et al. 2010; Michel et al. 1996), indicating that it can withstand the conditions in the alimentary canal and still retain its chemical structure. Fucoidan deacetylase has been identified for partial degradation of fucoidan. This enzyme can deacetylate fucoidan and, however, is not able to desulfate or degrade fucoidans into lower molecular weight or fractions. That is, the enzyme activity is specific to the acetyl and sulfate bonds on the fucoidan. This enzyme was identified in the marine bacteria which utilized fucoidan, Luteolibacter algae H18 (Nagao et al. 2017). Such enzyme specificity is desirable, where a fucoidan of specific degree of acetylation is required to achieve a specific bioactivity.

A number of fungi (Rodriguez-Jasso et al. 2010), bacteria (Chang 2010) and certain invertebrates (Bilan et al. 2005) are known to contain enzymes capable of degrading fucoidan. The level of activity and mode of activity vary for different enzymes. For instance, the marine bacteria species such as Formosa algae produce the enzyme fucoidanase which has the ability to hydrolyze fucoidan (Sichenko et al. 2013). This enzyme can optimally function in a wide range of pH value (6.5–9.1). The level of activity of fucoidanase has been shown to vary for different fucoidan structures. For example, fucoidan-utilizing bacteria of the Flavobacteriaceae family extracted from seawater showed a fucoidan utilization rate of 81.5% for fucoidan extracted from sea cucumber (Chang 2010), while the enzymes from terrestrial fungi, *Aspergillus niger,* can degrade fucoidan from the brown algae Laminaria japonica, that from *Fucus evanescens* is hydrolyzed by another type of enzyme from marine bacteria. Deacetylated fucoidan has been found to be hydrolyzed more readily compared to desulfated fucoidan. This is attributed to the specificity of the enzymes to the 1 → 4 bonds of the polysaccharide chains, specifically for sulfated alpha-L-fucopyranose. Such degradation by fucoidanase is significant toward producing the immunomodulatory-active sulfated fuco-oligosaccharide.

With the aim to achieve improved pharmacological bioactivity by producing lower molecular weight fucoidan from higher molecular weight ones, Lahrsen et al. (2018) degraded fucoidan from a molecular weight of 4.9–38.2 kDa using hydrogen peroxide. However, the lower molecular weight fucoidans produced, lost their antioxidant and antiproliferative activities. It is, therefore, important that the activities of fucoidan extracts can be tailored to meet desired bioactivities by optimizing the right combination and conditions of enzyme activities.

Certain microbes, mainly marine bacteria or mollusks, contain endo- and exo-enzymes which can break down fucoidans (Kusaykin et al. 2001). These are important for either understanding the breakdown of the fucoidan-based products when exposed to the environment at the end of use or in the body or as source of enzymes used to modify fucoidan into other forms using enzymes. An example of such enzyme is fucoidan hydrolase, alpha-L-fucosidase. This can be used in combination with a desulfating enzyme arylsulfatase to break down the sulfated carbohydrate fucoidan structure (Silchenko et al. 2013). This results in the formation of sulfated oligosaccharides as a result of cleavage of the fucoidan chain into shorter chains. Fucose is also produced in the process. Interestingly, some enzymes which are capable of cleaving or hydrolyzing fucoidans from some species are not able to do the same for fucoidans from other species. This is due to the diverse structure of fucoidans as they vary for different sources. For example, the enzyme extracted from marine bacteria which degrade fucoidan from *Fucus evanescens* and *Fucus vesiculosus* does not hydrolyze fucoidan from another species *Saccharina cichorioides* (Rodriguez-Jasso et al. 2010). This difference in the degradation process limits the large-scale modification of fucoidans to obtain more standard batches, for example, to control the chain length or degree of sulfation.

5.4 Availability of Raw Materials

Brown algae, one of the two sources of fucoidan, are dominant in many coastal regions and are already readily available as a food product in many Asian cuisines. It is also gaining popularity in Western and other economies for its health benefits. The chapter on alginate has discussed the global availability of brown algae. Taking an example crude yield of 4.02% fucoidan by dry weight of brown algae (Sinurat et al. 2015) and using Norway, the highest producer in Europe, as an example brown algae producing country where 154,230 tonnes of brown algae is harvested from wild stocks annually (Monagail et al. 2017), this means 6200 tonnes of fucoidan is attainable from brown algae resource in Norway annually. As we will discuss in the latter sections of this chapter, the yield, structure and bioactivity of fucoidans are very species specific. This variation means that while biopolymers such as alginate from brown algae could be mass extracted from a mixture of species of brown algae, fucoidans need to be extracted in species-specific batches to obtain more specific bioactivity.

Echinoderms, which comprised of species such as starfish, sea urchins and sea cucumbers, are aquatic animals which also are a source of fucoidans. Some echinoderms such as sea cucumbers are consumed by humans as food. Sea cucumbers can be found in the Mediterranean Sea and the eastern parts of the Atlantic Ocean (Chang et al. 2015). Species which have been explored for fucoidan production include *Holothuria tubulosa*, *Stichopus japonicus*, *Apostichopus japonicus* and *Acaudina molpadioides* among several others. The reproductive rate of algae is much faster than that of sea cucumber which although is moderate compared to other invertebrates in the sea and is currently declining in population due to increasing interest as a functional food with bioactivities such as anticoagulant and antioxidant effects (Chang et al. 2015; Qin et al. 2018). Commercial fucoidan production should therefore focus more on extraction from brown algae or at least diversify production of fucoidan from more than a single source.

In the bid to achieve standardized fucoidan-based product with consistent and controllable chemical structures and hence bioactivity, some research work has gone into developing enzymes involved in fucoidan synthesis and development of cell cultures which express the genes to produce the desired fucoidan chemical structure (Kasai et al. 2015).

5.5 Extraction of Fucoidans

Various methods exist for extraction of fucoidan, and here we shall look at some of them. The type of extraction and purification method strongly determines the structure and bioactivity of fucoidan (Ponce et al. 2003).

5.5.1 Acid Extraction

The use of acid for extraction of fucoidan is the oldest known method. This involves solubilizing the fucoidan in an aqueous acid solution at high temperature. This is then followed by precipitation of the dissolved fucoidan with ethanol. This process is based on the fact that fucoidan is soluble in polar liquids at low pH and elevated temperature but insoluble in less polar solvents. Subsequent solubilization and precipitation can then be carried to further purify the extracted fucoidan (Fitton et al. 2015a, b). The proteins and alginate within the algae or sea cucumber will also be dissolved at low pH and high temperature. The alginates can be separated from the fucoidan by precipitation.

An example process is as follows (Sinurat et al. 2015; Lu et al. 2018): The brown seaweeds are first soaked in a mixture of methanol, chloroform and water in the ratio 4:2:1 over a duration of 12 h. This process removes the fats and depigments the brown algae. This is then followed by washing in acetone, and the defatted and depigmented algae are then air-dried. The dried biomass is then treated with 0.01M HCl at a pH of 4 with a weight-to-volume ratio of 1:10 (g:ml) under continuous stirring for 6 h. The liquid extract obtained after filtering contains fucoidan as well as acid-soluble proteins, alginates and smaller particles which get through the filter pore space. The alginate is separated by precipitation with 4M of calcium chloride ($CaCl_2$) under incubation for a period of 30 min after which the precipitates are filtered off. The precipitation is repeated to further remove more alginates. This is then followed by centrifugation at 1544 g for 15 min in order to remove the smaller particles. Crude fucoidan is then obtained from the solution by precipitation with ethanol and further purified by dialysis at a cutoff point of 10 kDa. This process is based on method reported for research studies which can also be scaled up for industrial fucoidan production. A flowchart of the extraction process is provided in Fig. 5.2.

5.5.2 Enzyme Extraction

Here an example extraction process of extraction of fucoidan from sea cucumber is used to explain the enzyme-based extraction of fucoidan. The enzyme extraction process involves breaking down of the proteins in the tissue using papain. The body of the sea cucumber which can be purchased from fishermen or from the market is dried and milled into smaller particle sizes. The pH is adjusted using hydrochloric acid, and the papain is added at a concentration of 0.15%. The extraction is carried out at 50 °C for a period of 2 h at a ph of 6 and a solid-to-liquid ratio of 1 g:16.26 mL— based on optimum conditions reported by Qin et al. (2018). This process results in the breaking down of the tissue proteins into peptides. The pH is then adjusted to a lower pH of 2.8 at a temperature of 4 °C over a period of 4 h. This allows the acidic proteins to be removed and separated by centrifugation (7441 × g for 20 min). The liquid supernatant obtained after the centrifugation contains a mixture of compounds. The

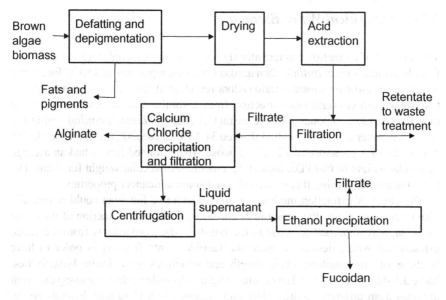

Fig. 5.2 Extraction of fucoidan

fucoidan is then separated by precipitation with alcohol followed by centrifugation at $4816 \times g$ for 15 min. The fucoidan obtained as the solid is then dried (Qin et al. 2018).

5.5.3 Microwave Extraction

Although the microwave extraction process allows for shorter extraction times and in some cases has demonstrated better control over structure, this process is limited to small-scale laboratory extractions. There are reports of microwave-extracted fucoidan having reduced anticancer activity compared to the conventional acid extraction (Rodriguez-Jasso et al. 2011), while other reports reported fucoidans with antioxidant property from *Ascophyllum nodosum* using the microwave-assisted extraction method (Yuan and Macquarrie 2015). As the bioactive properties of fucoidan vary from species to the other, it is not yet concluded that this incidence of reduced anticancer property using microwave extraction applies to all species. This process is nonetheless useful for high-value, low-quantity fucoidans with more controllable bioactive properties such as their antioxidant property.

5.5.4 Subcritical Water Extraction

This method of extraction has recently shown to result in fucoidan yield of 25.98% from *Nizamuddinia zanardinii* when used at optimal temperature of 150 °C for 29 min at a 21 g/mL biomass-to-water ratio (Alboofetileh et al. 2019). This is much higher that yield from conventional extraction from a similar species. Evaluation of the monosaccharide components of the extract of *Nizamuddinia zanardinii* using the subcritical water extraction method showed 34.13% fucose, 30.70% mannose, 9.35% xylose, 23.19% galactose and 2.65% glucose. The extracted fucose had an average molecular weight of 694 kDa, indicating a medium molecular weight fucoidan. The fucoidan extracted using this method showed some anticancer properties.

The effect of extraction methods on the structure of fucoidan could potentially give more control over the structure of fucoidan. Following extraction of the crude fucoidan, it is important to separate the non-fucoidan components from the crude extracts and where desired separate the fucoidans into fraction in order to have fractions of more uniform chain length and structures with similar bioactivities. Table 5.1 shows the yield and molecular weight of fucoidan extraction using different methods from different studies. Note that the extraction yield also depends on the efficiency of the individual process and the purity of the final product; therefore, the values presented within the table are not absolute values for the particular method or source.

Table 5.1 Yield and molecular weight of fucoidan from different extraction methods

Fucoidan source	Yield % dry weight	Molecular weight	Extraction method	References
Sea cucumber	6.83	40–400 kDa	Enzyme	Qin et al. (2018)
Nizamuddinia zanardinii	25.98	694 kDa	Subcritical water	Alboofetileh et al. (2019)
Brown algae (*Undaria pinnatifida*)	NR	2–440 kDa	Hot water	Lu et al. (2018)
Brown algae (U. pinnatifida)	12.9	NR	Hot water	Zhao et al. (2018)
Sea cucumber (*Holothuria tubulosa*)	2.5%	1567.6 ± 34.1 kDa	Enzyme	Chang et al. (2015)
Brown algae (*Sargassum* sp.)	4.02	NR	Acid	Sinurat et al. (2015)
Brown algae	18.2%	NR	Microwave-assisted	Rodriguez-Jasso et al. (2011)

NR Not reported in study

5.6 Environmental Implications

Like many biopolymers, the production of fucoidan, from the process of harvesting the raw material to the final product, is not completely benign. Here we explore some of the environmental issues associated with the extraction of fucoidan from brown algae.

5.6.1 Resource Utilization of Brown Algae and Sea Cucumber

Brown algae play a significant role in the aquatic environment, serve as hosts to some microbes with which it has a symbiotic relationship, are a source of food to animals and humans, fix nitrogen and CO_2 and also play a key role in water remediation. Removing such from the environment results in an imbalance in the ecosystem, and this must be weighed against the economic and environmental significance of harvesting brown algae for fucoidan production. Already the heavy harvesting of brown algae is causing some environmental concerns in parts of the world. On the other hand, algae cultivation and harvesting at a sustainable rate could have some benefits on the environment.

Furthermore, brown algae as an alternative source of fucoidan could potentially reduce overfishing of sea cucumber which have recently suffered from severe decline in population is some seas in area such as Madagascar and China. In these regions, they are seen as highly priced for their medicinal purpose. Fishermen often free dive into the sea to collect seaweeds and sea cucumber within the sunlight zone of the sea. The brown algae and sea cucumber are sold in local markets and restaurants as food and as an important source of income to the coastal communities. Cultivation of both seaweed and sea cucumber should only be carried out at a rate balanced by the rate of their reproduction in the wild; otherwise, these aquatic resources considered as renewable resources will become depleting resources. Harvesting brown algae and sea cucumber for large-scale extraction of products such as fucoidan particularly where yields could be as low as 4% requires a strain on these natural resources. Alternatives to reduce the environmental impact of overheating of these resources from the wild include genetically modified brown algae strains which can be cultivated in controlled environment to yield desired structure of fucoidan and cultivation of sea cucumber in aquaculture (Jour et al. 2012). Limitation of aquaculture of sea cucumber and brown algae includes costs of facilities and the risk of losing whole batches to infection outbreak.

5.6.2 Use of Mineral Acids

HCl is used at a relatively low concentration of 0.01M in the acid extraction method. Since the acid is not chemically altered, it is washed out at the end of the process and goes to the wastewater treatment. Even where it is neutralized, this eventually results in increasing the mineral content of the aquatic environment where it does not lead to direct acidification.

One way of eliminating or minimizing the need for acids in the process of the extraction is to use enzymes. Enzyme extraction has been more commonly reported for extraction of fucoidan from sea cucumber. This is due to the fact that sea cucumber involves mainly proteolytic enzymes to dissolve the proteins within the tissue. Papain is commonly used at a concentration of 15% (Qin et al. 2018). Enzyme extraction of fucoidan from brown algae is less feasible as this would require the use of a variety of enzymes such as alginate lyase enzymes to break down the alginate, proteolytic enzymes to break down the protein and then cellulases to remove the cellulose.

5.6.3 Energy Consumption

The amount of energy consumed varies with extraction method. The enzyme method using papain can be carried out at 50 °C which is lower than the temperature required for the hot water extraction at 70–80 °C for longer time period. Meanwhile, the acid extraction process can be carried out at room temperature (Sinurat et al. 2015). Further energy is consumed in centrifugation process which for the enzyme process is required at two different stages at $7441 \times g$ for 20 min and at $4816 \times g$ for 15 min.

The amount of energy and materials consumed for a typical acid extraction of fucoidan from brown algae is given in Table 5.2. The values are based on values

Table 5.2 Typical consumptions for fucoidan extraction from brown algae using acid extraction

Consumption	Quantity
Brown algae	24.87 g/g fucoidan
Methanol[a]	5.71 ml/g fucoidan
Chloroform[a]	2.86 ml/g fucoidan
Water[a]	1.43 ml/g fucoidan
HCl	10 ml 0.01M per gram fucoidan
Calcium chloride[a]	4M 20 ml
Ethanol[a]	10 ml 95%
Energy	
• Stirring	200 rpm 4 h
• Centrifuge	1544 g for 15 min

[a]indicates estimated based on typical values

from studies carried out at laboratory scale (Sinurat et al. 2015). When the precise values are not provided in the particular study, estimates are used based on typical values.

5.7 Applications of Fucoidan

Fucoidans have been identified with a range of bioactive properties which give them potential applicability in food, cosmetics, pharmaceutical and biomedical industries. Presently, clinical applications of fucoidans are limited as they are yet to be approved by regulatory organizations such as FDA for such use. Part of the reason for this is that fucoidans vary significantly from batch to batch, and this variation affects the bioactivity significantly. For instance, while several fucoidan-producing species demonstrate ability to inhibit cancer cell adhesion to platelets (Cumashi et al. 2007), which in turn inhibits their ability to metastasize, other fucoidan-producing species like the *Cladosiphon okamuranus* do not demonstrate this property, in fact fucoidans extracted from the latter showed increased tumor growth (Azuma et al. 2012). Therefore, approving fucoidans, in general, for specific applications could mean approving forms of fucoidans which might not be suitable for the same applications and vary significantly in effectiveness and safety. There needs to be set standards for extraction, characterization and processing of fucoidans which guides their approval for therapeutic applications. Furthermore, although there are many studies pointing to the bioactive properties of fucoidan which makes it suitable for many applications such as anticancer agent, the exact mechanism of much of these bioactivities is yet unknown. Nonetheless, fucoidan's broad range of bioactivities open up great potentials for this biopolymer to be of significant and economic impact.

5.7.1 Biomaterials in Biomedicine and Tissue Engineering

Fucoidans either in the neat form or as a composite with other materials such as chitosan, alginates, hydroxyapatite and polycaprolactone have been tested in various biomedical applications. Fucoidan–polycaprolactone composites have been used as macroporous sutures which in cellular mineralization, similarly fucoidan–chitosan–alginate and fucoidan–hydroxyapatite composites aid in cellular mineralization (Lee et al. 2012; Venkatesan et al. 2014; Jeong et al. 2013). Scaffold formed from fucoidan is able to inhibit the activities of osteoclasts (Kim et al. 2014a, b), while promoting the activities of osteoblast cells (Park et al. 2012; Pereira et al. 2014) such that the breakdown of new cells forming in the scaffold is prevented while new cell growth is promoted, therefore leading to faster tissue repair. Fucoidan-based scaffolds also promote growth and cell differentiation of mesenchymal cells based on results from studies in vitro (Han et al. 2015) and in vivo using laboratory mice (Huang and Liu

2012). These various reports indicate that fucoidan has robust structural compatibility to form functional composites with other polymers and non-polymers.

5.7.2 Anticancer Therapies

Studies of coculture cells and animals have confirmed fucoidan has some anticancer properties. The key issues requiring further investigation include sufficient bioavailability to deliver the potential anticancer effect at physiologically relevant rate and reproducibility of the anticancer effect in different fucoidan extracts across different species. Fucoidan from *Sargassum muticum* with molecular weights ranging from 5 to 100 kDa showed a range of anticancer properties. While fractions with molecular weight above 100 kDa containing the highest level of sulfates and phenolics showed the highest antiradical activity, the fraction with molecular weight in the range 10–30 kDa had the highest cytotoxic effect on cervical cancer cells (Alvarez-Vinas et al. 2019). Fucoidans from different sources can therefore be used as a multiple mechanisms for cancer therapies since different fractions from the same source show different forms of antitumoral activities.

The use of fucoidan in treatment of cancer is so far focusing on orally delivered fucoidan as an adjunct either to enhance the effectiveness of conventional cancer therapies, prevent side effects or to act as an alternative medication for cancers where no known medications exist (Fitton et al. 2015a, b). It is hypothesized that fucoidan takes multiple routes to act against cancer; this includes prevention of blood vessel formation by cancer cells (angiogenesis), prevention of metastasis (spread of cancer cells beyond the point of origin), boosting the body's immune response to cancer cells, making cancer cells more vulnerable, antioxidant activity and interfering with the cancer cell metabolism pathways (Kwak 2014). For example, some study has found that cancer cell apoptosis could be induced in breast cancer and colon cancer using fucoidan. The mechanism through which fucoidan achieved this is thought to be through modulation of endoplasmic reticulum stress cascades (Chen et al. 2014).

Fucoidan ingested orally can have some anticancer effect within the gastrointestinal tract before absorption into the bloodstream and before reaching the stomach enzyme, while much of the fucoidan remains intact. This could be in different preparations such as food supplements or processed foods with fucoidan added as a functional food ingredient. This would, however, be use of fucoidan as a preventive approach to cancer through its many pathways such as prevention of blood vessel formation and mobilization of the body's immune cells, thus preventing the formation of cancerous cells in the first place. There are already some natural products which are used in this manner (Esmaeelian et al. 2014).

Fucoidan can be used to minimize or prevent the side effects attached to cancer therapy such as chemotherapy. A study which showed this used low molecular weight fucoidan extracted from *Acaudina mòlpadioides*, a sea cucumber. When administered to mice with cyclophosphamide-induced intestinal mucositis, these mice showed restored levels of immunoglobulin A in the mucosa and moderated cytokine levels.

The side effects associated with chemotherapy drug cisplatin, such as weight loss and changes in levels of hormones gastrin and serotonin, were prevented with similar effectiveness as drugs which are used for such purpose (Zuo et al. 2015), proving that fucoidan can act as an adjunct to anticancer drugs for reducing or eliminating the side effects associated with such drugs. This will contribute significantly to improving the patients' quality of life during such treatments.

As an adjunct therapy, fucoidan could work in synergy with conventional chemotherapy and radiotherapy for improved treatment effectiveness (Zhang et al. 2014a). However, this is still under investigation as other studies also suggest that fucoidan could have antagonistic effects on conventional cancer therapy (Oh et al. 2014). This could be attributed to the wide variation in the structure, hence bioactivities of fucoidans for different sources. Therefore, further investigations are required to validate this concept of fucoidan as an adjunct cancer therapy.

At a recommended dose of 4.0 g per day, human subjects ingesting fucoidan from the *Cladosiphon okamuranus* species during chemotherapy treatment of unresectable colon cancer showed improved tolerance of repeated rounds of chemotherapy compared to those who were not taking fucoidan. The patients taking fucoidan also showed less chemotherapy-related fatigue as it is common in cancer therapy (Azuma et al. 2012; Ikeguchi et al. 2011). Studies on mice, given the equivalent dosage of fucoidan, also showed reduced fatigue during exercise. This was attributed to increased levels of glucose in the serum for energy and decreased levels of ammonia, lactate and triglycerides which could cause stress in the cells (Chen et al. 2015). The relationship between reduced fatigue and fucoidan ingestion still requires further research to understand the mechanism and extent.

Fucoidan, therefore, shows some bioactivities either as a sole anticancer therapeutic agent, as an adjunct to use along with conventional chemotherapy or radiotherapy or to abate the side effects associated with cancer therapy, or as a preventive agent against cancer initiation or proliferation. These have been shown in studies in cell cultures, in animal models and in humans against various cancers such as breast, prostate, lung and liver cancers. While these are very promising, the limitations lie in the broad variation in structure and bioactivities of fucoidan from different sources as well as the variation in the different types of cancers. Such that, the successful application of fucoidan for cancer awaits further development in both understanding of cancer and fucoidan variation from species to species.

5.7.3 Drug Delivery Agent

As a biopolymeric material, fucoidans have demonstrated potential for use as drug delivery compounds. Doxorubicin, an anticancer drug, showed better treatment against multidrug resistance cancer cells when loaded onto fucoidan nanoparticles compared with the drug in its free form (Lee et al. 2013a, b). The heterogeneous polysaccharide structure of fucoidan makes it potentially robust such that various fractions can show compatibility with other materials, thus widening the range of possible biocomposite drug delivery biomaterials attainable from fucoidans.

5.7.4 Immune Modulation

Ingestion of fruits and vegetables has been known to boost the immune system, thereby preventing the onset of diseases (Gibson et al. 2012). One of the problems with complaints to this is the large portions required and in some cases food shortage. In, for example, regions suffering from draft or where human conflicts have disrupted farming and food production activities, it is important to provide the necessary quality of nutrient intake to maintain a healthy immune system. Fucoidan provides immune modulation when used in a much lower quantity than a larger amount of fruits and vegetables required for the same level of immune modulation (Negishi et al. 2013).

An example of such, fucoidan extracted from *Undaria pinnatifida* induced improved antibody response to vaccine in elderly human subjects who were above retirement age. Such findings are very important as it could reduce the quantity of vaccine required to achieve the same immune response. This would reduce the cost of vaccines and the amount of energy and manpower required to produce them.

The immune modulation activity of fucoidan at a lower gram requirement compared to dietary consumption of fruits and vegetables is also important in providing treatment and care for individuals who are not capable of ingesting large amounts of food. Fucoidan intake as low as 1 g per day has been shown to have sufficient immune modulation activity.

A series of in vivo and in vitro studies show that fucoidans from different species such as *Fucus vesiculosus* and *Ascophyllum nodosum* promote both growth and activities of the dendritic cells of the body responsible for coordination of immune response (Jin et al. 2014; Zhang et al. 2014a, b). These include recognition of pathogens, antigen presentation and initiation of immune response.

It is understood that one of the ways fucoidan acts in immune modulation is through binding to specific receptors and such interactions resulting in activation or mediations of specific immune responses in a manner that is not toxic to the cells. Such interactions result in production of pathogen-destroying biochemicals such as cytokines and chemokines. Such immune bioactivity has been demonstrated by fucoidans extracted from species such as *Laminaria japonica*, *Laminaria cichorioideae* and *Fucus evanescens* (Makarenkova et al. 2012). Immune modulation is one of the current commercial claims of fucoidan-based food supplements. However fucoidan is not commercially useed as a therapeutic as there is a need for more studies in this area to confirm clinical applicability.

5.7.5 Antipathogenic Agent

Fucoidans have shown to act against a number of pathogens which include Leishmania parasite (Sharma et al. 2014), influenza virus (Synytsya et al. 2014), canine distemper virus (trejo-Avila et al. 2014) and new castle virus (Elizondo-Gonzalez et al. 2012). The mode of action includes inhibition of entry of these pathogens into the cell and interfering with the pathogen's defences against immune response of the host. Due to the complex structure of fucoidan, various fractions of the polymer could be responsible for different antipathogenic activities. Although much is yet to be known of the specific mode of action, there are indications of what fractions are active against certain pathogens, for example, low molecular weight sulfated O-acetyl fucogalactan fraction of fucoidan with average molecular weight of 9 kDa from the species *Undaria pinnatifida* orally administered to laboratory mice acted against different forms of influenza virus infection (Hayashi et al. 2013; Synytsya et al. 2014).

Fucoidans have shown antipathogenic response to a variety of diseases causing organisms in a variety of organisms including humans, canines and birds with no toxic effect on the body's own cells. These findings are consistent across different research groups.

Although fucoidans have not been identified to directly act against bacteria, their antibacterial effect lies in their ability to boost the effectiveness of antibiotic agents (Lee et al. 2013a, b). Such that, they can be used as an adjunct to antibacterial therapeutics or simply consumed as preventive care against bacterial infection. As much of the known bacteria strains have developed one form of immunity against antibacterial drugs, the search for new measures against bacteria remains active. Fucoidan could potentially have a significant economic impact as a versatile and readily available preventive and therapeutic biopolymer for bacterial infection. Tests carried out on pseudomonas culture showed that fucoidan extracted from *Ascophyllum nodosum* resulted in upregulation of the genes responsible for immune response in the organism, while the genes for metabolism, sensing and survival in the pathogenic bacteria were downregulated, showing that the particular fucoidan had, to great extent, improved the resistance of the organism to bacterial infection (Kandasamy et al. 2015). Fucoidan has also shown to protect the body against damage by endotoxins when taking either orally or injected subcutaneously (Kuznetsova et al. 2014).

The potential economic impact of such antipathogenic impact demonstrated by fucoidan cuts across the human and animal healthcare industry and commercial poultry, increasing the quality of life and compliance with an antipathogenic remedy which can be sourced from oral fucoidan preparations in the form of food supplements.

5.7.6 Antiinflammatory

Fucoidan has potential application as an antiinflammatory agent in skin, gut and aortic. These antiinflammatory effects have been demonstrated in cells and tissues in humans and other organisms. Various antiinflammatory effects have been demonstrated by fucoidan when delivered orally, topically or parenterally. For topical applications, this will depend on the range of molecular weight fucoidans that are able to permeate the skin layers for bioavailability. Fucoidans from species such as *Fucus vesiculosus* (Carvalho et al. 2014) and *Undaria pinnatifida* (Fitton 2011) demonstrate these antiinflammatory properties.

By inhibiting the action of inflammatory compound, selectin, through prevention of inflammatory cell entry into tissue and prevention of platelet adhesion to inflammatory cells (Fitton 2011), treatment of inflammatory-related conditions such as pancreatitis (Kambhampati et al. 2014), aortic aneurysm (Alsac et al. 2013), atopic and allergic dermatitis (Fitton et al. 2015a, b) and colitis (Lean et al. 2015) could potentially be addressed with fucoidans as indicated by various reports. Fucoidan could serve as a replacement for some antiinflammatory drugs with unpleasant or chronic side effects.

5.7.7 Renal and Hepatic Disease Treatment

The renal or hepatic effect of fucoidan is demonstrated by the manner in which it prevents the accumulation of compounds which lead to renal or hepatic damage. Accumulation of fat in the liver and increase in overall body weight, which lead to chronic liver diseases, are reduced by orally ingested fucoidan included in the diet of the laboratory mice (Kim et al. 2014a, b). Laminaria japonica-sourced fucoidan when orally administered to diabetes-induced test animals showed lowered blood sugar, decreased blood urea nitrogen levels and preserved renal nitrogen excretion (Wang et al. 2014). This shows a potential for use as a less toxic diabetes treatment for diabetes-related kidney disease as fucoidan showed lower toxicity compared to more toxic available treatments for diabetes. Although at very early stage, there have been research studies into potential use of fucoidan in the treatment of hepatitis C. Clinical study of fucoidan from *Cladosiphon okamuranus* using fifteen chronic hepatitis C patients who were orally administered fucoidan over a period of 12 months showed some improvements in patients treated with fucoidan. Such results point to potential use of fucoidan as a therapeutic agent in the treatment of hepatitis C (Mori et al. 2012). As such research advances, this would have an impact in the area of improving patient care for a disease which presently has no vaccine and existing treatment has poor level of effectiveness.

5.7.8 Blood Anticoagulant

Although highly effective anticoagulant drugs already exist in the market, fucoidan shows anticoagulant activities similar to heparin with potential to overcome the limitation of heparin. One limitation to heparin's antithrombotic effect is the risk of hemorrhage when administered at levels required to attain antithrombotic effect. Fucoidan extracted from *Undaria pinnatifida* achieved antithrombotic effect at safe levels without similar risk of delayed blood clot when injected intravenously in laboratory mice. However, as is known of the variation in the bioactivity of fucoidan, that from another species, *Fucus vesiculosus* showed some increase in clotting time (Min et al. 2011).

Studies on the anticoagulant property of fucoidans have found some correlation between the structural properties of fucoidan from *Fucus vesiculosus* and the anticoagulant property (Zhang et al. 2014a, b). This promises a step closer to standardized fucoidan formulations which can be reproduced on a large scale. Minimum charge density of 0.5 sulfates for every sugar unit and a degree of polymerization of 70 have been shown to be the required parameters for fucoidan from *Fucus vesiculosus* with pro-coagulant property. Anticoagulant property of fucoidan from Laminaria japonica showed molecular weight dependency as well as dependency on fucose to galactose ratio within the polymer chain (Jin et al. 2014). Platelet aggregation activation by fucoidans has also been reported in different studies (Manne et al. 2013; Dürig et al. 1997).

Fucoidans, therefore, have multiple bioactivities in terms of blood clotting and coagulation; they can act as antithrombotic, anticoagulant and pro-coagulant. This makes then applicable for the treatment of conditions such as hemophilia and air-travel-induced deep vein thrombosis. Fucoidans delivered orally and intravenously from a variety of species have shown these properties. This further adds to the robust applicability of fucoidan.

Other recent potential applications of fucoidans include application in the treatment of type 2 diabetes through inhibiting the breakdown of starch into sugar in the body by inhibiting the action of the enzymes, amylase and glucosidase, which are responsible for catalyzing the hydrolysis of starch into glucose (Senthil et al. 2019) or through other mechanisms like improving effectiveness of insulin, as demonstrated in experiments with mice (Sim et al. 2019). Fucoidans are also being investigated for treatment of Alzheimer's disease by acting as a neuroprotector (Alghazwi et al. 2019). These applications vary for fucoidan fractions from different sources.

In general, the bioactive properties of fucoidan, some of which are listed in Table 5.3, occur in a wide range of species. Applications of fucoidan for these bioactivities range from anticancer drug or adjuvant to antipathogenic agents. Successful commercialization of fucoidan in medicinal applications poses significant economic and social impact in terms of providing an easily accessible raw material for such fucoidan-based therapeutics, a natural source which is less likely to result in undesirable or serious side effects and less costly drug development compared to synthetic-sourced alternatives.

Table 5.3 Some reported bioactivities of fucoidans from various sources

Fucoidan source	Bioactivity	References
Sea cucumber	Modulation of metabolic syndromes and gut *Microbiota dysbiosis* (in vitro)	Shan et al. (2019)
Kjellmaniella crassifolia	Antioxidant activity and CCl4-induced liver injury (in vitro)	Liu et al. (2018)
Nizamuddinia zanardinii	Antioxidant and anticancer properties (in vitro)	Alboofetileh et al. (2019)
Undaria pinnatifida	Breast cancer cell inhibition (in vitro)	Lu et al. (2018)
Turbinaria conoides	Antiangiogenesis in cancer cells	Matsubara et al. (2005)

5.8 Commercial Production

Fucoidans of specified characterizations such as molecular weight and source are commercially available as research chemicals by suppliers such as Sigma Aldrich. Despite its broad spectrum of bioactive properties, fucoidan is yet to be FDA approved for any of the clinical applications. The fact that fucoidans are not the major biopolymers of economic importance contained in brown algae (Hahn et al. 2016) is one of the limiting factors of their commercial exploration. The inconsistency in the structural and bioactive properties that varies from species to species, harvest period and extraction technique also further limits the commercial production of fucoidan. The inconsistency in the chemical properties, extraction, purification and production methods for different forms of fucoidan means it does not meet the good manufacturing practice as set out for pharmaceutical products by the world health organization (WHO 2014), for example, the inherent contamination with other polymers and phenols due to the biological source and the low bioavailability and broad variation in the chemical structure of fucoidans.

Although fucoidans still await regulatory approval for therapeutic applications, it, however, is approved for use as food and food supplements since there are sufficient evidence to support its safety and bioavailability when consumed as a food or a food supplement (Fitton et al. 2015a, b). Fucoidan is detected in the serum and urine following oral administration although more efforts are being directed toward increasing bioavailability and having a better understanding of the mechanism of absorption into the body. Proposed methods for improving bioavailability following oral consumption of fucoidan include use of nanoparticles and liposome-based formulations (Pinheiro et al. 2015; Lee et al. 2013a, b; Kimura et al. 2013). Orally consumed fucoidan generally has a bioavailability of around 2% w/w or just within detectable limits (Fitton et al. 2015a, b). This needs to be improved in order to utilize the bioactive potential of fucoidan. While a food product being safe is a minimal requirement, inability to fully absorb and utilize the bioactives within the food product when consumed is a form of wastage in itself. The oral route has been known to lead to loss of bioavailability due to the first-pass metabolism, where much of the component of the ingested substance is exposed to degrading enzymes and the harsh conditions in the stomach. Such that very little or no part of the active component gets into the bloodstream. Therefore, to make more use of the benefits of fucoidan as a bioactive compound, the applications need to go beyond oral administration.

A number of patents exist for use of fucoidans in biomedical applications. One of such is for use as scaffolds in tissue repair (Le Visage et al. 2015), and another existing patent covers the formulation and use of a fucoidan-based treatment for bleeding disorders (Dockal et al. 2014). However, obtaining a patent for a product only protects the right of the inventor but does not grant the approval to commercial production.

Impurities significantly affect the bioactivity of fucoidans; therefore, in addition to the requirement for clinical application, to study the structure and bioactivity of fucoidan extracts, the crude fucoidan must be purified in order to understand the true structure and bioactivity of fucoidan in its pure form. Purification methods for fucoidans such as ion-exchange chromatography (Isnansetyo et al. 2017), gel permeation chromatography and biological affinity purification (Zayed and Ulber 2019), add significant cost to the production of pure fucoidans. Because of the symbiotic relationship marine algae have with marine microbes, some of these microbes are deeply attached to the cell walls of the algae and there usually exist different types of microbes (Nambisan 1999), such that a broad spectrum sterilization technique is required to ensure these contaminants do not get to the fucoidan extract.

5.9 Conclusion

Fucoidans are presently commercially available as food supplements. Their bioactive properties cannot be standardized and validated due to inconsistency in structure and bioavailability in oral form. The production of brown algae for fucoidan competes with use of brown algae as food and for production of other products from brown algae such as alginates. With potential applications of fucoidan as a low-cost treatment for diseases such as colon cancer and type 2 diabetes, these fucose sulfated polysaccharides remain of significant research and commercial interests.

References

Alboofetileh M, Rezaei M, Tabarsa M, You S, Mariatti F, Cravotto G (2019) Subcritical water extraction as an efficient technique to isolate biologically-active fucoidans from *Nizamuddinia zanardinii*. Int J Biol Macromol 128:244–253

Alghazwi M, Smid S, Karpiniec S, Zhang W (2019) Comparative study on neuroprotective activities of fucoidans from *Fucus vesiculosus* and *Undaria pinnatifida*. Int J Biol Macromol 122:255–264

Alsac JM, Delbosc S, Rouer M, Journe C, Louedec L, Meilhac O, Michel JB (2013) Fucoidan interferes with *Porphyromonas gingivalis*-induced aneurysm enlargement by decreasing neutrophil activation. J Vasc Surg 57:796–805

Alvarez-Vinas M, Florez-Fernandez N, Gonzalez-Munoz JM, Dominguez H (2019) Influence of molecular weight on the properties of *Sargassum muticum* fucoidan. Algal Res 38 (Article101393)

Azuma K, Ishihara T, Nakamoto H, Amaha T, Osaki T, Tsuka T, Imagawa T, Minami S, Takashima O, Ifuku S et al (2012) Effects of oral administration of fucoidan extracted from *Cladosiphon okamuranus* on tumor growth and survival time in a tumor-bearing mouse model. Mar Drugs 10:2337–2348

Berteau O, Mulloy B (2003) Sulfated fucans, fresh perspectives: structures, functions, and biological properties of sulfated fucans and an overview of enzymes active toward this class of polysaccharides. Glycobiology 13:29–40

Bilan MI, Kusaykin MI, Grachev AA, Tsvetkova EA, Zvyagintseva TN, Nifantiev NE, Usov AI (2005) Effect of enzyme preparation from the marine mollusk *Littorina kurila* on fucoidan from the brown alga *Fucus distichus*. Biochemistry (Moscow) 70:1321–1326

Carvalho AC, Sousa RB, Franco AX, Costa JV, Neves LM, Ribeiro RA, Sutton R, Criddle DN, Soares PM, de Souza MH (2014) Protective effects of fucoidan, a p- and l-selectin inhibitor, in murine acute pancreatitis. Pancreas 43:82–87

Chang Y (2010) Isolation and characterization of sea cucumber fucoidan utilizing marine bacterium. Lett Appl Microbiol 50(3):301–307

Chang Y, Hu Y, Yu L, McClements DJ, Xu X, Liu G, Xue C (2015) Primary structure and chain conformation of fucoidan extracted from sea cucumber *Holothuria tubulosa*. Carbohyd Polym. https://doi.org/10.1016/j.carbpol.2015.10.016

Chen S, Zhao Y, Zhang Y, Zhang D (2014) Fucoidan induces cancer cell apoptosis by modulating the endoplasmic reticulum stress cascades. PLoS ONE 9:e108157

Chen YM, Tsai YH, Tsai TY, Chiu YS, Wei L, Chen WC, Huang CC (2015) Fucoidan supplementation improves exercise performance and exhibits anti-fatigue action in mice. Nutrients 7:239–252

Cho ML, Lee BY, You S (2011) Relationship between oversulfation and conformation of low and high molecular weight fucoidans and evaluation of their in vitro anticancer activity. Molecules 16:291–297

Cumashi A, Ushakova NA, Preobrazhenskaya ME, D'Incecco A, Piccoli A, Totani L, Tinari N, Morozevich GE, Berman AE, Bilan MI (2007) A comparative study of the anti-inflammatory, anticoagulant, antiangiogenic, and antiadhesive activities of nine different fucoidans from brown seaweeds. Glycobiology 17:541–552

Deniaud-Bouët E, Hardouin K, Potin P, Kloareg B, Hervé C (2017) A review about brown algal cell walls and fucose-containing sulfated polysaccharides: cell wall context, biomedical properties and key research challenges. Carbohyd Polym 175:395–408. https://doi.org/10.1016/j.carbpol.2017.07.082

Dockal M, Ehrlich H, Scheiflinger F (2014) Methods and compositions for treating bleeding disorders. U.S. Patent 8,632,991

Dürig J, Bruhn T, Zurborn K-H, Gutensohn K, Bruhn HD, Béress L (1997) Anticoagulant fucoidan fractions from *Fucus vesiculosus* induce platelet activation in vitro. Thromb Res 85:479–491

Elizondo-Gonzalez R, Cruz-Suarez LE, Ricque-Marie D, Mendoza-Gamboa E, Rodriguez-Padilla C, Trejo-Avila LM (2012) In vitro characterization of the antiviral activity of fucoidan from *Cladosiphon okamuranus* against newcastle disease virus. Virol J 9:307. https://doi.org/10.1186/1743-422X-9-307

Esmaeelian B, Abbott CA, Le Leu RK, Benkendorff K (2014) 6-bromoisatin found in muricid mollusc extracts inhibits colon cancer cell proliferation and induces apoptosis, preventing early stage tumor formation in a colorectal cancer rodent model. Mar Drugs 12:17–35

Fitton J (2011) Therapies from fucoidan; multifunctional marine polymers. Mar Drugs 9:1731–1760

Fitton JH, Dell'Acqua G, Gardiner VA, Karpiniec SS, Stringer DN, Davis E (2015a) Topical benefits of two fucoidan-rich extracts from marine macroalgae. Cosmetics 2:66–81

Fitton JH, Stringer DN, Karpiniec SS (2015b) Therapies from fucoidan: an update. Mar Drugs 13:5920–5946

Fletcher HR, Biller P, Ross AB, Adams JMM (2017) The seasonal variation of fucoidan within three species of brown macroalgae. Algal Res-Biomass Biofuels Bioprod 22:79–86

Gibson A, Edgar JD, Neville CE, Gilchrist SE, McKinley MC, Patterson CC, Young IS, Woodside JV (2012) Effect of fruit and vegetable consumption on immune function in older people: a randomized controlled trial. Am J Clin Nutr 96:1429–1436

Hahn T, Schulz M, Stadtmüller R, Zayed A, Muffler K, Lang S, Ulber R (2016) A cationic dye for the specific determination of sulfated polysaccharides. Anal Lett 49(12):1948–1962

Han YS, Lee JH, Jung JS, Noh H, Baek MJ, Ryu JM, Yoon YM, Han HJ, Lee SH (2015) Fucoidan protects mesenchymal stem cells against oxidative stress and enhances vascular regeneration in a murine *Hindlimb ischemia* model. Int J Cardiol 198:187–195

Hayashi K, Lee JB, Nakano T, Hayashi T (2013) Anti-influenza a virus characteristics of a fucoidan from sporophyll of *Undaria pinnatifida* in mice with normal and compromised immunity. Microbes Infect/Inst Pasteur 15:302–309

Huang YC, Liu TJ (2012) Mobilization of mesenchymal stem cells by stromal cell-derived factor-1 released from chitosan/tripolyphosphate/fucoidan nanoparticles. Acta Biomater 8:1048–1056

Hwang PA, Yan MD, Kuo KL, Phan NN, Lin YC (2017) A mechanism of low molecular weight fucoidans degraded by enzymatic and acidic hydrolysis for the prevention of UVB damage. J Appl Phycol 29:521–529

Ikeguchi M, Yamamoto M, Arai Y, Maeta Y, Ashida K, Katano K, Miki Y, Kimura T (2011) Fucoidan reduces the toxicities of chemotherapy for patients with unresectable advanced or recurrent colorectal cancer. Oncol Lett 2:319–322

Isnansetyo A, Lutfia LNF, Nursid M, Susidarti RA (2017) Cytotoxicity of fucoidan from three tropical brown algae against breast and colon cancer cell lines. Pharmacogn J 9(1):14–20

Jeong HS, Venkatesan J, Kim SK (2013) Hydroxyapatite-fucoidan nanocomposites for bone tissue engineering. Int J Biol Macromol 57(138–141):123

Jin JO, Zhang W, Du JY, Wong KW, Oda T, Yu Q (2014) Fucoidan can function as an adjuvant in vivo to enhance dendritic cell maturation and function and promote antigen-specific t cell immune responses. PLoS ONE 9:e99396

Jour TY, Purcell S, Hair C, Mills D (2012) Sea cucumber culture, farming and sea ranching in the tropics: progress, problems and opportunities 368(369):68–81

Kambhampati S, Park W, Habtezion A (2014) Pharmacologic therapy for acute pancreatitis. World J Gastroenterol 20:16868–16880

Kandasamy S, Khan W, Kulshreshtha G, Evans F, Critchley AT, Fitton JH, Stringer DN, Gardiner VA, Prithiviraj B (2015) The fucose containing polymer (fcp) rich fraction of *Ascophyllum nodosum* (l.) le jol. Protects caenorhabditis elegans against *Pseudomonas aeruginosa* by triggering innate immune signaling pathways and suppression of pathogen virulence factors. Algae 30:147–161

Kannan RR, Arumugam R, Anantharaman P (2013) Pharmaceutical potential of a fucoidan-like sulphated polysaccharide isolated from *Halodule pinifolia*. Int J Biol Macromol 62:30–34

Kasai A, Arafuka S, Koshiba N, Takahashi D, Toshima K (2015) Systematic synthesis of low-molecular weight fucoidan derivatives and their effect on cancer cells. Org Biomol Chem 13(42):10556–10568

Kim MJ, Jeon J, Lee JS (2014a) Fucoidan prevents high-fat diet-induced obesity in animals by suppression of fat accumulation. Phytother Res 28:137–143

Kim YW, Baek SH, Lee SH, Kim TH, Kim SY (2014b) Fucoidan, a sulfated polysaccharide, inhibits osteoclast differentiation and function by modulating rankl signaling. Int J Mol Sci 15:18840–18855

Kimura R, Rokkaku T, Takeda S, Senba M, Mori N (2013) Cytotoxic effects of fucoidan nanoparticles against osteosarcoma. Mar Drugs 11:4267–4278

Kusaykin MI, Burtseva YV, Svetasheva TG, Sova VV, Zvyagintseva TN (2001) Distribution of o-glycosyl hydrolases in marine invertebrates. Enzymes of the marine mollusk *Littorina kurila* that catalyze fucoidan transformation. Biochemistry (Mosc) 68:317–324

Kuznetsova TA, Besednova NN, Somova LM, Plekhova NG (2014) Fucoidan extracted from Fucus evanescens prevents endotoxin-induced damage in a mouse model of endotoxemia. Marine Drugs 12(2):886–898. https://doi.org/10.3390/md12020886

Kwak JY (2014) Fucoidan as a marine anticancer agent in preclinical development. Mar Drugs 12:851–870

Kylin H (1913) Zur Biochemie der meeresalgen. Z Für Physiol Chemie 83:171–197

Lahrsen E, Liewert I, Alban S (2018) Gradual degradation of fucoidan from fucus vesiculosus and its effect on structures, antioxidant and antiproliferative activities. Carbohyd Polym 192:208–216

Lean QY, Eri RD, Fitton JH, Patel RP, Gueven N (2015) Fucoidan extracts ameliorate acute colitis. PLoS ONE 10:e0128453

Lee JS, Jin GH, Yeo MG, Jang CH, Lee H, Kim GH (2012) Fabrication of electrospun biocomposites comprising polycaprolactone/fucoidan for tissue regeneration. Carbohydr Polym 90:181–188

Lee KW, Jeong D, Na K (2013a) Doxorubicin loading fucoidan acetate nanoparticles for immune and chemotherapy in cancer treatment. Carbohydr Polym 94:850–856

Lee KY, Jeong MR, Choi SM, Na SS, Cha JD (2013b) Synergistic effect of fucoidan with antibiotics against oral pathogenic bacteria. Arch Oral Biol 58:482–492

Le Visage C, Chaubet DLF, Autissier A (2015) Method for preparing porous scaffold for tissue engineering. U.S. Patent 9,028,857

Liu S, Wang Q, Song Y, He Y, Ren D, Cong H, Wu L (2018) Studies on the hepatoprotective effect of fucoidan from brown algae *Kjellmaniella crassifolia*. Carbohyd Polym 193:298–306

Lu J, Shi KK, Chen S, Wang J, Hassouna A, White LN, Merien F, Xie M, Kong Q, Li J, Ying T, White LW, Nie S (2018) Fucoidan extracted from the New Zealand *Undaria pinnatifida*-physicochemical comparison against five other fucoidans: unique low molecular weight fraction bioactivity in breast cancer cell lines. Mar Drugs 16(461):1–25

Manne BK, Getz TM, Hughes CE, Alshehri O, Dangelmaier C, Naik UP, Watson SP, Kunapuli SP (2013) Fucoidan is a novel platelet agonist for the c-type lectin-like receptor 2 (clec-2). J Biol Chem 288:7717–7726

Matsubara K, Xue C, Zhao X, Mori M, Sugawara T, Hirata T (2005) Effects of middle molecular weight fucoidans on in vitro and ex vivo angiogenesis of endothelial cells. Int J Mol Med 15:695–699

Michel C, Lahaye M, Bonnet C, Mabeau S, Barry JL (1996) In vitro fermentation by human faecal bacteria of total and purified dietary fibres from brown seaweeds. Br J Nutr 75:263–280

Min SK, Kwon OC, Lee S, Park KH, Kim JK (2011) An antithrombotic fucoidan, unlike heparin, does not prolong bleeding time in a murine arterial thrombosis model: a comparative study of *Undaria pinnatifida* sporophylls and *Fucus vesiculosus*. Phytother Res 26:752–757

Miyazaki Y, Iwaihara Y, Nakamizo M, Takeuchi S, Takeuchi H, Tachikawa D (2018) Potentiating effects of high-molecular weight fucoidan-agaricus mix (CUA) feeding on tumor vaccination. J Immunol 200(181):22–34

Monagail MM, Cornish L, Morrison L, Araujo R, Critchley AT (2017) Sustainable harvesting of wild seaweed resources. Eur J Phycol 52(4):371–390

Mori N1, Nakasone K, Tomimori K, Ishikawa C (2012) Beneficial effects of fucoidan in patients with chronic hepatitis C virus infection. World J Gastroenterol 18(18):2225–2230. https://doi.org/10.3748/wjg.v18.i18.2225

Nagao T, Kumabe A, Komatsu F, Yagi H, Suzuki H, Ohshiro T (2017) Gene identification and characterization of fucoidan deacetylase for potential application to fucoidan degradation and diversification. J Biosci Bioeng 124(3):277–282

Nambisan P (1999) Seaweed biotechnology. Cyanobacterial and algal metabolism and environmental biotechnology, 236–246

Negishi H, Mori M, Mori H, Yamori Y (2013) Supplementation of elderly Japanese men and women with fucoidan from seaweed increases immune responses to seasonal influenza vaccination. J Nutr 143:1794–1798

Oh R, Kim J, Lu W, Rosenthal D (2014) Anticancer effect of Fucoidan in combination with Tyrosine Kinase Inhibitor Lapatinib. Evid Based Complement Alternat Med. 865375. https://doi.org/10.1155/2014/865375

Park SJ, Lee KW, Lim DS, Lee S (2012) The sulfated polysaccharide fucoidan stimulates osteogenic differentiation of human adipose-derived stem cells. Stem Cells Dev 21:2204–2211

Peng Y, Wang Y, Wang Q, Luo X, He Y, Song Y (2018) Hypolipidermic effects of sulfated fucoidan from *Kjellmaniella crassifolia* through modulating the cholesterol and aliphatic metabolic pathways. J Funct Foods 51:8–15

Pereira J, Portron S, Dizier B, Vinatier C, Masson M, Sourice S, Galy-Fauroux I, Corre P, Weiss P, Fischer AM et al (2014) The in vitro and in vivo effects of a low-molecular-weight fucoidan on the osteogenic capacity of human adipose-derived stromal cells. Tissue Eng Part A 20:275–284

Pinheiro AC, Bourbon AI, Cerqueira MA, Maricato E, Nunes C, Coimbra MA, Vicente AA (2015) Chitosan/fucoidan multilayer nanocapsules as a vehicle for controlled release of bioactive compounds. Carbohydr Polym 115:1–9

Ponce NMA, Pujol CA, Damonte EB, Flores ML, Stortz CA (2003) Fucoidans from the brown seaweed *Adenocystis utricularis*: extraction methods, antiviral activity and structural studies. Carbohyd Res 338:153–165

Qin Y, Yuan Q, Zhang Y, Li J, Zhu X, Zhao L, Wen J, Liu J, Zhao L, Zhao J (2018) Enzyme-assisted extraction optimization, characterization and antioxidant activity of polysaccharides from sea cucumber *Phyllophorus proteus*. Molecules 23(590):1–19

Rodriguez-Jasso RM, Mussatto SI, Pastrana L, Aguilar CN, Teixeira JA (2010) Fucoidan-degrading fungal strains: screening, morphometric evaluation, and influence of medium composition. Appl Biochem Biotechnol 162:2177–2188

Rodriguez-Jasso RM, Mussatto SI, Pastrana L, Aguilar CN, Teixeira JA (2011) Microwave-assisted extraction of sulfated polysaccharides (fucoidan) from brown seaweed. Carbohyd Polym 86:1137–1144

Sanjeewa KKA, Lee JS, Kim WS, Jeon YJ (2017) The potential of brown-algae polysaccharides for the development of anticancer agents: an update on anticancer effects reported for fucoidan and laminaran. Carbohydr Polym 177:451–459

Senthil L, Raghu C, Arjun HA, Anantharaman P (2019) In vitro and in silico inhibition properties of fucoidan against alpha-amylase and alpha-D-glucosidase with relevance to type 2 diabetes mellitus. Carbohyd Polym 209:350–355

Shan L, Li J, Mao G, Yan L, Hu Y, Ye X, Tian D, Linhardt RJ, Chen S (2019) Effect of sulfation pattern of sea cucumber-derived fucoidan oligosaccharides on modulating metabolic syndromes and gut microbiota dysbiosis caused by HFD in mice. J Funct Foods 55:193–210

Sharma G, Kar S, Basu Ball W, Ghosh K, Das PK (2014) The curative effect of fucoidan on visceral leishmaniasis is mediated by activation of map kinases through specific protein kinase c isoforms. Cell Mol Immunol 11:263–274

Silchenko AS, Kusaykin MI, Kurilenko VV, Zakharenko AM, Isakov VV, Zaporozhets TS, Gazha AK, Zvyagintseva TN (2013) Hydrolysis of fucoidan by fucoidanase isolated from the marine bacterium, formosa algae. Mar Drugs 11:2413–2430

Sim S, Shin Y, Kim H (2019) Fucoidan from *Undaria pinnatifida* has anti-diabetic effects by stimulation of glucose uptake and reduction of basal lipolysis in 3 t3-L1 adipocytes. Nutr Res (in press)

Sinurat E, Peranginangin R, Saepudin E (2015) Purification and characterization of fucoidan from the brown seaweed *Sargassum binderi* sonder. Squalen Bull Mar Fish Postharvest Biotechnol 10(2):79–87

Synytsya A, Bleha R, Synytsya A, Pohl R, Hayashi K, Yoshinaga K, Nakano T, Hayashi T (2014) Mekabu fucoidan: Structural complexity and defensive effects against avian influenza a viruses. Carbohydr Polym 111:633–644

Tokita Y, Nakajima K, Mochida H, Iha M, Nagamine T (2010) Development of a fucoidan-specific antibody and measurement of fucoidan in serum and urine by Sandwich ELISA. Biosci Biotechnol Biochem 74:350–357

Trejo-Avila LM, Morales-Martinez ME, Ricque-Marie D, Cruz-Suarez LE, Zapata-Benavides P, Moran-Santibanez K, Rodriguez-Padilla C (2014) In vitro anti-canine distemper virus activity of fucoidan extracted from the brown alga *Cladosiphon okamuranus*. VirusDisease 25:474–480

Venkatesan J, Bhatnagar I, Kim SK (2014) Chitosan-alginate biocomposite containing fucoidan for bone tissue engineering. Mar Drugs 12:300–316

Wang J, Liu H, Li N, Zhang Q, Zhang H (2014) The protective effect of fucoidan in rats with streptozotocin-induced diabetic nephropathy. Mar Drugs 12:3292–3306

Weelden G, Bobinski M, Okla K, Weelden WJ, Romano A, Pijnenborg JMA (2019) Fucoidan structure and activity in relation to anti-cancer mechanisms. Mar Drugs 17(32):1–30

Wei X, Cai L, Liu H, Tu H, Xu X, Zhou F, Zhang L (2019) Chain conformation and biological activities of hyperbranced fucoidan derived from brown algae and its desulfated derivative. Carbohyd Polym 208:86–96

WHO (2014) Good manufacturing practices for pharmaceutical products: main principles. WHO technical report series no. 986

Yuan Y, Macquarrie D (2015) Microwave assisted extraction of sulfated polysaccharides (fucoidan) from ascophyllum nodosum and its antioxidant activity. Carbohydr Polym 129:101–107

Zayed A, Ulber R (2019) Fucoidan production: approval key challenges and opportunities. Carbohyd Polym 211:289–297

Zhang W, Du JY, Jiang Z, Okimura T, Oda T, Yu Q, Jin JO (2014a) Ascophyllan purified from *Ascophyllum nodosum* induces th1 and tc1 immune responses by promoting dendritic cell maturation. Mar Drugs 12:4148–4164

Zhang Z, Till S, Jiang C, Knappe S, Reutterer S, Scheiflinger F, Szabo CM, Dockal M (2014b) Structure-activity relationship of the pro- and anticoagulant effects of *Fucus vesiculosus* fucoidan. Thromb Haemost 111:429–437

Zhao Y, Zheng Y, Wang J, Ma S, Yu Y, White WL, Yang S, Yang F, Lu J (2018) Fucoidan extracted from *Undaria pinnatifida*: source for nutraceuticals/functional foods. Mar Drugs 16(321):1–17

Zuo T, Li X, Chang Y, Duan G, Yu L, Zheng R, Xue C, Tang Q (2015) Dietary fucoidan of *Acaudina molpadioides* and its enzymatically degraded fragments could prevent intestinal mucositis induced by chemotherapy in mice. Food Funct 6:415–422

Chapter 6
Carrageenans

Abstract Carrageenans are the sulfated polysaccharides that are obtained from red algae. They are most commonly applied as gelling agents. They are made up of alternating disaccharide units of 1, 3 linked beta galactose linked to either 1,4 alpha galactopyranose or 3,6 anhydrogalactose. They are classified as either λ, κ, ι, ε, μ, depending on the degree of sulfation. The extraction process makes use of alkali, acids and salts and requires energy for heating and additional processes to recover and purify carrageenan from seaweed biomass. These are considered in evaluating the environmental impact of carrageenan production. Presently, carrageenan is more commonly used for its rheological properties as a gelling agent in food and other consumer goods. Some studies have also presented potential application as a bioactive compound and in renewable energy system components.

Keywords Carrageenan · Polymer · Polysaccharide · Phycocolloids · Algae

6.1 Introduction

Red algae are a source of polysaccharides. The polysaccharides serve as structural as well as functional compounds within the red algae cell wall. Carrageenans are the sulfated polysaccharides present in the cell walls of red algae where they play mainly structural roles. They are one of the aquatic sourced polymers which are exclusive to the aquatic ecosystem. The only known source of carrageenan is the red algae which grow mainly in the epipelagic zones of the sea, and more rarely some species grow in littoral zones of freshwater with moderate flow rate (Sheath and Vis 2015). The harvesting of feedstock for carrageenan production is therefore from within these regions of the aquatic ecosystem or simulation of these environments in aquaculture for the cultivation of the feedstock.

Familiar consumer products within which carrageenans are commonly applied include toothpastes, air fresheners and pet food. The important role carrageenan plays in these fast moving consumer goods puts it at a value of around 300 million USD (McHugh 2003). The value of the aquatic environment is extended to not only the market value of the resources directly obtained from it but also the products obtainable from these resources. While the present commercial value of carrageenan

© Springer International Publishing 2020   121
O. Olatunji, *Aquatic Biopolymers*, Springer Series on Polymer and Composite Materials,
https://doi.org/10.1007/978-3-030-34709-3_6

is mainly attributed to its use for its rheological properties, further modifications of carrageenan result in the form of carrageenans with important bioactivities thus increasing the potential market value of carrageenans to beyond just food additives.

Carrageenans are classified as galactans which are generally characterized by the galactose repeating units. It is one of three hydrophilic phycocolloids obtained from algae, and it is a sulfated polysaccharide found in the cell wall of red algae species such as *Kappaphycus alvarezii* (Manuhara et al. 2016), *Hypnea musciformis* (Souza et al. 2018) and *Palisada flagellifera* (Ferreira et al. 2012). Carrageenan has relatively well-established application in the food industry where it is used for texture modification and stabilization.

The subsequent sections in this chapter discuss the occurrence of carrageenan in nature, the abundance of the raw material for production of carrageenan and the present state of these resources as well as past trends. A couple of extraction methods are reviewed to give the reader an idea of what is required to produce this aquatic biopolymer. As naturally sourced products received increasing attention from consumers, researchers and producers, it is important to review the environmental implications of these products beyond just the fact that they are biodegradable and renewable. A section in this chapter is therefore dedicated to the environmental implications of the process of producing carrageenans and discusses some key aspects.

6.2 Occurrence in Nature

Carrageenan is present in nature as one of the sulfated polysaccharides of red algae (Rhodophyta). It is a component of the cell wall where it plays a structural role alongside the other cell wall polysaccharides. Red algae species which contain carrageenan include species such as *K. alvarezii* (Manuhara et al. 2016), *H. musciformis* (Souza et al. 2018), *P. flagellifera* (Ferreira et al. 2012), *Gigartina skottsbergii* (Gonçalves et al. 2005), *Eucheuma spinosum* (Ghani et al. 2019). Prior to carrageenan being known as the sulfated polysaccharide phycocolloid of red algae, carrageenan has been used for its gelling properties in its whole form as seaweed as far back as 1905 where the *Chondrus crispus* commonly referred to as Irish moss is used as a gelling agent in recipes such as blancmange (Smith 1905). Carrageenans have been identified as the gelatinous extract obtained from boiling of the red algae species *C. crispus* in hot water (Stanford 1862). Carrageenan-producing species of red algae of genus Eucheuma have documented use for medicinal, food and adhesive purposes in the far east regions dating back to the early 1950s (Eisses 1952; Zaneveld 1959; Stanley 1987).

In the wild, red algae such as *K. alvarezii* are abundant in places such as Karimun Jawa Islands in town of Jepara located in Central Java, Indonesia, and in Ubatuba, São Paulo in Brazil (Gonçalves et al. 2005) and species such as *H. musciformis* are abundant in regions like Flecheiras Beach in Ceara, Brazil (Souza et al. 2018). *H. musciformis* is valued for its kappa carrageenan content and is one of the species of economic importance in the northeast coast of Brazil (Souza et al. 2018). Red algae grow in the epipelagic (sunlight) zones of the sea. Some freshwater red algae also exist (Sheath and Vis 2015), albeit rare.

Until the 1970s, carragenophytes, mainly *K. alvarezii* and *Eucheuma denticulatum*, were obtained only from wild stocks. As the technology to cultivate carragenophytes began in the Philippines and Indonesia where they were first cultivated, carragenophyte cultivation spread to other regions of the world where it has become a significant source of livelihood for many seaweed farmers in countries such as Tanzania and Vietnam. Today, some carragenophytes are still harvested from the wild. For example, the *Betaphycus gelatinum* is harvested in regions in the Philippines China, Hainan Island and Taiwan Province of China, and *G. skottsbergii*, *Sarcothalia crispata* and *Mazzaella laminaroides* are harvested in some parts of Chile. *Gigartina skottsbergii* also grows in Argentina. However, over 90% of the carragenophytes are cultivated. In other regions such as the Prince Edward Island in Canada, USA and France, carragenophyte commonly grown here is the *C. crispus*. *Gigartina canaliculata* and *H. musciformis* grow in Mexico and Brazil, respectively. In some cases, the occurrence of the wild-growing species can be sporabic. Chile is one example where all of the carragenophytes produced in this region are sourced from natural stock. Despite this reliant on naturally existing carragenophytes, there is no report of overharvesting and sufficient aquatic space exists for further expansion of carragenophyte cultivation.

Carrageenans yield obtained from bench-scale extractions are around 34.3% (Manuhara et al. 2016) although the actual content of carrageenan in red algae could be higher since the yield of extraction is highly dependent on the efficiency and method of extraction. Seaweeds from which carrageenans are obtained are referred to as carragenophytes. The types of red algae from which carrageenans are sourced vary with the locations within which they exist or conditions within which they can be cultivated. Main factors for growth include temperature, salinity, water current, nutrients composition and water depth which in turn affect light intensity. Most algae exist in the epipelagic zones of the water (see Chap. 2 on aquatic ecosystem classification) where sufficient light penetrates for photosynthesis.

One of the red algae species which is the most common source of carrageenan is *K. alvarezii*. This species grows in marine water where the temperature is ~21 °C and there is sufficient bright light for growth. This species grows well in water exposed to slow to moderate tides and rocks, sand and coral substrates. Another species which grows in this region is the *E. denticulatum* prefers moderate to strong tides and *B. gelatinum* grows in stronger tides close to the reef edge.

Species which grow at warmer temperatures include chondrus crispus which grows in the littoral fringe with moderate light and large rock areas. In colder climates, it grows best in the summer periods and late spring with minimal growth in winter. *G. skottsbergii* also grows in warmer seasons in the sublittoral region on the sea. *M. laminaroides* and sarcothalia, *G. canaliculata* and *H. musciformis* all grow along the seashore line while *S. crispata* grows in the epipelagic zone. Cultivations in tanks and seaweed farms require mimicking these growth conditions for optimal yield of the particular species of interest.

6.3   Chemistry of Carrageenan

The structure of carrageenan is the sulfated polysaccharide polymer chain made up
of alternating disaccharide units of 1, 3 linked beta galactose linked to either 1, 4
alpha galactopyranose or 3,6 anhydrogalactose. It can therefore be said to be an
alternating copolymer of these two units. Carrageenans are linear polymer structures characterized by the monomeric units and also importantly sulfate ester groups
attached to some of the repeating units. Carrageenans can be differentiated from the
other sulfated polysaccharide present in red algae (agarans) by the stereochemistry
of the 1-4 linked alpha galactopyranose which is D form in carrageenan but L form
in agarans (Jiao et al. 2011). They are classified as either λ, κ, ι, ε, μ, depending on
the degree of sulfation which could range from 22 to 35%. The most common types
are λ, κ and ι, while κ is the most used of the three types. While k carrageenan has
alternating units of 1, 3 beta-D-galactose with a sulfate group at C4 and 1-4 linked
anhydrogalactose unit, iota carrageenan has a similar structure but with further sulfation on the C2 carbon of the anhydrogalactose unit such that the iota carrageenan
has two sulfate groups on the repeating disaccharide unit. The lambda carrageenan is
further sulfated by having a third sulfate group attached to the C6 carbon of the alpha
1-4 linked galactose unit, such that it has 3 sulfate groups per disaccharide unit. The
lambda carrageenan further differs from the other forms by having no 3,6 anhydride
bridge on the 1-4 linked galactopyranose (Funami et al. 2007). The types present
vary in different species of red alga; for example, *K. alvarezii* is the most common
source of k carrageenan and lambda carrageenan is more abundant in species such
as Gigartina and *Chondrus genera* (Zhou et al. 2006). The presence of the sulfate
group makes carrageenans anionic polymers. Extraction methods have significant
effect on the properties of the carrageenan obtained. The degree of sulfation in turn
affects properties such as the gel strength.

Other than the natural form in which they occur within different species, carrageenan of different forms can be obtained by chemically processing the original
forms. For example, theta carrageenan can be obtained from lambda carrageenan
through alkali treatment to form anhydride bridges between the units. This alkali
treatment usually follows the extraction process in order to convert the extracted
from into the desired form of carrageenan (Doyle et al. 2010). In nature, carrageenans
occur as a mix or hybrid of different forms of carrageenan. In addition to this, the
carrageenan structure can be further complicated by the presence of methyl groups,
pyruvic acetal or other sugars attached to the main chain (Yu et al. 2010). Figure 6.1
shows the different forms of carrageenan and how they can be transformed from one
form to the other.

In FTIR spectrometry, carrageenan is characterized by absorption peaks at
1234 cm^{-1} which is indicative of the presence of sulfate ester ($S = 0$) functional
group, 3,6, anhydrogalactose at 926 cm^{-1} and glycosidic linkages at 1072 cm^{-1}.
The band occurs around 840–850 cm^{-1} which indicates the presence of sulfated
anhydrogalactose for *k* type carrageenan. The vibration of the O–H hydroxyl groups

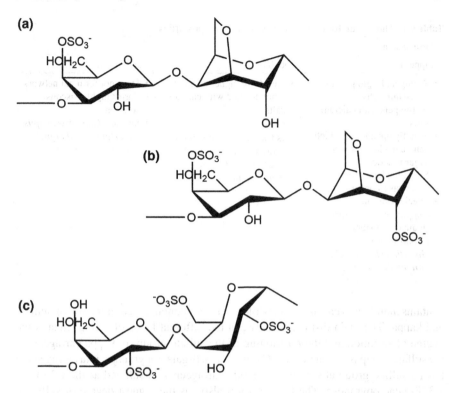

Fig. 6.1 Disaccharide structures of λ, κ and ι forms of carrageenans

and the C–H alkyl group is indicated at 3441 and 2934 cm^{-1}, respectively (Souza et al. 2018; Manuhara et al. 2016).

Carrageenans are soluble in water at high pH where they can form either viscous solutions or thermoreversible gels. The temperature at which carrageenans dissolve in water is increased as the level of sulfation decreases (Ghani et al. 2019). The interaction with water is dependent on the type of carrageenan, while iota and kappa form thermoreversible gels, and lambda carrageenan forms viscous solutions (Williams and Phillips 2003). Hence why, some forms of carrageenan are more suitable as viscosity enhancers and others are suitable as gelling agents. Carrageenans form gel by unraveling of their random coil structure to form helical secondary structures. These helixes then form networks in water to form thermoreversible gels. The solubility and nature of solution or gel formed also depend on the types of electrolytes present. The properties of these carrageenans vary significantly and consequently so do their applicabilities. For example, k carrageenan finds application in food industry as a thickening and gelling agent, in acetic acid production (Iglauer et al. 2011) and in industrial effluent treatment (Necas and Bartosikova 2013).

Occurence of carrageenan varies in different species of red algae. For example, *Eucheuma cottonii* and *K. alvarezii* contain mainly kappa carrageenan, *E. spinosum*

Table 6.1 Three main forms of carrageenan and their properties

Carrageenan		
Kappa (κ)	Iota (ι)	Lambda (λ)
• Strong and rigid gels in potassium salts • Brittle gels with calcium salts • Slightly opaque gel which can be made clear when sugar is added • Slight tendency for syneresis to occur • Species present: *Kappaphycus alvarezii* (main component), *Chondrus crispus*, *Gigartina skottsbergii*, *Sarcothalia crispata*	• Gels have higher elasticity when formed with calcium salts • Clear gels • Is less prone to syneresis • More freeze/thaw stable • Species: *Eucheuma denticulatum*	• Does not form gel network • Forms highly viscous solution • Species: *Chondrus crispus*, *Gigartina skottsbergii*, *Sarcothalia crispata*

contains mainly iota carrageenan, while *C. crispus* contains a combination of lambda and kappa. The main distinction between the different types of carrageenan is the degree of sulfation and the positioning of the sulfate group. In kappa carrageenan, the sulfate group is attached to the C4 of the 1-3, β-galactopyranose, iota carrageenan has its sulfate group at C2, while lambda carrageenan is sulfated as the C2 of the 1,3, β-galactopyranose. The lambda form also has the highest degree of sulfation of 70% (Williams and Phillips 2003). Table 6.1 summarizes the different types of carrageenan and their properties.

The molecular weight of carrageenan has significant effect on its rheological, thermal and mechanical properties as well as the bioactivity (Souza et al. 2011). Although the effect of the molecular weight on the properties of carrageenan varies for different types of carrageenan which in turn vary in the degree and pattern of sulfation and sequence of repeating units, some deductions can be made from reported studies to relate molecular weight to some specific parameters. For instance, lower molecular weight carrageenan shows higher gel elasticity than higher molecular weight ones (Souza et al. 2011). This was observed for a mixture of k and l carrageenan extracted from *Mastocarpus stellatus* red algae. The solubility and dissolution temperature as well as the conformation upon interaction with water are affected by the molecular weight. Generally, polymers' molecular weight is related to the solution viscosity such that the viscosity average molecular weight can be obtained from the intrinsic viscosity using the Mark–Houwink equation. Molecular weight of carrageenan of commercial grade is between 200 and 800 kDa (Weiner et al. 2017).

6.4 Availability of Raw Materials

Carrageenan is sourced from red algae (Rhodophyta), which is one of the three main types of seaweeds. Seaweeds are grown commercially in 35 countries of the world; as of 2003, the seaweed industry was valued at an estimated 5.5–6 billion USD annually (McHugh 2003). Of this estimated value, about 1 billion occurs from hydrocolloids production from seaweed with carrageenan being one of such. The majority of seaweed production still goes toward direct consumption by humans as food, while a smaller portion goes toward other miscellaneous uses such as fertilizers and additives in animal feed. Species such as *Undaria pinnatifida*, *Caulerpa* spp. and *Porphyra* spp. which are dominant in Eastern and Southeastern Asia are produced mainly for direct consumption as food (FAO 2018).

As of 1995, 13.5 million tonnes of seaweed was reportedly produced globally (McHugh 2003). This rose to over 30 million tonnes by 2016 (FAO 2018). The seaweed production output is currently led by China and Indonesia. China's seaweed production rose from near 11 million tonnes in 2010 to over 14 million tonnes by 2016. Indonesia's seaweed production rose from 4 million tonnes in 2010 to well over 11 million tonnes by 2016. Indonesia produces mainly *K. alvarezii* and *Eucheuma* spp. which are the main sources for carrageenan production. Globally, *Eucheuma* spp. is currently the highest produced seaweed species with output rising from 3.5 million tonnes in 2010 to 10.5 million tonnes in 2016. *K. alvarezii* comes in as the fifth highest output increasing from 1.2 million tonnes in 2010 to 1.5 million tonnes in 2016 (FAO 2018).

In 2018, FAO lists China, Indonesia, Philippines, Republic of Korea, Democratic People's Republic of Korea, Japan, Malaysia, Tanzania, Madagascar, Chile, Solomon Islands, Vietnam, Papua New Guinea, Kiribati and India, as the major seaweed producers in the world (FAO 2018). However, there exists a large gap between the output quantities, ranging from over 14 million tonnes produced by the top producer, China, to 3 thousand tonnes produced by India, the lowest producer. Seaweed production by far predominates over aquatic plant production globally. This can be attributed to their significant role as highly nutritious food, potential medicinal applications and increasing interest in cultivating them for biofuel production.

Seaweeds grow in a variety of conditions from cold to warm such that they can grow in different parts of the world from northern to Southern Hemisphere. About 7.5–8 million tonnes of seaweed is harvested annually on wet basis. These are either obtained from wild-growing stocks or cultivated in seaweed farms (McHugh 2003). Using artificial growth systems which replicate the sea conditions, seaweeds are being commercially grown even in the middle of the desert.

The earliest red seaweed species which served as a source of carrageenan is the *C. crispus* which is commonly referred to as Irish moss. It naturally occurs in France, Portugal, Ireland, Spain and eastern coast of Canada. Other species from which carrageenan is sourced are Irideae, Gigartina and Eucheuma which grow in Chile, Spain and the Philippines, respectively, as additional species to source carrageenan became necessary as demand for carrageenan grew. Today, the red seaweed species *K. alvarezii* and *E. denticulatum* have become the more common raw material for carrageenan production (McHugh 2003). For more diverse range of applications, there are continuous efforts to explore the variety of structural forms of carrageenan and this involves uses of a broad range of species which serve as sources of various forms of carrageenan. These include *H. musciformis, M. laminaroides, S. crispata, B. gelatinum, C. crispus, E. denticulatum, Eucheuma gelatinase, E. spinosum, G. canaliculata* and *G. skottsbergii* (McHugh 2003).

Red seaweed, which serves as the source of carrageenan, is primarily used as food as it makes up a sizeable part of Asian cuisine which is adopted across different parts of the world. In cold waters, red seaweed species of economic importance such as *Palmaria palmata* can be sourced from cold regions such as Canada, Iceland, Ireland and southern Chile, while in warmer waters they can be found in areas such as Morocco, Portugal, the Philippines, and Indonesia.

Most common species of red seaweed that is harvested for food is the Porphyra which is more commonly called nori or laver. It is commonly used in preparation for sushi, a Japanese delicacy which comprises raw fish and boiled rice wrapped in red seaweed which has been processed into thin sheets. The seaweed plays a structural role as an edible food wrap and also has significant nutritional value.

A large proportion of the red seaweed harvested globally is from cultivation of red seaweed as seaweed growing naturally in the wild is insufficient to meet increasing demands. The cost of harvesting from wild is also higher, and in some cases access is limiting. Cultivation in tanks requires a good understanding of the life cycle and required growth conditions of the particular species. Republic of Korea, China and Japan are some of the major large-scale producers of red seaweed. Red seaweed nori is one of the most highly priced seaweeds which is valued at 1200 USD per wet tonne, relative to, for example, brown seaweeds which are valued at 610–530 USD per wet tonne.

The type of carrageenan present in each species of red algae varies. Some species contain almost only one form of carrageenan, while some species contain a mixture of two different types of carrageenan. These mixtures are referred to as hybrids. A species of red seaweed known as *K. alvarezii*, for example, contains almost entirely kappa carrageenan, and another species called *E. denticulatum* (*E. spinosum*) contains almost entirely iota carrageenan. Species which contain hybrids include Chondrus which contain a mixture of kappa and lambda. A large fraction of the red seaweeds used for carrageenan production are sourced from the Philippines. Improvement in seaweed harvesting technology has led to red seaweed cultivation spreading to warmer regions such as Tanzania and Indonesia. Seaweeds which can be grown vegetatively are preferred for commercial carrageenan production as those requiring sexual reproductive cycles are more cost intensive making it non-economical as raw materials for carrageenan production. For seaweeds sold as food, the cost can be recovered more easily since very little further value addition occurs in the product preparation. However for use as raw materials for carrageenan production, further cost is involved in the extraction and purification of carrageenan such that it is paramount to minimize the cost of raw material, thus making vegetative cultivation a more favorable option.

According to FAO reports, production of some red algae species has increased in the past 10 years while some have dropped. *K. alvarezii*, for example, showed annual increase in production from 1285 MT in 2005 to 1963 in 2012. This was then followed by a decline to 1726 MT in 2013 to 1527 in 2016 (FAO 2018). On the other hand, Porphyra has shown a general annual rise from 703,000 tonnes in 2005 to 1,353,000 tonnes in 2016. This excludes a drop from 1,072,000 to 1,123,000 tonnes between 2010 and 2012. On this basis, the future availability of red algae is quite unpredictable.

The factors which may pose a threat to the availability of red algae are inter-woven with the factors which pose a threat to the aquatic environment as a whole. Exploring renewable sources of polymers used in products such as processing agents, packaging, energy storage, printing, cosmetics and textiles is a route to abating the threat of depleting fossil fuel reserves. The other issue is the impact of the process of extraction of these resources on the environment. The next section explores the environmental impact of producing carrageenan from red algae.

6.5 Extraction of Carrageenan

Carrageenan is soluble in aqueous alkali solution at high temperature and insoluble in alcohol and certain salt solutions (Williams and Phillips 2003), and this property serves as a basis for its extraction from red seaweed. Alkalis commonly used include sodium hydroxide, potassium hydroxide and calcium hydroxide. Here, we use a typical chemical extraction process as used in the literature (Manuhara et al. 2016) and an enzyme extraction method used by Souza et al (2018). Variations in method include change in the choice of alkali, use of enzymes and extraction conditions such as extraction pH, temperature and time.

6.5.1 Chemical Extraction

In the chemical extraction method by Manuhara et al. (2016), the collected red algae biomass is washed with water and then soaked in water for 24 h to soften. The wet biomass is then pulverized until a pulp is obtained. Extraction is carried out in alkali solution of calcium hydroxide at a pH of 9. The system is heated to 90 °C and the extraction carried out for 2 h under continuous stirring. The alkali serves to decrease the degree of sulfation and also increase the number of 3,6 anhydrous carrageenan units. This makes for stronger gels (McHugh 2003). The mass is then separated by filtration. The liquid filtrate which contains the carrageenan is retained for further purification. Hydrochloric acid is added until the pH reduces to 7. The neutralized liquid filtrate is then heated at 60 °C for 30 min. At this point, the filtrate contains other soluble molecules; therefore, the carrageenan needs to be precipitated out of the solution. This is done using potassium chloride (KCl) at a concentration which ranges between 1 and 3.5% in the volume ratio 1:1. This variation was done in this particular study to investigate the effect of KCl concentration on the properties of the carrageenan formed, and 2.5% was found to be the optimum concentration (Manuhara et al. 2016). A duration of 15 min is allowed for the precipitation. The carrageenan precipitates out as a gel which is separated from the liquid by filtration. This is then followed by washing in 96% alcohol, drying at 70 °C in a cabinet dryer for 24 h. The carrageenan is then milled into desired particle size. Figure 6.2 shows a flowchart summarizing the extraction process. Here, it is shown that two routes can be followed after filtration and initial concentration which is done by evaporation of the solvent. The filtrate can be precipitated either with a salt or with alcohol. When precipitated with alcohol, a solid precipitate is formed which can then be dried and then taken through a repeated precipitation stage to improve purity. When precipitated with potassium chloride, a gel is formed. This only happens for kappa carrageenan which has the ability to form stronger gel potassium chloride and hence is more suitable for gel pressing prior to drying.

After initial filtration, the filtrate usually contains about 1–2% w/v carrageenan, following further filtration which could be done using vacuum distillation or ultrafiltration, and the final concentrate contains 2–3% carrageenan (w/v) (McHugh 2003). The rest of the moisture then needs to be removed by drying to obtain a dried carrageenan which can then be milled for more effective storage and packaging.

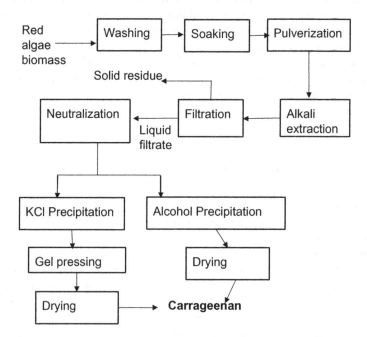

Fig. 6.2 Flowchart summarizing extraction process for carrageenan from red seaweed

6.5.2 Enzyme Extraction

Carrageenan can also be isolated from red algae using enzymatic digestion. This involves breaking down the proteins within the structure to release the polysaccharides. The sulfated polysaccharide, which in red algae is carrageenan, is then purified through precipitation. This method has been used by Souza et al. (2018) to extract carrageenan from the *H. musciformis* red algae. In the said study, the harvested biomass was cleaned and washed to remove epiphytes and other unwanted matter. The process of extraction then begins with proteolytic digestion using the protease enzyme, papain from the papaya fruit (*Carica papaya*). This breaks down the protein into smaller units of peptides and amino acids. The digestion was carried out at 60 °C in a sodium acetate buffer at pH 5 containing 5 mM of EDTA and cysteine. The digestion was allowed a period of 6 h. The next stage is then to isolate the freed sulfated polysaccharide from the solution. This was done by precipitation with cetylpyridinium chloride. This method resulted in a carrageenan yield of 28% carrageenan per gram of dry red algae biomass. The carrageenan extracted showed 17.3% free sulfate content and a very polydisperse extract with peak molar mass of 519.1 kDa measured using gel permeation chromatography.

6.5.3 Semi-refined Carrageenan

The semi-refined form of carrageenan also finds some use. This form of carrageenan is one, whereby other water-soluble components of the carrageenan are removed by dissolving them of at lower water temperature such that the carrageenan alongside the salt still remains behind in the seaweed. This process requires much lower cost of processing and is used where the presence of cellulose alongside the carrageenan does not affect the application. The seaweed is washed to get rid of debris. It is then treated in a solution of potassium hydroxide at a moderately high temperature. This allows the reduction in degree of sulfation as well as the hardening of the gel structure, while the carrageenan is still within the seaweed biomass. The heat allows the aqueous solution of alkali to penetrate the cell walls allowing interaction between the OH- ions and the sulfate groups and the K+ ions and the gel networks resulting in a stronger gel which does not dissolve in the solution. The other components of the seaweed such as proteins, water-soluble carbohydrates and other smaller molecules, however, are removed in this process. This leaves behind a solid mass which is mostly carrageenan and cellulose. This can be heat sterilized and used as low-grade gelling agent in canned pet food where a cheaper alternative to refined carrageenan is acceptable. Alternatively, it can be sold to carrageen manufacturers as feedstock for producing refined carrageenan. The latter option enables value addition to seaweed from place of origin and reduces the cost of transportation and waste treatment for the manufacturer. When semi-refined carrageenan is produced using more stringent processes which ensures a clean sterile product containing carrageenan and cellulose, it can be sold as carrageenan suitable for human consumption (McHugh 2003).

The hardening of the carrageenan gel only occurs in kappa and to a lesser extent in iota carrageenan; therefore, the gel formation-based extraction can only be used for seaweeds containing these forms of carrageenan. The other main form of carrageenan, lambda carrageenan, does not form gels and therefore can only be precipitated using alcohol (McHugh 2003). The end use of the carrageenan must therefore be considered prior to determining a suitable extraction method.

6.6   Environmental Implications

Carrageenan extraction is dependent on elevated temperature in aqueous alkaline conditions. In the main extraction process of the crude carrageenan, the concerns here are energy consumption, water usage and use of alkali. Other factors to consider include land use changes, additional energy used in separation, milling, drying and purification as well as other components such as use of salts for precipitation and alcohol for washing and drying.

Table 6.2 gives a summary for a typical process of extraction of carrageenan from red algae using chemical extraction method, and some of these are discussed further in the following subsections. This method is chosen as an example since it is more commonly used in present industry. The values here are simply the basic consumptions for a bench-scale extraction.

Table 6.2 Typical consumptions for extraction of carrageenan

Consumption	Quantity
Red algae biomass	2.92 g per gram carrageenan
Calcium hydroxide[a]	15 ml at pH 9 per gram carrageenan
HCl	Neutralize 15 ml CaOH from pH 9 to 7
Potassium chloride	15 ml of 2.5% per gram carrageenan
Ethanol	15 ml 96% per gram carrageenan
Energy	
• Heat	90 °C for 2 h
• Drying	60 °C for 3 min
• Stirring	70 °C for 24 h
	200 rpm for 2 h

6.6.1 Cultivation and Harvest of Red Algae

An estimated 1180 kg of seaweed biomass is required to produce 1 kL of the carrageenan containing say which can then be further processed to produce carrageenan. In addition to the benefits of growing algae as discussed in Chap. 5 and this chapter on alginate and fucoidan, red algae compared to the brown algae are easier to cultivate on a large scale since they have a vegetative reproductive cycle which is less complex than that of brown algae. A particular type of the Rhodophyta known as coralline algae has played a significant role in the formation of coral reefs for billions of years and still continues to do so (Moreira-Gonzalez et al. 2019). Harvesting of red algae from the wilds or cultivating in aquaculture or open seas ensures that the conditions in the water are equally conducive for the growth and functioning of these coral-forming red algae. Hence, red algae cultivation for alginate production encourages the maintenance of marine biodiversity.

6.6.2 Energy Consumption

The drying stage consumes much of the energy in the production process. The filtrate from the extraction only contains about 2–3% carrageenan, and the final product needs to be dried to at least 4% moisture content for storage and packaging. In this dry state, the growth of microbes is minimized hence allowing a longer shelf life of the carrageenan. Where there is intense sunlight and the weather is sufficiently hot in the day, drying can be done outdoors to save cost from using electrical dryers.

Energy is also consumed in transportation of the cultivated biomass to the factory for further processing. In the life cycle assessment carried out by Gosh et al. (2015), transportation of the cultivated biomass accounted for 13% of the environmental impact in the cultivation stage. In the processing of biomass into carrageenan and sap, the high-density polyethylene production for packaging shed and electricity made up 97.3% of the carbon footprint, with electricity consumption accounting for 25.2% and plastic packaging accounting for 54.2% of this.

When transportation by sea, road and rail was compared, transportation by sea for conveying biomass to factory for processing proved to be the least energy-consuming method while road consumed the most energy. Climate impact by road, rail and sea was 138.5%, 51.8% and 141%, respectively (Ghosh et al. 2015). In some cases where transportation by road is the only option, the most efficient means of transportation needs to be sought. For instance where the place of cultivation (usually shore) is at a relatively close distance to the processing factory, tricycle carts could be considered as an option; however where quantities are on a large scale, this is not practical.

Energy consumption can be minimized by using more fuel efficient machinery and transportation systems. Even where manual labor is used, it is important to have the most efficient process in place, to make optimal use of manpower. Transportation by sea to distances of up to 7200 km to other countries is even less energy-consuming that transportation by road within the same country by distances of about 1500 km (Gosh et al. 2015).

6.6.3 CO_2 Emission

The 118.6 kg of CO_2 equivalents is produced per kL of sap produced from *K. alvarezii* (Ghosh et al. 2015). To put this in perspective, cultivation of maize on one hectare of land results in production of 599 kg of CO_2. The 142.5 L of sap which is equivalent to a hectare produces 25.65 kg of CO_2 (Gosh et al. 2015). The process for producing the sap results in the production of semi-refined carrageenan which can then be further refined to produce refined carrageenan of higher grade. This process was based on using the same cultivation process for carrageenan production, and the process of extraction of sap is part of the process of carrageenan production. This quantity of CO_2 includes the entire process of sap production from cultivation to the factory gate. Following the process of separation of the sap from carrageenan, the two processes then separate such that this estimated CO_2 emission is not the same as that for carrageenan production as the process of refining to final packaging differs. Therefore, the CO_2 emission for carrageenan production is expected to be greater than 118.6 kg.

The environmental impact of the CO_2 emission from the process is weighed against the benefits to the environment of carrageenan and its by-product. For example if sap is coproduced along with sap which is in turn sprayed on crops, this substitutes the chemical fertilizers providing 21 g per l of potassium per sap. The 1000 L of sap substitutes 25.3 kg of chemical fertilizers (Ghosh et al. 2015).

6.6.4 Water Consumptions

Water serves the key role as the extraction solvent. It is also used in the washing of the raw material and solvent for the salt for precipitation. Gosh et al. (2015) estimate 262.205 m^3 of water used in the production of 1 kl of unrefined carrageenan (sap). In the reported experimental extractions (Manuhara et al. 2016), if we include the water used in washing and soaking and as solvent and continuous media for the process, an estimated 75 ml of water is used per gram of red algae processed. Although water is recycled, the release of mineral acids and alkali and salts into the water cycle results in gradual acidification and alterations of water salinity as the water is returned into the sea. The cultivation of red algae could be integrated into the wastewater treatment from the extraction process to achieve a net zero release of minerals into the water. This will require modification of the process such that the acids, alkali, salts and alcohols used can be broken down into minerals which are consumed by algae such as nitrates and sulfates.

6.6.5 Use of Plastics

In a life cycle assessment carried out by Ghosh et al. (2015), the economic and environmental impact of cultivation of *K. alvarezii* for the production of biostimulant sap alongside semi-refined carrageenan was assessed. It was deduced that 83.7% of the impact on environment in the cultivation stage came from the use of plastics. These include the polypropylene and high-density polyethylene ropes used for the rafts and the use of plastic nets. The ropes and nets were assumed to have an average life span of ten cultivation cycles. Implementing plastic reuse within the value chain in, for example, packaging significantly reduced the carbon footprint of the process as a whole by about 28% (Ghosh et al. 2015).

6.7 Applications

Much of the application of carrageenan is reliant on its rheological properties, gelling property and viscosity-enhancing properties. Most common commercial application of carrageenan is in food. However, its established and potential applications expand to other industries which include pharmaceutical, cosmetics, printing and textiles. Thus far, it seems all sulfated polysaccharides possess common properties which can be attributed to the sulfated pyranose structure. Each sulfated polysaccharide, however, distinguishes itself through the mechanism of action that results from the variation in the monosaccharide units and sequence of monosaccharides occurring in the polymer chain. This makes some sulfated polysaccharides more predominant in some applications than others. Carrageenan finds its key relevance in food industry as a thickener or gelling agent. However, other applications exist and either are already being commercially explored or have recently come to light and are being investigated for commercial use. In most applications in foods, carrageenans are used commercially at concentrations ranging between 0.005 and 0.5% (Davidson 1980; Graham 1977). Such that, they are high-value additives used in relatively low amounts.

6.7.1 Milk Stabilizer

Carrageenan is commonly added to liquid milk products as a stabilizer. The stabilizing properties of carrageenan are attributed to its ability to interact with the proteins in milk and also with itself to form a weak gel network which prevents formation of sediments and creaming of the milk product and by so doing prolongs the shelf life of the milk product. For such purposes, it is important to have a balance between the interaction the carrageenan polymer chains have with each other and the interactions they have with the protein polymers in the milk.

In milk preservation, the three types of carrageenans, λ, κ and ι, are usually mixed together. Carrageenan is usually used in concentration ranging between 0.01 and 0.05 wt% (Towle 1973; Syrbe et al. 1998; Tijssen et al. 2007). Carrageenan is particularly of importance in ultrahigh temperature (UHT) treated milk in which the milk protein has been slightly denatured due to high-temperature treatment.

Carrageenan stabilizes milk by forming a secondary helical structure and then bridges forming at different junctions of the helices. The κ- and ι-type carrageenans usually form these helical secondary structures resulting in gelation. Gelation is affected by the presence of certain ions, particularly calcium (Ca+) ions and potassium (K+) ions. The high amount of sulfate groups present in the λ-type carrageenan prevents it from gelation. This gelation is then further combined with complexation which occurs as a result of electrostatic interactions between the carrageenan and the protein. This is a result of the positive charged regions on the surface of the milk protein and the sulfate groups of the carrageenan. In some cases, this may be further aided by cations such as Ca+ forming bridges between the carboxyl group of the proteins and the sulfate groups of carrageenan (Tijssen et al. 2007).

Carrageenan is also used in other dairy products such as cheese, ice creams, chocolate milk and coffee creamers for the separation of whey and fat. 33% of the carrageenan market is from use in dairy products (McHugh 2003). This makes up the market where much of carrageenan is used.

6.7.2   Gelling Agent

In the presence of electrolytes such as rubidium, potassium and cesium, kappa carrageenan can form gels at much lower concentration (Williams and Phillips 2003). Such that, less carrageenan is required to obtain the required gel texture. This is important for both reducing cost of food processing and also minimizing the additive content. Hence why, kappa form of carrageenan is much preferred as a gelling agent compared to other forms. Gels formed in the presence of potassium chloride (KCl) show superior gel strength (William and Phillips). Iota carrageenan does not show this same increased gel strength in the presence of potassium chloride. Although kappa forms strong gels, they are rather brittle and more prone to syneresis in the absence of enhancers. Konjac mannan and locust bean gum are used to improve the gel properties and prevent syneresis (Williams and Phillips 2003).

In water-based jelly, a combination of kappa and iota carrageenan can be used as a replacement for pectin where a little or no calorie alternative is needed. In nondairy alternatives to ice cream such as sorbet, the creamy mouthfeel is achieved using a mixture of iota and kappa carrageenan. For this purpose, locust bean or pectin is also added to enhance the texture. A mixture of kappa and iota carrageenan is also used in stabilizing oil in water emulsions in mayonnaise with reduced oil content. Xanthan gum is also added here to aid the activity of carrageenan (McHugh 2003).

6.7.3 Meat and Poultry

Precooked meat and poultry are desirable as they minimize preparation time and in some cases provide added flavor in cases such as smoked thin sliced ham or smoked chicken. However, the cooking process often results in loss of some of the food mass and protein as some of it is lost in the heated water. This otherwise lost protein can be retained in the product using carrageenans. Due to their ability to bind and interact with water molecules, carrageenans are used by the food processing industries to control texture and retain structure of precooked meat and poultry products. The challenge here is that once dissolved in water, carrageenan tends to increase the viscosity of water which makes it difficult for the carrageenan to penetrate the meat and then interact with the free water and protein within it in order to have the desired effect. To overcome this challenge, the carrageenan is introduced in the water after the brine has been dissolved in it. The high salt content prevents the carrageenan from dissolving before cooking such that the carrageenan only begins to dissolve as the meat cooks hence allowing for better penetration. In the process, the carrageenan also acts to improve the texture of the meat or poultry product.

Carrageenans are also used to improve the otherwise dry texture of meat with reduced fat. The fat in meat contributes a juicy and tender mouthfeel, and it also contributes to flavor. When the fat is reduced, the meat tends to lose this appeal. Carrageenans have been used in some cases to recover this texture and flavor in low-fat meat (McHugh 2003).

6.7.4 Toothpastes

The key components of toothpastes are mild abrasives such as chalk, flavor, detergent and water. These ingredients are then held together by a thickening agent which ensures the product does not separate in storage, easily flows out of the toothpaste tube and also can be easily laid on the toothbrush during usage. Iota carrageenan at about 1% concentration is commonly used for this application.

6.7.5 Pet Food

Since animals have a more tolerant digestive system than humans, the requirement for pet food-grade carrageenan is less than that for humans. Pets also have less demand for food appearance than humans do so the slightly less clear seaweed powder which is a more crude form of carrageenan can be successfully used in pet food. In pet food, they are used as thickening agent for the meat gravy or as gelling agent for the flavored jelly. They also help bind the meat pieces and provide texture since much of the meat used in pet food is scraps from human meat cutting and processing.

For such application, kappa carrageenan containing seaweed flour is often used in combination with locust bean. The less refined form of carrageenan costs only a fraction of the cost of refined carrageenan (one-fourth). Over 5 thousand tonnes of seaweed powder is used for such applications.

6.7.6   Air Freshener Gels

An estimated 200 tonnes of lower-grade carrageenan in the form of seaweed flour finds non-food applications as air freshener gels annually (McHugh 2003). The air freshener perfume is mixed in potassium salt carrageenan and water. The gel is sealed to prevent evaporation over its storage shelf life. Once opened, the perfume and water begin to evaporate against the concentration gradient with the environment. Eventually, the water and perfume completely evaporate and what remains is a hard dry gel at the end of the product usage life.

6.7.7   Immobilized Cells and Biocatalysts

For different applications such as cell biology or tissue culture preparation, cells need to be stored in inactive dry forms such that they are protected from the environment and can also be easily stored, transferred and transported. The ability of kappa carrageenan to form hard gels makes them applicable for such purpose. When required for use, these immobilized biocatalysts and cells can then be rehydrated and activated. In one of the early studies on the use of carrageenan for such purpose, the ethanol-producing bacterium *Zymomonas mobilis* has been entrapped in kappa carrageenan (Luong 1985). Kappa carrageenan is preferred as it allows immobilization of the biological component under mild conditions without damage (Chibata et al. 1987). More recently, enzymes such as cellulase and pollunase have been immobilized using carrageenan-based polymer complexes (Yasin et al. 2019; Long et al. 2016). Ability to immobilize cells and enzymes makes more effective processes where the release and activity of the enzyme can be better managed. Enzyme immobilization is further discussed in this book in the chapter on Enzymes.

6.7.8   Antimicrobial Properties

Although not its main application, carrageenan extracted from red algae strains such as *H. musciformis* has shown antimicrobial activity. Kappa carrageenan from *H. musciformis* had antibacterial effect against *Staphylococcus aureus* and antifungal effect against *Candida albicans* (Souza et al. 2018). The carrageenan extracted from this study had a molecular weight of 519 kDa. Carrageenan molecular weight varies between 200 and 800 kDa; therefore, this has a molecular weight in the midrange. When tested on gram-negative bacteria *S. enteritidis*, *E. coli* and *P. aeruginosa*, the carrageenan extracted in this study did not present any antimicrobial activity against these gram-negative bacteria. However, growth inhibition was observed when tested against gram-positive bacteria *S. aureus* and *C. albicans* (Souza et al. 2018).

6.7.9 Antioxidant

In the long polymeric form, carrageenan has little or no reported antioxidant activity; however when processed into derivative forms with shorter chain, varied degree of sulfation, acetylation and phosphorylation, these derivatives have significant antioxidant activity (Yuan et al. 2005). For instance, short-chain oligosaccharides of carrageenan obtained from hydrochloric acid hydrolysis of the polysaccharide form which were then oversulfated and acetylated showed antioxidant properties in vitro. This antioxidant property was evident as scavenging effect on hydroxy oxides and superoxides and DPPH radicals (Yuan et al. 2005).

Although carrageenan extracted from *H. musciformis* showed no cytotoxic effects on cancer cells investigated, it demonstrated anticancer effect through the inhibition of cancer cell proliferation (Souza et al. 2018). Taking results from different studies leads to the conclusion that if at all, carrageenan only shows very low antioxidant property which is dependent on the chemical structure of carrageenan in terms of sulfation pattern. Much of the anticancer property of carrageenan seems to be due to its ability to enhance innate immunity of the body in, for example, enhancing the activity and production of killer cells.

6.7.10 Neuroprotective

However, very few studies have investigated the neuroprotective property of carrageenan. Results from some studies show carrageenan activity against 6-hydroxydopamine, a neurotoxicant (Souza et al. 2018). The mechanism of action is demonstrated to be through having an inhibitory effect on caspase enzyme activity and affecting the mitochondrial transmembrane potential. Although these studies are early stage studies on cell cultures, it does point to some potential neuroprotective activity of carrageenan.

6.7.11 Immunomodulatory Activity

In cases where immune response leads to adverse reactions such as inflammations, carrageenans have shown potential immunomodulatory effect. It is thought to do this by chemically attaching to phagocytic cells such as macrophages by binding to Toll-like receptors or pattern recognition cells. This has been demonstrated in mouse T cells and in vivo experiments in laboratory mice. Carrageenan showed potential use in ameliorating the allergic reaction from immune response. When given an oral dosage of lambda carrageenan mice that have been treated with ovalbumin to trigger an allergic reaction, the reduction in serum histamine release and ovalbumin-specific IgE indicates a reduction in allergic reaction (Maruyama et al. 2005).

Immune modulation is also required where the immune activity needs to be enhanced, for example, in the case of vaccination. Evidence also indicates that carrageenans could potentially aid innate immune response against cancer cells by enhancing the cytotoxic effect of the immune system's lymphocytes and macrophages such that the anticancer effect of carrageenan can be attributed to its immunomodulatory effect. In studies using oligosaccharides of kappa carrageenan of carrageenan extracted from the red algae *Kappaphycus striatum* administered to mice which had been induced with cancer (sarcoma S180), carrageenan enhanced the activity of killer cells, inhibited cancer cell growth and increased macrophage phagocytosis (Yuan et al. 2006). It should be noted that in the aforementioned study, the short-chain kappa carrageenans used are quite different from the high molecular weight carrageenans which are commercially used in food industry. Sulfated polysaccharides have generally shown structure-dependent bioactivity such that the difference in molecular weight means higher molecular weight carrageenans are not likely to show the same immunomodulatory activity as oligosaccharide carrageenans. These oligosaccharides are prepared from the acid hydrolysis of the longer-chain polysaccharide extracted from the red algae (Yuan et al. 2005), therefore making the polysaccharide carrageenans relevant in understanding the immunomodulatory activity of the oligosaccharides.

6.7.12 Antiviral Activity

Carrageenans show relatively weak antiviral activity compared to other aquatic polymers which have been investigated for their potential antiviral activity. When tested against denge viva virus DENV-1, DENV-2, DENVV-3 and DENV-4 in vero cells, carrageenans only had antiviral effect of DENV-2. Carrageenan acted by inhibiting the multiplication of the DENV-2 virus by preventing cell absorption and entry. This effect was much less potent in DENV-3 and DENV-4 and completely ineffective in DENV-1 (Talarico and Damonte 2007). However, even the activity against DENV-2 is dependent on the virus taking the normal route into the cell and being introduced into the host cell at the same time as the virus or shortly after, once the normal entry is bypassed or the virus is introduced at a longer time before the carrageenan, it then has no antiviral infect and the DENV-2 virus is able to successfully infect the host.

6.7.13 Polymeric Electrolytes

Biopolymer-based electrolytes are favoured as energy storage devices for their bio-compatibility, renewable source, relative abundance, ease of processing and relative low cost. Iota carrageenan has been investigated for use as a polymeric electrolyte. This particular form of carrageenan shows good prospects in this field of application due to its level of sulfation and 25–30% anhydrogalactose component which results in a medium gel strength with a moderate level of amorphous structure leaving a sufficient amount of functional groups available for interactions with charge carriers, making it more suitable as a polymer electrolyte (Ghani et al. 2019). Carrageenan can be dissolved in water to form a solid-state electrolyte. This makes it possible for low-temperature processing which is desirable in lowering processing costs in such applications. Purification of the iota carrageenan significantly affects the electrochemical stability. In the purified form, iota carrageenan attained a conductivity of 1.57×10^{-5} S cm^{-1} while the unpurified form was 1.65×10^{-6} S cm^{-1} (Ghani et al. 2019).

From all the applications of carrageenans discussed so far, applications based on the rheological properties and viscosity- and gel-forming properties are the most well established. The bioactive properties such as antimicrobial and anticancer properties need further investigations to understand and possibly extend the effectiveness and mechanism of action. Particularly important is to understand how these activities are affected by variables such as extraction conditions, molecular weight and species.

It is indeed desirable to relate the different bioactivities of carrageenans with specific chemical properties or structure. However, there is much yet to be known about the structural activity relationships of sulfated polysaccharides. In attempts to understand the structural relationship with the bioactivity, one approach is to make inferences from similar but simpler structures. For example, studying more linear simpler structures and relating the bioactivities of such to similar more complex branched structure, such has been done for some sulfated polysaccharides such as fucoidan (Jiao et al. 2011).

6.8 Commercial Production

Carrageenan is one of the relatively well-explored sulfated polysaccharides of algae. It mainly finds commercial applications in food as the largest market for carrageenan. However, its potential applications also extend beyond food industry to others such as cosmetics and nutraceuticals and possibly in the future pharmaceuticals. The total market for carrageenan is valued at over 300 million USD (McHugh 2003). Demand for carrageenan is further boosted by the need for a substitute for a replacement for animal-sourced gelatin with the outbreak of mad cow disease (bovine Spongiform encephalopathy). Its thermostable property makes it form more stable gels in hotter environments compared to gelatine.

Commercially, carrageenan is available either as refined carrageenan which is assigned the E number $E = 407$ or as semi-refined carrageenan produced to high standards which is referred to as PES assigned the E number E-407a (McHugh 2003). Weiner et al. (2017) emphasized the difference between commercial carrageenan used in food additives and other forms of carrageenan which have no commercial applications in food but are used in other applications such as medical. The group further highlighted that the difference in the molecular weight had significant effect on the physicochemical and rheological properties of carrageenan and should therefore not be used interchangeably. Commercial carrageenan used as a food additive is sulfated polysaccharide that has a average molecular weight between 200 and 800 kDa with alternating disaccharide repeating units of 1–3 alpha-D-galactose and 1–4 linked beta-D-galactose or 3,6 anhydrogalactose with sulfate groups attached to some of the units at C2 or C4 depending on the type of galactan (3 main ones Iota, kappa and lambda). The low molecular weight carrageenan which is extracted at high temperature under low pH and has a molecular weight of less than 20–40 kDa is referred to as degraded carrageenan of poligeenan. The significantly shorter chain length and change in configuration due to low pH and high temperature of extraction result in a structure which does not share the typical attributes that characterize carrageenan.

As of 2001, 42,930 tonnes of carrageenan was produced by the top 3 producers of carrageenan in the world (McHugh 2003). Asia-Pacific contributed 2018 tonnes of this, 9900 of this being PES grade and the rest being either gel or alcohol-precipitated carrageenan. Europe contributed a total carrageenan production of 13,500 tonnes with 500 tonnes of this being PES and the rest refined carrageenan. The Americas produced 9150 tonnes; 1100 of this was PES grade, while the rest was refined carrageenan. About half of the carrageenan produced is therefore PES grade with most of it coming from the Asia-Pacific region.

Some of the companies producing carrageenans in different parts of the world include Kelcco, Ingredients Solutions Inc, Shemberg Biotech Corporation, Marcel Carrageenan Corporation, FMC Biopolymer, Degussa Texturant Systems, Danisco Cultor, Rhodis Food, Gelymar S.A., CEAMSA, Hispanagar S A, Ina food Industry Co. Ltd., Myeong Shin Chemical Ind. Co. Ltd., Soriano SA, Chuo Food Materials Co. Ltd., Quest International Philippines Corp., FMC Corporation, Geltech Hayco Inc., TBK Manufacturing Corp., Iberagar S.A., PT Gumindo Perkasa Industri, CV Cahaya Cemerlang, PT Surya Indoalgas, P.T. Asia Sumber Laut Indonesia (McHugh 2003).

6.9 Conclusion

Carrageenan is a relatively well-explored aquatic biopolymer primarily sourced from the aquatic environment with no reported terrestrial source; therefore, its availability is highly reliant on the availability of red algae in the aquatic ecosystem. With a potential yield of ~30% per dry mass of red algae, it can be considered a major algal resource since it makes up a relatively large proportion of the red algae composition. Although the majority of commercial applications of carrageenan mainly relies on its rheological properties, carrageenan has some promising bioactivities which are being investigated. Carrageenans with bioactive properties have a promising future in developing multifunctional products such as food additives which also have health benefits. While carrageenan itself as a polymer is a renewable resource produced from red algae and the product itself has no toxic effect on the environment, the process of production requires fossil energy consumption and use of chemicals which might have less benign effects. Some of these environmental impacts of carrageenan production process can be minimized by using milder and higher yielding processes such as enzyme-assisted extraction and integration of algae cultivation process in the wastewater treatment.

References

Chibata I, Tosa T, Sato T, Takata I (1987) Immobilization of cells in carrageenan. Methods Enzymol 135:189–198

Davidson RL (ed) (1980) Handbook of water-soluble gums. New York McGraw-Hill Book Co.

Doyle JP, Giannouli P, Rudolph B, Morris ER (2010) Preparation, authentication, rheology and conformation of theta carrageenan. Carbohyd Polym 80:648–654

Eisses J (1952) The research of gelatinous substances in Indonesian seaweeds at the laboratory for chemical research. Bogor J Sci Res Indon 1:44–49

FAO (2018) The state of world fisheries and aquaculture 2018—Meeting the sustainable development goals. Rome. CC BY-NC-SA 3.0 IGO

Ferreira LG, Noseda MD, Goncalves AG, Ducati DRB, Fujii MT, Duarte MER (2012) Chemical structure of the complex pyruvylated and sulfated agaran from the red seaweed *Palisada flagellifera* (Ceramiales, Rhodophyta). Carbohyd Res 347:83–94

Funami T, Hiroe M, Noda S, Asai I, Ikeda S, Nishinari K (2007) Influence of molecular structure imaged with atomic force microscopy on the rheological behavior of carrageenan aqueous systems in the presence or absence of cations. Food Hydrocolloids 21:617–629

Ghani NAA, Othaman R, Ahmad A, Anuar FH, Hassan NH (2019) Impact of purification on iota carrageenan as solid polymer electrolyte. Arab J Chem 12:370–376

Ghosh A, Anand VKG, Seth A (2015) Life cycle impact assessment of seaweed based biostimulant production from onshore cultivated *Kappaphycus alvarezii* (Doty) Doty ex Silva—is it environmentally sustainable? Algal Res 12:513–521

Gonçalves AG, Ducatti DRB, Paranha RG, Duarte MER, Noseda MD (2005) Positional isomers of sulfated oligosaccharides obtained from agarans and carrageenans: preparation and capillary electrophoresis separation. Carbohydr Res 340:2123–2134

Graham HD (1977) Food colloids. AVI Publishing Co., Inc., Westport, Connecticut

Iglauer S, Wu Y, Schuler P, Tang Y (2011) Goddard III WA. Dilute iota- and Kappa-Carrageenan Solutions with high viscosities in high salinity brines. J Petrol Sci Eng 75:304–311

Jiao G, Yu G, Zhang J, Ewart HS (2011) Chemical structures and bioactivities of sulfated polysaccharides from marine algae. Mar Drugs 9:196–223

Long J, Xu E, Xingfei, Wu Z, Wang F, Xu X, Jin Z, Jiao A, Zhan X (2016) Effect of chitosan molecular weight on the formation of chitosan- pullulanase soluble complexes and their application in the immobilization of pullulanase onto Fe_3O_4-k-carrageenan nanoparticles 202:49–58

Luong JH (1985) Cell Immobilization in kappa carrageenan for ethanol production. Biotechnol Bioeng 27(12):1651–1661

Manuhara GJ, Praseptiangga D, Riyanto RA (2016) Extraction and characterization of refined K-carrageenan of red algae [*Kappaphycus alvarezii* (Doty ex P.C. Silva, 1996)] Originated from Karimun Jawa Islands. Aquat Procedia 7:106–111

Maruyama H, Tamauchi H, Hashimoto M, Nakano T (2005) Suppression of Th2 immune responses by mekabu fucoidan from *Undaria pinnatifida* sporophylls. Int Arch Allergy Immunol 137:289–294

McHugh DJ (2003) A guide to seaweed industry FAO fisheries technical paper 441. Rome. Downloaded 5/1/2019, pp 1–6

Moreira-Gonzalez AR, Fernandez-Garces R, Batista MG, Leon-Perez AR, Caballero YC, Garcia-Moya A, Fujii MT, Suarez-Alfonso AM (2019) Marine red algae from central-southern coast of Cuba. Reg Stud Mar Sci 25(100450):1–9

Necas J, Bartosikova L (2013) Carrageenan: a review. Veterinarni Medicina 58(4):187–205

Sheath RG, Vis LM (2015) Red algae. Freshwater Algae of North America (Second Edition). Ecology and Classification pp 237–264

Smith HM (1905) The utilization of seaweeds in the United States. Bull US Bur Fish 24:169–171

Souza HKS, Hilliou L, Bastos M, Goncalves MP (2011) Effect of molecular weight and chemical structure on thermal and rheological properties of gelling k/l-hybrid carrageenan solution. Carbohyd Polym 85(2):429–438

Souza RB, Frota AF, Silva J, Alves C, Neugebauer AZ, Pinteus S, Rodrigues JAG, Cordeiro EMS, de Almeida RR, Pedrosa R, Benevides NMB (2018) In vitro activities of kappa-carrageenan isolated from red marine alga *Hypnea musciformis*: antimicrobial, anticancer and neuroprotective potential. Int J Biol Macromol 112:1248–1256

Stanley N (1987) Production, properties and uses of carrageenan. FAO report, 1987, Rome

Syrbe A, Bauer WJ, Klostermeyer H (1998) Polymer science concepts in dairy systems: an overview of milk protein and food hydrocolloid interaction. Int Dairy J 8:179–193

Talarico LB, Damonte EB (2007) Interference in dengue virus adsorption and uncoating by carrageenans. Virology 363:473–485

Tijssen RLM, Canabadv-Rochelle LS, Mellema M (2007) Gelation upon long storage of milk drinks with carrageenan. J Dairy Sci 90:2604–2611

Towle GA (1973) Carrageenan. In: Whistler RL (ed) Industrial gums, polysaccharides and their derivatives. Academic Press, New York, pp 83–114

Weiner ML, McKim JM, Blakemore WR (2017) Addendum to Weiner ML (2016) Parameters and pitfalls to consider in the conduct of food additive research, carrageenan as a case study. Food Chem Toxicol 107:208–214

Williams PA, Phillips GO (2003) GUMS: properties of individual gums. In: Caballero B (ed) Encyclopedia of food sciences and nutrition. Academic Press

Yasin MA, Gad AAM, Ghanem AF, Rehim MHA (2019) Green synthesis of cellulose nanofibers using immobilized cellulase. Carbohydr Polym 205:255–260

Yuan H, Song J, Li X, Li N, Dai J (2006) Immunomodulation and antitumor activity of kappa-carrageenan oligosaccharides. Cancer Lett 243(2):228–234

Yuan H, Zhang W, Li X, Lu X, Li N, Gao X, Song J (2005) Preparation and in vitro antioxidant activity of kappa-carrageenan oligosaccharides and their oversulfated, acetylated and phosphorylated derivatives. Carbohyd Res 340(4):685–692

Yu G, Hu Y, Yang B, Zhao X, Wang P, Ji G, Wu J, Guan H (2010) Extraction, isolation and structural characterization of polysaccharides from a red alga *Gloiopeltis furcata*. J Ocean Univ China Nat Sci 9:193–197

Zaneveld JS (1959) The utilization of marine algae in tropical south and east Asia. Econ Bot 13(2):89–131

Zhou G, Sheng W, Yao W, Wang C (2006) Effect of low molecular [lambda]-carrageenan from *Chondrus ocellatus* on antitumor H-22 activity of 5-Fu. Pharmacol Res 53:129–134

Chapter 7
Agar

Abstract Agar is obtained from red algae. It occurs as a mixture of agarose and agaropectin with agarose having the more desirable properties. Its applications are mainly attributed to its rheological properties. Such applications extend to the food, biotechnology and pharmaceutical industries where it is used as a thermoreversible gelling agent, stabilizer, texture modifier and thickener. The ability to serve as a substitute additive for the animal-sourced gelatin also contributes to its economic value. In this chapter, the extraction, chemistry, occurrence in nature and applications are discussed. In the process, the economic and environmental impact of agar production is evaluated.

Keywords Agar · Agarose · Agaropectin · Polymers · Gels

7.1 Introduction

A range of polysaccharides can be sourced from the aquatic environment, and some of these polysaccharides have well-established process technology and market presence. Agar is one of such well-explored polysaccharides. It is present exclusively in red algae; there are yet to be other non-aquatic sources of agar to date. The most common form of agar is agarose which is familiar with almost every biology student and researcher. This partially sulfated polysaccharide of the red algae is most valuable in its gel form for its rheological properties, which has earned it a potential global annual market valued at around 172.8 million USD (Bixler and Porse 2011). It can therefore be considered a highly valuable aquatic-derived resource.

Agar is one of the three main phycocolloids obtained from the aquatic environment, the other two being carrageenan and alginate. Agar is comprised of agarose and agaropectin, where the composition of each within the organism and this in turn affects the properties and applications of the agar (Armisen and Gaiatas 2009). Agar finds applications in food, biotechnology and pharmaceutical industries where it is used as a thermoreversible gelling agent, stabilizer, texture modifier and thickener. It is often used as a substitute additive for the animal-sourced gelatin (Marcus 2014) for reasons of ethical, health or dietary preferences where animal-derived gelling agents are not desirable.

© Springer International Publishing 2020   145
O. Olatunji, *Aquatic Biopolymers*, Springer Series on Polymer and Composite Materials,
https://doi.org/10.1007/978-3-030-34709-3_7

Global production of agar is estimated at 10,600 tonnes annually with the major producing countries in the Asia-Pacific regions (Rhein-Knudsen et al. 2015). Agar has well-established large-scale production process technology, and several companies across the world are involved in agar production. Examples of such companies are MSC in Korea and Huey Shyang Seaweed Industrial Company in China. The species of the genus Gracilaria remain the most preferred source for commercial agar production.

In this chapter, we discuss the environmental and economic impact of agar as an aquatic biopolymer. In doing so, we look at the red algae as a source, the occurrence of red algae in nature and the process of extracting agar from red algae and the impact of these processes on the environment. A review of the chemistry of agar and how it relates to its rheological properties explain how agar differs from other polymers. The applications of agar in several products are also explored to give an understanding of the commercial value and the versatility of agar. Some of the progress and limitations of commercialization of agar are also explored.

7.2   Occurrence in Nature

Agar has been in use as far back as 1658 in Japan (Armisen and Gaiatas 2009). It occurs naturally in cell wall and within the intracellular spaces of red algae, mainly those belonging to the genus Gracilaria, Gelidium and Gelidiella (Rhein-Knudsen et al. 2015). These include species such as *Gracilaria tikvahiae* (Rocha et al. 2019). *Gelidium sesquipedale* (Martinez-Sanz et al. 2019), Pterocladia, Ancatkopeltis and *Ceramium genera* (BeMiller 2019; Sudha et al. 2014), *Gracilaria cliftonii* (Kumar and Fotedar 2009), *Gracilaria asiatica* (Li et al. 2008) and *Gracilaria lemaneiformis* (Li et al. 2008). Algae which serve as a source of agar are referred to as agarophytes (Armisen and Gaiatas 2009).

Red algae species belonging to other genera are also explored for agar extraction such as the *Pyropia yezoensis* (Sasuga et al. 2017) which, although more commonly used for food as nori, is also a source of agar. The rhodophytes can be found growing in marine, brackish and less commonly some freshwater. Although conditions of the sea can be replicated in artificial growth environments for cultivation of these algae as biomass source for production of agar, certain species grow naturally in several regions across the world. For example, the *Gracilaria tikvahiae* is endemic to Western North Atlantic region of the USA (Rocha et al. 2019), while the *Gracilaria lemaneiformis* grows in China along the coast of Liaodong Peninsula (Li et al. 2008). They are cultivated in ponds and estuaries and also occur naturally in larger water bodies.

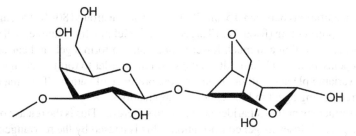

Fig. 7.1 Basic structure of agar repeating dimer unit

7.3 Chemistry of Agar

7.3.1 Polymeric Structure

Agar is a mixture of two polymers, agarose and agaropectin. These can be further fractionated and characterized upon extraction by methods such as gel permeation, ultracentrifugation or SDS-PAGE since they vary by molecular weight. Agarose is a neutral linear copolymer comprising of 3-O substitute β-d-galactopyranosyl and 3,6-anhydro-α-l-galactopyranosyl repeating units (Sudha et al. 2014), while agaropectin is the charged branched chain comprising of the same repeating unit (BeMiller 2019) as shown in Fig. 7.1.

Compared to other phycocolloids from red algae, agar has relatively lower hydrophilicity and water solubility. This is in part due to the presence of some methyl ether groups along the agaropectin polymer chain which may also include some small amounts of sulfation (BeMiller 2019). Agarose forms gels, while agaropectin does not form gels. The gel properties of the agar depend on the amount of agarose and agaropectin presents. The sulfate groups are mainly present in the agaropectin, and this has an effect on the syneresis generated in the gel (Mizrahi 2010).

7.3.2 Gel Formation

One of the desirable properties of agar is the ability to form a gel at relatively low concentration. This means less of the product is needed as an additive; this reduces cost and chances of altering other properties of the product. Gelation occurs in agar as a result of hydrogen bonds between the functional groups of the polymer chains forming bridges (Armisen and Gaiatas 2009). Because these bridges are as a result of secondary bonds which can be easily broken as conformation of the polymer chain varies at different temperatures, this makes this gelation a reversible one. Gelation is enhanced by the formation of 3,6-anhydrogalactose which occurs during alkali extraction where the sulfated galactose is converted to 3,6-anhydrogalactose (Bixler and Porse 2011).

At concentrations between 0.5 and 2% w/v in water at around 80–100 °C, agar will form a gel upon cooling (Rioux and Turgeon 2015). Gelatin, for example, will require a higher concentration and much lower temperature to form a gel, and carrageenan requires potassium or calcium salts in the solution in order to form a gel; agar forms a gel at variable pH and without requiring the presence of cations. This makes agar a preferred gelling agent in biotechnology applications.

Gels formed by polysaccharides are prone to syneresis. This is the release of water from the gel as it loses its gel conformation. This is caused by the rearrangement of the polymer chains due to different factors (Mizrahi 2010). The relationship between degree of syneresis and agar concentration is quantified in Eq. (7.1) (Mizrahi 2010).

$$\text{Degree of Syneresis} = 1/\text{Concentration}^2 \tag{7.1}$$

The gel strength is affected by the growth method and growth conditions. For example, a gel strength of 505 g cm^{-2} was measured for agar extracted from Gracilaria grown from tissue-cultured seedlings using the broadcast method of seaweed cultivation, while 201 g cm^{-2} was obtained from the same species grown using the off-bottom method from tissue-cultured seedlings (Rejeki et al. 2018).

The lower sulfate content in agar compared to the other more sulfated seaweed polysaccharides gives it a higher gel strength and gel melting point. Compared to carrageenan, for example, which has a melting point between 50 and 70 °C, agar has a melting point between 85 and 95 °C and while gel strength of agar varies between 700 and 1000 g/cm^2 for a 1.5% w/v concentration, that of carrageenan is between 100 and 350 g/cm^2 (Rhein-Knudsen et al. 2015).

7.3.3 Viscosity

Viscosity of polymers is related to the molecular weight. This relationship is quantified by the Mark–Houwink equation (Eq. 7.2), where [η] is the intrinsic viscosity, M is the average molecular weight and K and a are constants which vary for different polymers.

$$[\eta] = K M^a \tag{7.2}$$

The constants K and a are obtained experimentally by plotting the values of intrinsic viscosity against molecular weight and fitting to the equation (Wang et al. 1997). The viscosity is therefore an important parameter which is used in the characterization of polymers. Agar has a lower viscosity (10–100 centipoise at 1.5%, 60 °C) than carrageen (30–300 centipoise) Rhein-Knudsen et al. 2015. This lower viscosity is attributed to the lower molecular weight of agar compared to carrageenan. While number average molecular weight of agar is between 36 and 1144 kDa, that of carrageenan is usually at the higher end from 200 to 800 kDa (Weiner et al. 2017).

This lower viscosity and higher gel strength make processing of the gels easier by allowing easier flow during processing of the product prior to gel formation.

7.3.4 Interaction with Sugars

Sugars are often used alongside agar-based products such as candies, icings and creams. Other than acting as a sweetener, sugar also contributes to the physical properties of the food. When agar gel interacts with sugar in a fluid gel system, this has an effect on the rheological properties of the resultant product and consequently the nature of gels and foams formed in such systems. This property-related interaction with sugar could be used as a means of reducing sugar contents of processed food while giving consideration to the change in properties such as the texture, syneresis and gel strength. These properties are important in food products, while the consumer and health agencies demand a lower sugar content for health reasons, and they also demand appealing taste, texture and appearance. The producer therefore needs to understand the agar–sugar interaction and how this affects the property of the product. Recent studies which analyzed the effect of concentrations of different sugars: glucose, sucrose and fructose which are commonly used in food products, have significant effect on the viscosity, gel strength and foam stability. At concentrations above 50% w/v of sugar (either fructose, sucrose or glucose), the gel formed is more elastic; at lower concentration below 50% w/v of sugar, the shear viscosity and modulus of elasticity increase, thus indicating a stronger more rigid gel formed. Increasing sugar concentration also increased the foam stability measured as foam half-life (Ellis et al. 2019). Earlier studies using agarose extracted from *Gelidium amansii* have established that stronger agarose gels with higher melting points are formed when sugar is added to the agarose gel solution up to a critical concentration (Watase et al. 1990). This relationship between the sugar concentration and physical properties of the resulting gel of agar is attributed to the interactions between the functional groups of the sugars and those of the agar. These interactions in turn affect the conformations during gelation. In the same study, it was observed that urea and guanidine hydrochloride resulted in a decreased young modulus.

7.3.5 Thermal Degradation of Agar

For a product to be considered safe, the products upon degradation should pose no harm to health while in storage or in use. One of the advantages of agar as a phycocolloid is its ability to withstand higher processing temperatures. This makes it suitable for heat-sterilized food products such as canned fish and meat. However, recent studies point to potential toxic effect of the degradation of agar within a food product when heated (Ouyang et al. 2018). Stronger agar gels tend to have higher thermal degradation temperature such that the chances of having toxic degradation products during heat treatment of agar-containing foods could be reduced by using stronger gels.

Upon thermal degradation, agar forms 3,6-anhydropyran galactopyranose and galactopyranose. These then further degrade 3,6-anhydropyranose galactopyranose units break down into furyl hydroxymethyl ketone, a potentially toxic compound to humans. The galactopyranose breaks down into 3,4-atrosan D-allose and two other potentially toxic compounds, furfural and 5-(hydroxymethyl)-2-furancarboxaldehyde (Ouyang et al. 2018). Therefore, although generally considered a safe and approved product, care should be taken during the process of food treatment to avoid the formation of toxic compounds either from partial or complete degradation, especially during sterilization.

The initial melting of agar-based gel varies with factors such as concentration, presence of compounds such as sugars and urea and the agarose/agaropectin content. Earlier studies have shown that at different agarose concentration, the melting temperature of the gel remains constant at 75 °C. However, the enthalpy of transformation from gel to sol increases as the concentration of agarose decreases from 12 w/w to 2 w/w (Watase et al. 1990).

7.3.6 Biological Degradation of Agar

Biological degradation is a relatively economic degradation method, whereby a substance is broken down into its smaller units by naturally existing organisms which produce the enzymes for digesting, hence degrading the substances which they can then use for their metabolism and energy production. These enzymes can be extracted from such organism for controlled degradation of the substance without the contamination of microbes. The understanding of degradation of biopolymers is important in determining their safety and better understanding of their bioactivity and applications.

Agar is naturally present in the cell walls and intracellular structures of red algae which occur in the aquatic environment; an ideal source of the organisms which degrade agar would therefore be in the same aquatic environment. These microbes either exist freely in the environment or within the bodies of other bigger organisms which feed on the red algae as these organisms will require the enzymes which break down all components of the red algae, which include the agarose, into smaller products which can then be used for metabolism.

Agar-degrading bacteria include Pseudomonas, Cellulophaga, Acinetobacter, Agarivorans, Microbulbifer, Pseudoalteromonas, Saccharophagus, Bacillus, Paenibacillus, Streptomyces and Zobellia (Kwon et al. 2019). There are more marine-based bacteria which degrade agar than the freshwater-based ones. This is expected since most of the red algae are found mostly in the marine environment; freshwater red algae are rarer. The enzyme which degrades agarose is called agarase. These break down agarose into cooligomers of 2–4 repeating units of galactose and anhydrogalactose; further cleavage of the glycosidic bonds breaks down these neoagarose units into neoagarobiose, dimers of agar which is consisting of one unit of galactose and

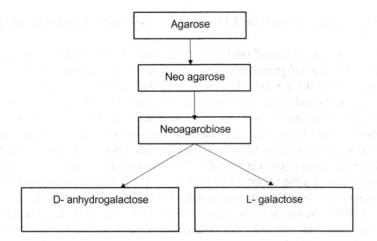

Fig. 7.2 Degradation of agarose into its monomer units

one unit of anhydrogalactose. This is then finally broken down into one D-galactose and one 3,6, anhydro-L-galactose (Kwon et al. 2019). These monosaccharides can then be utilized for production of energy and carbon. Figure 7.2 summarizes the degradation process of agarose.

Agarase activities are bond specific. Some α agarases cleave the α-1,3 glycosidic bonds between the monomer units, while the beta agarases cleave the β-(1-4) glycosidic bonds. Therefore, the products of degradation depend on the mix of the enzymes being used. This has commercial relevance where, for example, neoagarobiose is required for specific bioactivities and the right enzymes are required for optimal degradation of the agarose to obtain a pure product.

7.4 Availability of Raw Material

Agar is produced from red algae. One of the most abundant species which are commercially grown for agar production is the *Gracilaria verrucosa* (Rejeki et al. 2018). The Gracilaria red algae are largely grown in regions such as Indonesia. This genus is more commonly used in food applications. The Gelidium is also a well-cultivated genus used more in pharmaceuticals and in biological applications. These red algae can either be grown in aquaculture or be harvested from natural stocks, although the majority of commercial agar production is from algae cultivated in aquaculture where aquaculture makes up 96.5% of aquatic plants produced globally (FAO 2018). The Gracilaria species which serves as a major commercial source for agar production ranks third in world aquaculture production volume of aquatic plants. Quantity of Gracilaria algae grown in aquaculture rose from 933 thousand tonnes in 2005 to

3.88 million tonnes in 2015 and 4.15 million tonnes in 2016 (weight of the live plant) (FAO 2018).

The yield of algae toward production of agar can be improved by modifying the growth conditions and growth methods. The algae can grow at rates between 0.5 and 1.3% per day on the low side (Rejeki et al. 2018) depending on the types of seeds and the growth conditions. The Gracilaria red algae can contain around 9.6% agar per weight. This particular strain of red algae contributes significantly to the export commodity of producing countries such as Indonesia where annual export is valued at an estimated 280 thousand USD annually from an export of 20 thousand tonnes of Gracilaria red algae (Rejeki et al. 2018). The red algae from which agar is sourced can be grown using the off-bottom, longline or broadcast method. Method of algae cultivation is discussed in the chapter dedicated to aquatic environment.

As of 2010, escalation in the price of seaweed was reported. This escalation in price was attributed to the emerging markets entering the seaweed market at a larger scale and increase in demand for seaweed products which makes their cost as raw materials for production of phycocolloids such as agar much higher. The value of seaweed in its whole form was competitive enough for producers to want to focus on a product requiring less energy and production cost, hence focusing on cultivation and quality such as nutrient value and appearance.

For many years, Gracilaria has been the main source of agar produced globally. The Gracilaria species are endemic to a few parts of the world, and this often results in companies either having to outsource production or move factories to locations nearer to the resource at some additional cost compared to if factory was located in home country. This therefore makes a compelling case to explore alternative red algae species to serve as sources of commercial production of agar. To this end, research efforts have been directed at seeking alternative sources of agar with yield and properties comparable to the agar from Gracilaria species. *Pyropia yezoensis,* for example, yields agar which forms gels which are suitable for bacterial culture media and DNA electrophoresis, two common applications of commercial agar (Sasuga et al. 2017). Having a more diverse range of red algae species which can serve as reliable sources of agar can improve the availability of raw material in more regions of the world and limiting the chances or resource scarcity.

7.5 Extraction of Agar from Red Seaweed

Agar can be extracted from red seaweed using aqueous extraction which generally involves heat treatment at boiling point temperature in water. The variable conditions of extractions include the pH and pressure. Here we look first at the conventional and then the more contemporary extraction methods.

7.5.1 Alkali Extraction

Commercial agar extraction is more commonly carried out using alkali treatment. The alkali treatment is more commonly used for the Gracilaria species to obtain strong gels (Bixler and Porse 2011). The goal of the alkali treatment is the disruption of the cell walls of the red algae biomass and dissolving off of the non-agar impurities such as proteins, minerals and polyphenols. An example of such method involved treating dried algae Gracilaria in 80% v/v NaOH at 27 °C (±3 °C) for 12 h at a mass-to-volume ratio of 1:10 (Rejeki et al. 2018). Under this high pH and temperature, the proteins and other smaller compounds are removed and the cell wall is much more pervious. After the alkali treatment, the biomass is washed with water followed by water extraction by heating at 90 °C in water for 1 h under continuous stirring. At this point, the agar is released into the aqueous media. The agar solution is then filtered followed by gel formation upon cooling. The agar can also be recovered using alcohol precipitation. The next stages involve water removal. This is commonly achieved using gel pressing using high-pressure membrane presses followed by drying to obtain agar in powdered form which is better for storage and packaging. Another example of optimum condition for alkali extraction of agar from the red algae *Gracilaria cliftonii* is: pretreatment time of 1 h at 30 °C using an alkali concentration of 5% with a mass-to-volume ratio of 1:150, extraction time of 3 h in water at 100 °C (Kumar and Fotedar 2009).

Alkali extraction could result in the degradation of part of the agar which could reduce the yield. However, alkali extraction method could result in agar with higher molecular weight, crystallinity and purity which are desirable properties for commercial agar application (Martinez-Sanz et al. 2019). When alkali method was used for the extraction of agar from *Gelidium sesquipedale* using hot water and alkali in combination with sonication, the yield using only hot water and sonication reduced from 10–12% to 2–3% when alkali was used (Martinez-Sanz et al. 2019). Figure 7.3 summarizes the extraction process in a flow chart.

7.5.2 Acid Extraction

Here we look at a typical extraction process used in the extraction of agar from the red seaweed of the Gelidium genera (Hernandez-Carmona et al. 2013). This method is commonly used for Gelidium red algae to obtain agar with superior gel strength (Bixler and Porse 2011). In this method, the alga is cooked at 100 °C in acidic water at pH between 6.3 and 6.5. The time of extraction varies depending on the conditions and the red algae used. Under these conditions, the agar in the red algae biomass is dissolved in water. The liquid containing the dissolved agar is then separated from the solid residue by filtration. Upon cooling the agar dissolved in water forms a gel. This gel contains 99% water and must be dehydrated for easy storage and longer shelf life. Dehydration can be achieved by thawing the gel and drying in oven or through high

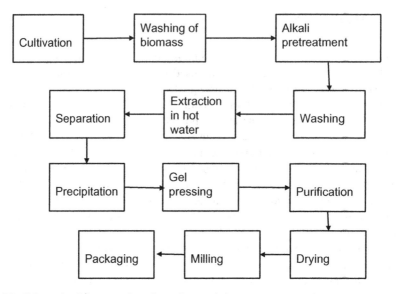

Fig. 7.3 Schematic of the extraction of agar from red algae

pressure pressing to separate the water from the solid. This can then be followed by further drying through heating to eliminate the remaining water. Further processing of the extract includes bleaching with sodium hypochlorite and subsequent washing to remove the bleach (Hernande-Carmona et al. 2013). Other studies have reported higher extraction temperature, for example, in earlier studies. Watase et al. (1990) reported extraction of agarose with a molecular weight of 78,000 ($\pm$320) kDa from the red seaweed *Gelidium amansii* at a temperature of 130 °C.

Sulfate groups, although present in relatively lower amounts, can be removed by treatment with sodium hydroxide. The extraction can be carried out at elevated pressure for better yield and/or faster extraction. Increasing the pressure allows more effective disruption of the cell walls, thus allowing for better isolation of the agarose from the cell walls. A compromise often needs to be made between extraction yield and quality of agarose obtained. The increased pressure and pH result in not only the disruption of the cell walls but also the partial degradation of the agar extracted (Hernandez-Carmona et al. 2013). The high-pressure extraction process therefore needs to be optimized toward an acceptable agar quality and yield.

7.5.3   Hot Water Extraction

However, this form of extraction will yield an impure form of agar which contains proteins, polyphenols and minerals removed along with the agar. These have some beneficial properties in terms of bioactivity such as antioxidant property. An extraction yield of 10–12% is achievable using the hot water method, albeit with brownish color and impurities (Martinez-Sanz et al. 2019). This method is relatively simpler; it involves boiling the washed red algae biomass in hot water for an extended period of time. The liquid extract is then separated from the solid algae residue by filtration.

7.5.4 Enzyme Extraction

Although alkali extraction is more commonly used for extraction of agar, enzyme-assisted extraction is also attractive prospect for the potential of low energy requiring and more environmentally friendly extraction process. More recently, enzyme extraction of agar has been explored. Gel strength of 1521 g/cm^2, 29.2% 3.6-anhydro-l-galactose and 0.84% sulfate content were obtained for agar extracted using the enzyme-assisted method. The agarase enzyme is capable of breaking the glycosidic linkages in agar which makes it more readily dissolve in the aqueous medium of extraction (Xiao et al. 2019). Using the enzyme extraction method to extract agar from Gracilaria caudata, agar with an average molecular weight of 116.51 kDa and a degree of sulfation of 0.14% was obtained (Alencar et al. 2019). Although alkali extraction results in agar with superior properties than enzyme-based extraction of agar, when considering the environmental impact, the enzyme-based extraction provides a more eco-friendly alternative. The biodegradation of agarose can be used as a means of extracting agarose oligomers, dimers or monomers from red algae. Although it results in short-chain degraded agar, these may be useful, for example, the more bioactive oligomeric forms of agar in cosmetics and pharmaceutical applications.

7.5.5 Ultrasound-Assisted Extraction

Ultrasound is commercially used in the food industry in the extraction and processing of various food compounds in the form of sonication. The ultrasound acts by creating multiple cavities in the process of formation, expansion and bursting of these microcavities; they exert pressure on the cell walls causing its disruption. The combination of the alkali extraction method with sonication results in more effective extraction process. Although alkali and heat are still required, the time of extraction could be effectively reduced, hence resulting in optimal use of time by up to fourfold (Martinez-Sanz et al. 2019). It is also possible to reduce the amount of the alkali required by combining the process with sonication. Unlike in alkali extraction which results in degradation of some of the agar upon extraction thereby resulting in reduced yield, use of ultrasound does not alter the yield or properties of the agar obtained.

Other extraction methods which have been explored include extraction by photobleaching process (Li et al. 2008). Photobleaching extracts of agar from *Gracilaria asiatica* and *Gracilaria lemaneiformis* showed good gel strength of 1913 g cm^{-2} which is higher than that obtained from using the conventional methods. For example using microwave-assisted method, a gel strength of 1331 g/cm^2 was obtained (Sousa et al. 2010). A yield of 14.4%, sulfate content of 1.73% and 3.6-anhydro-l-galactose content of 39.4% are achievable using the microwave-assisted method from *Gracilaria vermiculophylla* (Sous et al. 2010).

7.6   Environmental Implications

Although agar is a biodegradable and biocompatible polymer, the process of extraction of agar requires energy and materials consumption, and the release of substances into the environment in the process. This therefore needs to be weighed against its advantages to the environment and commercial value. Agar degradation goes through two stages of degradation before finally degrading into its monomer which can then be broken down by microbes for use as energy or as carbon source. The product itself is therefore not toxic or a nuisance to the environment unlike the non-biodegradable petrochemical-derived plastics which over time have accumulated in the oceans. An estimated 10–20 million tonnes of plastics accumulate in the ocean (Ubanek et al. 2018) annually to the point that these form what is now known as plastic islands (Lebreton et al. 2018). In developing solutions and alternatives to the plastic management problem, it is important to evaluate the environmental implications of the process involved in the production of these biopolymers and identify areas where improvements are required to make these biopolymers as environmentally benign as possible.

7.6.1   Cultivation of Red Algae

Red algae are a renewable natural resource as they grow freely in the wild, and when harvested, they are replenished naturally within a number of days. Unlike fossil fuels which are formed over billions of years and are therefore constantly being depleted, demand for fossil products increases with increasing population. The growing demand for algae as food and as a feedstock for the production of phycocolloids can potentially reduce overfishing of some aquatic species being experienced in some parts of the world. By serving as an alternative source of income, it reduces the pressure on the fish resource as people in the fishing industry diversify their commercial interests.

Naturally, algae are able to extract nutrients from the environment for their growth and metabolism. This is advantageous where they can be used to clean polluted waters. While uncontrolled algae growth can have adverse impacts such as harmful algae blooms, well-controlled algae growth in a balanced ecosystem can be an efficient means of keeping the water bodies clean and safe. In some parts of the world, biological methods of cleaning water bodies include the use of aquatic plants and algae for biological nutrient uptake. The process involves an extraction system built into the flow channel which is comprised of aquatic plants and algae. As the water flows through the channel, it passes through the extraction system where the plants and algae extract the nutrients and the water exits the channel cleaner than it entered. This process of extraction of nutrients from the water can be integrated into the cultivation of biomass for the production of agar (Rocha et al. 2019). The limitation of such systems is controlling the nutrient level in the water coming from homes and industry to achieve the desired growth conditions for agar production and ensuring that the agar produced is not contaminated and safe for intended applications.

Another approach of integrating seaweed cultivation is integration of integrated multi-trophic aquaculture (IMTA). This process capitalizes on the synergy between fish growth and algae growth and nutrient requirement. The algae take up the nutrients produced by the fish and in turn produce the oxygen required by the fish. An example of an IMTA system cultivates Gracilaria alongside the milkfish (Bixler and Porse 2011).

Agar production therefore adds value to red algae beyond food. This is particularly important as outside of Asia; algae are not commonly consumed as food in other parts of the world. There are therefore less incentives for them to be cultivated or preserved as a natural resource. Creating awareness of the value of red algae encourages better preservation of the resource and diversifies interest from aquatic species which are currently being fished at unsustainable rates.

7.6.2 CO_2 Emission

While agar in itself is a biodegradable polymer with no direct adverse impact on the environment, the process of production and the chemicals used in the production can result in considerable generation of CO_2, use of fossil energy and the toxification of human and aquatic environment. The growth and cultivation of algae for the production of agar mitigate against some of these environmental impacts as algae remove CO_2 from the environment for the production of polysaccharides and other organic compounds which can then be used by humans, agar being one of such. Algae also remove nutrients from the water and are used for cleaning up polluted water as a biological treatment method.

The cultivation of red algae removes CO_2 from the environment and can therefore be considered as carbon negative. 260 thousand tonnes of carbon can potentially be converted into biomass by macroalgae annually, assuming that all of this carbon is generated from the CO_2 fixation from the environment (Chung et al. 2011; Sahho et al. 2012). The process of extraction does not involve direct release of CO_2 into the environment. However, other processes attached such as transportation, packaging and running of the plants are processes which result in CO_2 emission into the environment. It is therefore important that the CO_2 emissions in the processes involved in the extraction as well as other ancillary processes are minimized.

7.6.3   Water Consumption

Much of the water used in the extraction process can be recovered during dewatering using the gel press method. This water then needs to be treated to remove impurities. The process of water treatment requires further use of energy and resources. During the drying process, there is return of water to the environment, especially in less sophisticated small or medium-scale production processes where the evaporated water is not recondenced. The water used in the extraction varies for different process and choice of processor. An example process for extraction of agar uses 150 ml of water per gram of algae biomass in the alkali pretreatment stage (Rejeki et al. 2018).

7.6.4   Use of Acid and Alkali

In the acid extraction method, a low pH is required to aid the process of cell wall disruption, a slightly acidic solution of pH ~6 at a temperature of 90–100 °C. In the alkali extraction method, a NaOH concentration of 5% w/v is used at a lower temperature of 30 °C, where a weight-to-volume ratio of 1:150 is used, that means for every gram of agar produced from red algae, 7.5 g of NaOH is used (Table 7.1). Sodium hypochlorite is also used to improve the appearance and optical properties of the agar extracted by bleaching it. Depending on the purification process used, alcohol may also be used for precipitation. Ethanol used can be sourced from non-fossil source such as from corn. The production of these acids, alkali and other chemicals involves further consumption of energy and CO_2 emissions and release of toxins. For example, in a life cycle assessment of sodium hydroxide production, it was deduced that 3.5 MJ of energy was consumed, much of which is fossil based, and 0.6329 kg of CO_2 was emitted per gram of sodium hydroxide produced (Thannimalay

Table 7.1 A summary of estimates of consumption in agar extraction process from red algae

Consumption	Quantity
CO_2	260 kt year^{-1}
Energy consumption	100 °C for 3 h
Acids	pH 6.5
Alkali	7.5 g per gram algae
Water	150 ml per gram algae
Biomass	10 g

et al. 2013). The process also leads to release of toxins to human and aquatic life and acidification of the atmosphere, although these adverse effects are not directly linked to the production of sodium hydroxide, but to the consumption of electricity and energy required for the process. This should therefore be considered in the real cost of producing biopolymers to the environment. Table 7.1 gives a summary of the estimated consumption for agar production based on reviewed works thus far.

7.7 Applications

7.7.1 Food

As an additive agar is required in relatively low amounts, a teaspoon of agar powder can achieve thickening effect in approximately 250 ml of liquid food (Marcus 2014), and at a concentration of 0.03 g per ml, agar fluid gels achieve optimal stability and half-life of up to 6 days (Ellis et al. 2017). Agarose is used in icings and frosting. This particular application is possible due to its compatibility with sugar and its stability at the relatively high room storage temperature used for such products, especially during transportation. It is used as a substitute for low-fat products such as oil-free salad dressings, creams and low-fat yoghurts. Other applications include in canned meat and fish to retain texture, in pie fillings to improve mouthfeel and in candies for gel strength. Agar can remain stable at the heat sterilization temperatures; hence, this makes it applicable for processed foods requiring heat sterilization such as canned foods.

Colloidal systems such as emulsions, gels and foams are used widely in the food industry. Some food products are aerated to form foam; these are dispersion of air within a continuous phase. Foams are used in some food products such as whipped cream and aerated chocolates. This foam structure gives a desirable texture to these foods and also serves to reduce mass per volume and calorie since less product is required within a pack. Typically, foams require a stabilizer to ensure even dispersion of air within the continuous phase, hence maintaining stability. Such stabilizer could be in the form of a surfactant such as Tween 20 (Ellis et al. 2017). Foams are even much less stable since the dispersed phase is air with low viscosity and surface tension, hence a higher chance of coalescence. Agar fluid gels are used in the stabilization of foams. Fluid gels, particles of agar gels suspended in a continuous phase, can be used to form stable foams. These fluid gels can serve as substitutes for colloidal systems which are conventionally made using fat. Agar gels stabilize foams via two mechanisms: absorption of water from the foam cavities hence preventing the release of water leading to destabilization of the foam and by acting as a viscosity enhancer. This increased viscosity prevents separation of the phases through coalescence of the dispersed phase. The presence of the gel particles also serves as a barrier to fluid drainage from the foam (Ellis et al. 2019). Agar therefore has a significant role to play in the preparation of foam-based food as a food ingredient stabilizer and also providing a healthier option to fat-based foams.

7.7.2 Tissue Mimicry

The ability to either test products or procedures on artificial tissue saves much time and cost in the sciences and medical research and training. Use of real tissue for in vitro studies often requires freshly excised tissue and/or constant refrigeration and freezing of the tissue for preservation. For example, where porcine skin is used in transdermal studies, the neonatal porcine obtained from the abattoir once slaughtered are stored in the freezer until required. In less developed regions where electricity is limited and access to fresh cadaver or animal models is more challenging, artificial tissue systems are required for either training, research or product in vitro testing and trials. Gelatine is commonly used in tissue mimetics, for example, as artificial skin for friction testing (Dabrowska et al. 2017). In mimicking tissue for ultrasound procedure training, gelatine gel used requires refrigeration and this poses a challenge in less developed region. Agar can retain its gel structure at room temperature without need for refrigeration and also has a longer period before spoilage since agar is particularly more resistant to bacterial infection (Kwon et al. 2019), thus making it more suitable as a tissue model without need for refrigeration. Agar-based tissue models have been tested for use in training of ultrasound medical imaging procedure. Agar concentration at 7.5–10% mixed with flour can serve as effective tissue models for training of medical practitioners in the use of ultrasound tissue imaging (Earle et al. 2016).

Access to cost-effective training of medical professionals has significant economic impacts, particularly in regions where resources such as electricity are scarce. Such low-cost training materials ensure the development of technical capacity to use advanced medical facilities such as ultrasound which could significantly improve the quality of health care rendered to the community. The quality of health in turn affects the economic activities such as farming and education. Furthermore, trials and training carried out on real animal or human tissue might violate some ethics and religious beliefs. Therefore, a plant-based tissue mimetic is often preferred over real animal or human tissue and is also a step better than the animal-based gelatin where a tissue model completely free of animal use is required.

7.7.3 Biotechnology

The ability to form a strong gel and serve as a moist, relatively unreactive and clear gel environment, makes agar applicable in several aspects of biology and biotechnology. In the biology, agar, particularly agarose, is very commonly used in preparation of culture media for microorganisms. It can be easily prepared from the powdered form when mixed with water and dissolved at high temperature forming a gel upon cooling. Some novel agarose preparations can form gels at lower temperature below 100 °C (BeMiller 2019), allowing for even easier gel preparations. Agarose is also used in electrophoresis for DNA analysis and some chromatography techniques (Sasuga et al. 2017).

7.7.4 Preservation of Post-Hatchery Chicks

When chicks are hatched, they undergo a period when there is no access to feed or water for a period of about 24 h as they are transported to production facilities where they will continue their life and be processed into poultry products. The period after hatching is a critical stage where the moisture content has an effect on the future productivity of the chicks in terms of weight, growth rate and quality of egg produced. As a means to address this moisture issue, agar-based aqueous system has been introduced. Then, this aqueous agar which contains 95% water and 5% agar is applied to the chick after hatching; they show better productivity and quality. For example, post-hatched chicks which were packed with aqua agar had body weight of 710 g, while those that were not packed with aqua agar weighed 670 g 21 days after hatching (Incharoen et al. 2015). The aqueous agar covering maintains adequate moisture content for the chicks during the period such that they retain their viability at this important stage of their growth. This application of agar is significant toward contribution to food security as a way of optimizing the poultry food resource for optimal yield.

7.7.5 Cosmetics

Due to the ability to entrap water and retain moisture, agar is used in the cosmetics industry as a moisturizing agent in skin and hair products (Rasmussen and Morrissey 2007). More recently, melanin has been loaded into agar-based composite films with the aim of achieving antioxidant effect in skin care (Roy and Rhim 2019). Much earlier studies have presented the moisturizing effect of short-chain neoagarobiose which also had skin-whitening effect (Kobayashi et al. 1997). This moisturizing effect of agar is likely due to the ability of agar to absorb water; the water is then released into the skin. This short-chain agar is better able to penetrate the skin stratum corneum barrier, hence why the short-chain agar forms show this moisturizing effect. The viscosity-enhancing property of agar also makes it applicable as a thickener in cosmetics products giving the product a more desirable texture and ease of use.

7.7.6 Antioxidant

When degraded into short-chain polymers (oligomers) referred to as agaro-oligosaccharides and neoagaro-oligosaccharides, agar demonstrates antioxidant activities (Chen and Yan 2005). These antioxidant activities of the short-chain agar forms are attributed to the improved chance of the short chains to cross tissue barriers and also the presence of some sulfate groups on the chain. Most sulfated polysaccharides studied so far are attributed with some antioxidant properties. Although agar has a relatively lower degree of sulfation (e.g., 0.14% of agar extracted from *Gracilaria caudata* through enzyme extraction) (Alencar et al. 2019), more research is required to fully understand the mechanism of its antioxidant activity. Sulfated polysaccharides with agar-like chemical structure extracted from the red algae *Gracilaria caudata* show antioxidant properties which suggest that agar could have some promising application beyond its rheological properties. Agar-based oligomers also show other bioactivities such as antitumor, prebiotic, anti-inflammatory, antidiabetic and anti-obesity activities (Kwon et al. 2019).

7.7.7 Packaging

Currently, one of the biggest global challenges of modern times is the accumulation of non-biodegradable plastic packaging waste. These have gone on to cause serious environmental issues as plastics from landfill clog up drainage systems and end up in the sea where they pose fatal risk to aquatic organisms. Their floating on the water prevents penetration of light and oxygen, and they end up inside the digestive systems of aquatic animals where they pose potentially fatal health risks. These fossil-based packaging materials are used for short periods; however, they take very long time to degrade. Degradation processes such as pyrolysis and incineration either require additional energy input (temperatures of ~450 °C for pyrolysis) or they pose a health risk (such as release of carcinogens from incineration of plastics). Marine microorganisms from the arctic are being studied as potential candidates to degrade these fossil-derived plastics; however, this is yet to be actualized as it is primarily dependent on the microbes adapting to develop enzymes and means to degrade these plastics (Ubanek et al. 2018).

One of the factors which make these materials attractive as packaging materials is the fact that they are not degradable by microbes and they have superior mechanical properties and water resistance; therefore, they can be used to hygienically pack foods. Biodegradable packaging materials on the other hand are almost as biodegradable as the food they are packed in. Such that, several researches have been directed toward developing biodegradable packaging materials, which incorporate antimicrobial properties. The fact that agar is relatively more resistant to bacterial degradation (Kwon et al. 2019) could contribute to its applicability in packaging materials.

Agarose has thermoplastic properties as well as moderate water resistance (Makwana et al. 2018), and this has been explored in the production of packaging materials. The limitations of agar for use in packaging materials are the brittleness at low moisture content, inferior mechanical properties compared to non-biodegradable thermoplastics such as low-density polyethylene and insufficient solvent resistance. These properties of agar can be improved by combining with other materials and other polymers in the form of composites and blends. Blends of two different polymers can result in a material with superior properties than each individual polymer alone. Similarly, a composite which is made up of a dispersed phase within a polymer matrix can significantly improve the polymer in its neat form. The dispersed phase could be made up of particles such as silver nanoparticles or fibers such as cellulose nanofibers (Olatunji and Olsson 2015). An example of such enhancement of agar's properties using a combination of nanocomposite and blend is a blend of agar with carboxymethyl cellulose (CMC) which is then dispersed with silver nanoparticles to produce a packaging film with thermomechanical properties suitable for production of packaging films. The introduction of silver nanoparticles improved antibacterial properties and also the mechanical properties (Makwana et al. 2018). Nanocomposites can significantly boost the physical properties of polymers in nanocomposites; however, their effect will be undermined if aggregation occurs. To prevent the aggregation of the silver nanoparticles within the agar-CMC nanocomposite, the silver nanoparticles were contacted with montmorillonite clay to prevent their aggregation within the polymer matrix. At 3 and 5% loading, silver nanoparticle and montmorillonite clay nanocomposite in agar-CMC blend resulted in 45 and 50% increase in the tensile properties and increased the young modulus by 20 and 89%, respectively (Makwana et al. 2018). Therefore, modified forms on agar can act as effective packaging materials with properties comparable to those of petroleum-based packaging materials and thus serve as a more environmentally friendly packaging material.

Agar can also be cross-linked to improve its applicability as a biodegradable plastic packaging. Cross-linking increases the water resistance, tensile and thermal properties of agar. Cross-linkers such as diisocyanates can be used. Better cross-links are achieved with aromatic diisocyanates than aliphatic ones. Non-cross-linked agar has a water uptake value of 206%; when cross-linked with aromatic diisocyanate, this is reduced to 33.6%. The tensile stress of cross-linked agar is 45.3 MPa, while non-cross-linked agar has a tensile stress of 31.7 MPa (Sonker et al. 2018).

Recent patented technology of biodegradable thermoplastic packaging material incorporates agar as a modifying agent. One of the requirements for packaging materials is the ability to be heat processed. They are therefore required to have thermoplastic characteristics. This allows for large-scale production using the conventional continuous plastic processing techniques such as blown film extrusion. This contributes to reducing the cost of production of biodegradable plastics and also requiring less technological transition when the feed is changed from non-degradable to biodegradable plastics. Biodegradable thermoplastic is a mixture of PLA and mTPS (polylactic acid and modified thermoplastic starch). The mTPS comprises of agar, epoxide and gum arabic. For example, formulation of such thermoplastic starch comprises of 80% unmodified starch, 28.5% glycerol, 20% agar, 10% epoxidized vegetable oil and 0.5% gum arabic (Justyna et al. 2016).

7.7.8 Preservation of Stone Structures

Agar has also recently been investigated for use in the desalination of limestone as a preservative measure for stone heritage structures which overtime are destroyed by salt accumulation. Aqueous-based methods are sometimes counterproductive as the water may cause other forms of damages such as rusting and erosion. Agar gel can gradually release water to remove the salt in a sponge-like manner (Martins et al. 2017). This is a gentler method than more abrasive sponges. The water released during syneresis dissolves off the salt, while the pores within the gel absorb the solution from the surface of the structure. The gel can then be wiped off the surface.

Other applications of agar include use in capsules for pharmaceuticals, glucose-lowering effects, in dentistry for preparation of denture molds, in archeology where it is used in the reproduction of ancient remains and in forensics for fingerprinting.

7.8 Commercial Production

As demand increases for more plant-based products, agar has potentially ever-increasing demand in food as a plant-based replacement to the animal-sourced gelatin. The demands for the hydrocolloids sourced from algae are generally increasing annually. The largest producers, which are the Asia-Pacific regions, have reported 2–3% growth in the hydrocolloids market in recent years, and this includes agar (Rejeki et al. 2018). There seems to be a general long-term increase in production rate since earlier years; production rate between 1999 and 2009 had shown an estimated 2% increase in production volume (Bixler and Porse 2011). Although the market has witnessed considerable price fluctuations over the past decades, the phycocolloid and algae market has gone through different phases since the 1930s. The Asia-Pacific region remains the highest producer of agar and has been for many years.

Of the three seaweed-derived hydrocolloids, agar has the highest commercial value in terms of commodity prices, selling at ~18USD per kg as of 2009 compared to carrageenan and alginates priced at 10.5 USD/kg and 12 USD/kg, respectively. As of 2009, an estimated global sale of 9600 tonnes was estimated for agar, while carrageenan and alginate were 50,000 tonnes and 26,500 tonnes, respectively (Bixler and Porse 2011). As of 2014, global agar production of agar was an estimated 10,600 tonnes annually (Rhein-Knudsen et al. 2015). Agar is also sold for research and laboratory use by chemical supplies company; for example, 100 g agar powder for microbiology from Sigma-Aldrich is priced at 117 Eur as of June 5, 2019, while the same quantity and quality of carrageenan from the same supplier are priced at 108 Eur (Sigma-Aldrich).

At smaller scale, agar of different grades is produced for research purposes. For example, high purity agar for laboratory research studies is supplied by Sigma-Aldrich (Ellis et al. 2019). These sources of agar for research are important in developing new applications of agar and also as essential materials for experimental procedures such as preparation of cell culture and tissue mimetics.

In commercial application, agar is mostly used as agarose (Makwana et al. 2018) where much of the agaropectin is removed since agarose has more desirable properties. Some species produce more agarose content than others such that the isolation of high agarose-producing species could reduce the processing time and cost in deriving more agarose rich fractions. Agar is sold as dry powder, squares or strips. This is unlike carrageenan and alginates which are almost exclusively sold in powdered form (Bixler and Porse 2011).

Successful commercial production of agar requires a mixture of factors: reliable supply of low-cost and good-quality agar producing red algae, an efficient production process, steady demand and good sales and marketing. It is also important for these factors to be fine-tuned to specific agar application. Companies involved in commercial production of agar include Algas Marinas in Chile, Agarindo Bogatama in Indonesia, MSC in Korea, Ina Food Industry Co Ltd. in Japan, Hispanagar in Spain, Setexan in Morocco, Huey Shyang Seaweed Industrial Company in China and B & V in Italy (outsourcing production to Indonesia and Morocco).

Production of agar from mixed species of red algae does not yield a desirable product with uniform properties; therefore, for agar production process to be better controlled and optimized for yield and agar property, a single type of red algae is used in the production process. Therefore, commercial cultivation of red algae for agar production is more focused on Gracilaria. For instance, 80% of the agar produced globally as of 2009 was produced from the Gracilaria species of red algae (Bixler and Porse 2011). However, some research efforts are directed at exploring other red algae species for commercial agar production toward having more diversified sources of raw materials.

7.9 Conclusion

Agar is a well-established aquatic biopolymer commercially sourced mainly from the Gracilaria red algae; its diverse applications in biotechnology and food make it a highly valued biopolymer of the aquatic environment. It also evidently has a role in the future polymer industry as more recent applications are emerging. This also points to a potential rise in market value of agar as newer applications emerge. The raw material is relatively available with main problem with commercial production being the competitive price for seaweed in general due to rising competition with use of seaweed as food. Although enzyme-based extraction and hot water extraction without use of chemicals are possible, alkali extraction remains the most widely used process.

Therefore, more research should be focused on more advanced extraction techniques which bring down the production cost and also minimize the environmental impact of the production process without compromising quality or yield.

References

Alencar POC, Lima CG, Barros FCN, Costa LEC, Freitas ALP (2019) A novel antioxidant sulfated polysaccharide from the algae *Gracilaria caudata*: in vitro and in vivo activities. Food Hydrocolloids 90:28–34

Armisen R, Gaiatas F (2009) Agar in: handbook of hydrocolloids. In: Phillips GO, Williams PA (eds) Woodhead Publishing Series in Food Science Technology and Nutrition, pp 82–107

BeMiller JN (2019) Carrageenans in: carbohydrate chemistry for food scientists, 3rd edn. Woodhead Publishing and AACC International, pp 279–291

Bixler HJ, Pose H (2011) A decade of change in the seaweed hydrocolloids industry. J Appl Phycol 23:321–335

Chen HM, Yan XJ (2005) Antioxidant activities of agaro-oligosaccharides with different degrees of polymerization in cell-based system. Biochim Biophys Acta 1722:103–111

Chung IK, Beardall J, Mehta S, Sahoo D, Stojkovic S (2011) J Appl Phycol 23:877–886

Dabrowska A, Rotaru GM, Spano F, Affolter C, Fortunato G, Lehmann S, Derler S, Spencer ND, Rossi RM (2017) A water -responsive, gelatine-based human skin model. Tribol Int 113:316–322

Earle M, De Portu G, DeVos E (2016) Agar ultrasound phantoms for low-cost training without refrigeration. Afr J Emerg Med 6:18–23

Ellis AL, Norton AB, Mills TB, Norton IT (2017) Stabilization of foams by agar gel particles. Food Hydrocolloids 73:222–228

Ellis AL, Mills TB, Norton IT, Norton-Welch AB (2019) The effects of sugars on agar fluid gels and the stabilisation of their foams. Food Hydrocolloids 87:371–381

FAO (2018) The state of world fisheries and aquaculture 2018—meeting the sustainable development goals. Rome. Licence: CC BY-NC-SA 3.0 IGO. ISBN 978-92-5-130562-1

Hernandez-Carmona G, Freile-Pelegrin Y, Hernandez-Garibay E (2013) Conventional and alternative technologies for the extraction of algal polysaccharides. In: Functional ingredients from algae for foods and nutraceuticals. Woodhead Publishing Series in Food Science Technology and Nutrition, pp 475–516

Incharoen T, Jomjanyouang W, Preecha N (2015) Effects of aqua agar as water replacement for posthatch chicks during transportation on residual yolk-sac and growth performance of young broiler chickens. Anim Nutr 1:310–312

Justyna K, Macie S, Helena J, Andrzej S, Katarzyna B, Aneta L (2016) Biodegradable thermoplastic polymer composition, method for its manufacture and use thereof. EP3064542A1

Kobayashi R, Takisada M, Suzuki T, Kirimura K, Usami S (1997) Neoagarobiose as a novel moisturizer with whitening effect. Biosci Biotechnol Biochem 61:162–163

Kumar V, Fotedar R (2009) Agar extraction process for *Gracilaria cliftonii*. Carbohyd Polym 78(4):813–819

Kwon GH, Kwon MJ, Park JE, Kim YH (2019) Whole genome sequence of a freshwater agar-degrading bacterium Cellvibrio sp. KY-GH-1. Biotechnol Rep 23: e00346

Lebreton L, Slat B, Ferrari F, Sainte-Rose B, Aitken J, Marthouse R, Hajbane S, Consolo S, Schwarz A, Levivier A, Noble K, Debeljak P, Maral H, Schoeneich-Argent R, Brambini R, Reisser J (2018) Evidence that the great Pacific garbage patch is rapidly accumulating plastic. Sci Rep 8(1):4666

Li H, Yu X, Jin Y, Zhang W, Liu Y (2008) Development of an eco-friendly agar extraction technique from the red seaweed Gracilaria lemaneiformis. Bioresour Technol 99(8):3301–3305

Makwana D, Castano J, Somani RS, Bajaj HC (2018) Characterization of agar-CMC/Ag-MMT nanocomposite and evaluation of antibacterial and mechanical properties for packaging applications. Arab J Chem (in press). https://doi.org/10.1016/j.arabjc.2018.08.017

Marcus JB (2014). Food science basics: healthy cooking and baking demystified. Culinary nutrition. Academic Press, pp 51–97

Martinez-Sanz M, Gomez LG, Rubio AL (2019) Production of unpurified agar-based extracts from red seaweed *Gelidium sesquipedale* by means of simplified extraction protocols. Algal Res 38:101420

Martins J, Dionisio A, Neves O (2017) Agar gel for anca limestone desalination. Procedia Earth Planet Sci 17(2017):754–757

Mizrahi S (2010) Syneresis in food gels and its implications for food quality. In: Chemical deterioration and physical instability of food and beverages. Woodhead Publishing Series in Food Science, Technology and Nutrition, pp 324–348

Olatunji O, Olsson RT (2015) Microneedles from fishscale-nanocellulose blends using low temperature mechanical press method. Pharmaceutics 7:363–378

Ouyang Q, Hu Z, Li S, Quan W, Wen L, Yang Z, Li P (2018) Thermal degradation of agar: mechanism and toxicity of products. Food Chem 264:277–283

Rasmussen RS, Morrissey MT (2007) Marine biotechnology for production of food ingredients. Adv Food Nutr Res 52:237–292

Rejeki S, Ariyati RW, Widowati LL, Bosma RH (2018) The effects of three cultivation methods and two seedling types on growth, agar content and gel strength of *Gracilaria verrucosa*. Egypt J Aquat Res 44:65–70

Rhein-Knudsen N, Ale MT, Meyer AS (2015) Seaweed hydrocolloid production: an update on enzyme assisted extraction and modification technologies. Mar Drugs 13:3340–3359

Rioux LV, Turgeon SL (2015) Seaweed carbohydrates. In: Tiwari BK, Troy DJ (eds) Seaweed sustainability, pp 141–192

Rocha MRC, Sousa AMM, Kim JK, Magalhaes JMCS, Goncalves MP (2019) Characterization of agar from *Gracilaria tikvahiae* cultivated for nutrient bioextraction in open water farms. Food Hydrocolloids 89:260–271

Roy S, Rhim J (2019) Agar-based antioxidant composite films incorporated with melanin nanoparticles. Food Hydrocolloids 94:391–398

Sahho D, Elangbam G, Devi SS (2012) Using algae for carbon dioxide capture and biofuel production to combat climate change. Phykos 42(1):32–38

Sasuga K, Yamanashi T, Nakayama S, Ono S, Mikami K (2017) Algal Res 26:123–130

Sonker AK, Belay M, Rathmore K, Jahan K, Verma S, Ramanathan G, Verma V (2018) Crosslinking agar by diisocyanates. Carbohyd Polym 202:454–460

Sousa AMM, Alves VD, Delerue-Matos MC, Goncalves MP (2010) Agar extraction from integrated multi trophic aquacultured *Gracilaria vermiculophylla*: evaluation of a microwave assisted process using response surface methodology. Biores Technol 101(9):3258–3267

Sudha PN, Gomathi T, Vinodhini PA, Nasreen K, (2014) Marine carbohydrates of wastewater treatment. In: Kim SK (ed) Marine carbohydrates: fundamentals and applications, part B. Elsevier, pp 103–143

Thannimalay L, Yusoff S, Zawawi NZ (2013) Life cycle assessment of sodium hydroxide. Aust J Basic Appl Sci 7(2):421–431

Ubanek AK, Rymowicz W, Mironczuk AM (2018) Degradation of plastics and plastic-degrading bacteria in cold marine habitats 102: 7669–7678

Wang K, Huang H, Sheng J (1997) Determination of the Mark-Houwink equation parameters and their interrelationship. J Liq Chromatogr Relat Technol 21(10):1457–1470

Watase M, Nishinari K, Williams PA, Phillips GO (1990) Agarose gels: effect of sucrose, glucose, urea, and guanidine hydrochloride on the rheological and thermal properties. J Agric Food Chem 38(5):1181–1187

Weiner ML, McKim JM, Blakemore WR (2017) Addendum to Weiner ML (2016) Parameters and pitfalls to consider in the conduct of food additive research, carrageenan as a case study. Food Chem Toxicol 87:31–44. Food and Chem Toxicol 107:208–214

Xiao Q, Weng H, Xiao A (2019) Physicochemical and gel properties of agar extracted by enzyme and enzyme-assisted methods. Food Hydrocolloids 87:530–540

Chapter 8
Ulvans

Abstract Ulvans are sulphated polysaccharides present in the cell walls of green algae alongside other cell wall polysaccharides. Their polymer structure is characterized by repeating units of disaccharide of sulfated rhamnose linked to other units of either uronic acid, guluronic acids or xylose. They are soluble polysaccharides; therefore, their extraction process is relatively milder compared to other insoluble polysaccharides. Ulvans are extracted from the Ulva species of green algae. These species of green algae as raw materials for biopolymer production have the advantage of growing in more diverse habitats and having a rapid growth rate. Extraction of ulvans is one way of utilizing the excess green algae resource which often results in algae blooms, to produce high-value bioactive polymers.

Keywords Ulvans · Biopolymers · Algae · Chlorophyta · Sulfated · Polysaccharides

8.1 Introduction

Ulvans are one of the three sulphated polysaccharides found in algae. While the sulfated polysaccharide of red algae is carrageenan and that of brown algae is fucoidan, ulvans are produced by green algae. Ulvans are found in the cell walls of green algae alongside cellulose and insoluble polysaccharide and two other soluble cell wall polysaccharides, xyloglucan and glucuronan, which are present in lower amounts (Kidgell et al. 2019). Ulvans are water-soluble polysaccharides and are characterized by their copolymer repeating units of disaccharide of sulfated rhamnose linked to other units such as uronic acid. The presence of sulfates, multiple sugar units make ulvans more complex than simple polysaccharides such as cellulose and chitin.

Like the other sulphated polysaccharides of algae, ulvan has been shown to have some bioactive properties which make it potentially applicable in diverse industries including food, pharmaceutics, cosmetics and biotechnology. Such bioactive properties include immune modulatory (Berri et al. 2017), antibacterial (Berri et al. 2016) and antiinflammatory effects.

© Springer International Publishing 2020   169
O. Olatunji, *Aquatic Biopolymers*, Springer Series on Polymer and Composite Materials,
https://doi.org/10.1007/978-3-030-34709-3_8

Marine algae are of great economical significance as they are a source of valuable biomaterials: polyphenols, polyunsaturated fatty acids, proteins, polysaccharides and pigments. All of which are useful in the pharmaceutical, cosmetics, food, biotechnology and microbiology fields. Some of the advantages of polymers sourced from algae include the higher biomass yield of algae compared to terrestrial plants similarly used as biorefinery crops, the use of aquatic rather than land space for their cultivation, removal of CO_2 from the atmosphere and uptake of eutrophying nutrients from the water. On the other hand, issues such as competition with algae used for food and relatively high cost of algae as feedstock pose a challenge in commercial production of ulvan.

This chapter reviews ulvan as an aquatic sourced biopolymer, its source, extraction process, environmental issues surrounding its production and the present state of commercial ulvan production. In so doing it furthers the goal of the book to highlight the diverse range of biopolymers in the aquatic ecosystem, the present issues as well as their environmental and economic significance.

8.2 Occurrence in Nature

Green marine algae such as *Ulva amoricana* and *Ulva rigida* and *Ulva enteromorpha* are the known sources of ulvan which forms a part of the cell wall alongside other cell wall polysaccharides such as mannan, xylan and cellulose (Misurcova et al. 2012). These green algae contain around 38–54% Ulvan by dry weight (Michel and Czjzek 2014) (8–29% Lahaye and Robic 2007), while others report between 9 and 36% by dry mass in Ulva species (Kidgell et al. 2019). Ulvan producing species include *U. rigida*, *Ulva pertusa*, Enteromorpha, *Ulva intestinalis* (Tabarsa et al. 2018), *E. linza*, *E. clathrata* (Tabarsa et al. 2018), *Ulva ohnoi* (Fernández-Díaz et al. 2017). The green algae Ulva species are rapid-growing species to the extent that they can result in algae blooms under uncontrolled growth in the wild aquatic habitat (Kidgell et al. 2019). Figure 8.1 is a schematic representation of Ulva within the green algae cell wall. Ulvan exists in a matrix with other cell wall polymers such as cellulose proteins and glucuronan.

Sulfated polysaccharides are also found in animals, and however, plants do not produce sulfated polysaccharides. Examples of sulfated polysaccharides of animals are glycosaminoglycans (GAGs) and proteoglycans (Scharnweber et al. 2015). Much of the studies on bioactivity of ulvans have explored mimicking the functions of these animal sulfated polysaccharides as it is expected that the similarity in structure will manifest as similarity in bioactivity. The plant macromolecule rhamnogalacturonans and rhamnolipids in phytopathogenic bacterium (Varnier et al. 2009) are the components of these organisms which have the closest similarity with ulvans of green algae.

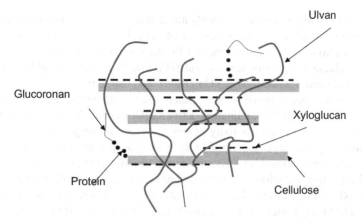

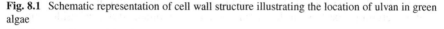

Fig. 8.1 Schematic representation of cell wall structure illustrating the location of ulvan in green algae

8.3 Chemistry of Ulvans

Ulvans are heteropolysaccharides made up of several sugar units mainly: sulfated rhamnose and xylose units, xylose, glucuronic acid and iduronic acid (Kim et al. 2011a). Figure 8.2 shows the structure of disaccharide units in ulvan. Ulvans are polydisperse and molecular weight of different fractions could vary, for example, Tabarsa et al. (2018) report weight average molecular weight (M_w) of fractions extracted from *U. intestinalis* ranging between 87,100 and 194,100 g/mol following extraction and fractionation of the crude ulvan into different molecular weights using DEAE-Sepharose Fast Flow column.

The ulvan polymeric chain is arranged as repeating unit disaccharides of either rhamnose and iduronic acid called the aldobiuronic acid (Lahaye and Robic 2007), sulfated rhamnose and glucuronic acid or sulfated rhamnose and xylose (Michel and Czjzek 2014). Arabinose and galactose are also said to be present in ulvans (Misurcova et al. 2012; Tabarsa et al. 2018). While the structure of the ulvan polymer could vary depending on factors like source, species and harvest period, it is established

Fig. 8.2 Structures of a disaccharide repeating unit in ulvan comprising of a glucuronic unit linked to a sulfated rhamnose

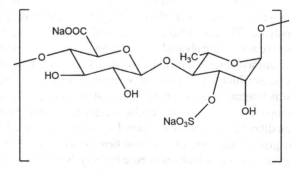

that the sulfate groups of ulvan generally attaches to either the second or third carbon of the rhamnose unit or the second carbon of the xylose unit (Tabarsa et al. 2018).

The bioactivity of ulvans is affected by the molecular weight. The molecular weight of ulvans varies between 100 and 8000 kDa (Lahaye 1998) although others report it between 150 and 2000 kDa (Rioux and Turgeon 2015). Lower molecular weight has been shown to favor interactions with immune cells which aid immunomodulatory and inflammatory activities (Chen et al. 2008; Jaswir and Monsur 2011). This superior bioactivity of lower molecular weight ulvan over higher molecular weight ulvan is attributed to the ease of access into the cells and ease of proton exchange as a result of lower molecular weight (Qi et al. 2005).

Each repeating unit within the ulvan structure plays a distinct role in the bioactivity of ulvan. Rhamnose can aid in hyaluronic and collagen production in human fibroblast dermal cells through interacting in biosynthetic pathways in the dermis (Adrien et al. 2017). The presence of functional groups such as sulfate and amine groups also confers certain properties on the ulvan polymer. The repeated disaccharide units of ulvans comprising mainly of uronic acid linked to sulfated rhamnose or xylose make them applicable for modulation of cellular activities carried out by polysaccharides in mammals. These properties make them applicable in areas such as wound healing, tissue repair and antiviral activities.

While each repeating unit has its contributory role to the physicochemical and bioactive properties of the ulvan, the configuration, order with which the repeating units are arranged within the polymeric structure, is also important. This in turn affects the conformation of the ulvan molecules and consequently the overall properties. Branching occurs usually at the 1.2 linked glucuronic acid linked to the rhamnose (Kidgell 2019; Lahaye and Robic 2007). As with most polymers the level of branching is related to properties such as crystallinity, thermal and mechanical properties of the polymer.

Ulvans are soluble polysaccharides that dissolve in water to form a gel-like solution. Like all polymers, ulvans will increase the viscosity of the solution within which it is dissolves in. The solution viscosity of ulvan depends on the pH and the presence of salts. At low to neutral pH, ulvan takes on a bead-like conformation which aggregates in the presence of salts such as sodium chloride. Such solutions of ulvan usually have low viscosity. In more alkaline environment, ulvan forms more viscous solutions with higher gel strength (Lahaye and Robic 2007).

The biochemistry of ulvans varies for different sources as a result of variation in the primary structure, molecular weight, branching and the polydisperse nature of the polymer. These variation correlate with their bioactive properties. For instance, low molecular weight ulvans have been shown to stimulate kidney macrophage better than ulvans with higher molecular weights (Fernández-Díaz et al. 2017). There have been studies which investigate the bioactive properties of crude ulvan extract and those which separate the ulvan into different molecular weight fractions to investigate the properties of specific molecular weight range, and other studies have extended to modification of ulvans to control specific properties. Chemical properties such as degree of sulfation can be modified by removal or addition of sulfate functional groups such as solvolysis or base hydrolysis (Qi et al. 2012). The chain length of the polymer can be altered by breaking specific glycosidic bonds using either chemical (Pengzhan et al. 2004) or enzyme hydrolysis (Reisky et al. 2018) of the ulvan polymer chain. The solution properties at different pH can be used to develop ulvan-based products with specific rheological properties. Ulvans therefore offer a broad range of tuneable chemical and rheological properties which presents a range of potential applicabilities.

8.3.1 Biodegradation of Ulvans

Understanding the biodegradation of ulvans is important in assessing their safe degradation when released into the environment and when consumed as food, cosmetics or therapeutics. It is also important in understanding the structure of ulvans. The biodegradation products and the enzymes which catalyze the biodegradation provide an insight into the structural configuration of the ulvan as well as an understanding of the biological activity of the microbes which degrade ulvans.

Ulvanolytic enzymes refer to enzymes which are capable of degrading ulvans. Microbes identified to be capable of degrading of ulvan polysaccharide such as gram-negative marine bacterium, ochrobactrum and flavobacterium produce ulvanolytic enzymes (Michel and Czjzek 2014). These enzymes are capable of cleaving specific glycosidic bonds, for example, a type of ulvanolytic enzyme called ulvan lyases catalyzes the cleavage of the bonds between sulfated rhamnose and glucuronic acid or that between rhamnose and iduronic acid (Nyvall Collen et al. 2011). Effectiveness of ulvanolytic enzymes therefore depends on the chain configuration of the specific ulvan and the repeating disaccharide units within the ulvan polymeric chain. Ulvan is not degraded by any enzyme produced by the human body, and it is also able to survive the gastrointestinal tract and colon undigested (Misurcova et al. 2012). This makes it a good option for dietary-soluble fiber.

8.4 Availability of Raw Material

Green algae are a renewable natural resource for biopolymer production. It also has the additional advantage of growing in more diverse habitat with varying abiotic factors. Compared to red or brown algae which primarily grow in marine waters, green algae grow in both marine and freshwater. The ulvan producing species grow in marine and freshwater. Example of marine species is *U. rigida* (Tabarsa et al. 2018) and that of freshwater is *Ulva thalli* (Rybak et al. 2012). This occurrence in diverse habitat means less limitation on region where the raw material for production of ulvans can be the sources. A readily available resource from multiple regions is always better for the economy than resources concentrated in only a few regions. This concentration of resources in limited regions has led to some of the socioeconomic problems of other natural resources such as crude oil and minerals.

Unlike brown algae with a relatively complex reproductive cycle which makes controlled cultivation in aquaculture more labor and cost intensive to be economically viable for commercial biopolymer production, the cultivation of green algae is relatively simpler. Due to its relatively rapid growth rate and simpler cultivation process, the ulvan species of green algae are more promising for the cultivation of monocultures which will result in more uniform ulvan extracts. This is important in developing standard formulations with specific bioactivities where factors such as molecular weight and degree of sulfation are within shorter ranges. Some Ulva green algae grow so easily that in situations where there is an environmental imbalance in the aquatic ecosystem such as increase in nutrient content of the water this could result in drastic population boom of green algae resulting in what is known as green tides (Kidgell et al. 2019). The Ulva species have been identified as one of the algae species which form algae blooms (Young and Gobler 2016). *Ulva prolifera*, a highly productive Ulva can increase by 28% in population within a period of 24 h (Zhang et al. 2019). In a well-managed controlled environment, this rapid reproductive rate is an advantage for generating feedstock for ulvan production.

Asian countries generally contribute a large fraction of global algae production. Algae are also produced in some countries in Europe and the USA. The FAO data on global commercial algae production makes no mention of the Ulva species. This could be classified under "other" (FAO 2018). However, the Ulva green algae are evidently abundant as they have resulted in algae blooms in areas such as the Yellow Sea in China off the coast of Jiangsu and Shandong Province (Zhang et al. 2019) and in Japan at Yatsu tidal flat in Tokyo Bay (Yabe et al. 2009). Green tides in the Yatsu tidal flat of Tokyo Bay could cover an area of water up to 27.1 ha (0.271 km^2) of a total area of 40 ha (0.4 km^2) at peak growth (Yabe et al. 2009). That is over 67% of the lagoon being covered by green tide. The green tide of the Yellow Sea in China covers an area of 20,000 km^2 and has been described as the largest ever experienced in the world (Zhang et al. 2019). Therefore, not only is the ulvan producing green algae available, they are presently causing environmental nuisance and finding ways to utilize these green algae is a priority.

8.5   Extraction of Ulvan

As a water-soluble fiber, ulvan can be extracted from green algae using aqueous extraction. Typical extraction temperature ranges between 80 and 90 °C. An extractant such as ammonium oxalate is often used at a 0.02M concentration (Robic et al. 2009) as a divalent cationic chelating agent to aid the extraction by making the biomass more susceptible to the solvent. The crude extract can be further purified using alcohol or quaternary ammonium salt precipitation (Lahaye and Robic 2007; Shao et al. 2011; Rioux and Turgeon 2015). High-temperature acidic extraction at a pH of 2 and a temperature of around 80 °C are also used to extract ulvan (Yaich et al. 2014; Rioux and Turgeon 2015).

Here, the extraction method used by Tabarsa et al. (2018) to extract ulvan from *Ulva intestinalis* is used as an example of extraction process. Algae strain *U. intestinalis* were sourced from the Iranian coast, Noor. The collected biomass was then washed using tap water followed by drying at a temperature of 60 °C. To reduce the particle size, the washed and dried algae mass was ground using a blender and sieved to particle size below 0.5 mm. To isolate crude polysaccharide, the alcohol-soluble pigments, lipids and low molecular weight compounds were removed from the ground seaweed by adding 80% ethanol at a mass-to-volume ratio of 1:10 (g:mL). This was left overnight at ambient conditions. Separation was carried out using a centrifuge at 10 °C at 8000 rpm for 10 min followed by decanting off of the supernatant. The solid residue which now contains a mix of polysaccharides and other macromolecules is then rinsed with acetone and dried at room temperature. The next stage which involved extraction of the water-soluble polysaccharides was carried out at 65 °C using distilled water in the mass-to-volume ratio of 1: 20 (g:ml). The extraction time allowed in the particular study was 2 h. Following the extraction period, the liquid extract was separated from the solid residue using centrifugation at 10 °C operating at 10,000 rpm for 10 min. The extraction was repeated with fresh distilled water to further remove soluble sulphated polysaccharides. The extracts from both runs were combined and evaporated under reduced pressure at a temperature of 60 °C. The ulvan was precipitated out using 90% ethanol (when added to liquid mixture concentration of ethanol decreases to 70%). Precipitation was allowed to take place overnight at 4 °C followed by centrifugation at 10,000 rpm for 10 min at 10 °C. The precipitated were further washed and dehydrated with acetone and 99% ethanol, respectively. The final product was dried at room temperature.

Following extraction of the crude ulvan, it is often desired to separate into different fractions. This could be for analysis or obtaining a less polydisperse product. The bioactivity of ulvans like other polysaccharides is affected by the molecular weight. Therefore, separating into fractions with specific range of molecular weight is important in analyzing and utilizing these bioactivities and physicochemical properties. Fractionation can be carried out using methods such as DEAE-Sepharose Fast Flow column (Tabarsa et al. 2018).

Co-extraction can be carried out to extract other polymers alongside the ulvan since the cell walls always contain other useful polysaccharides such as cellulose, xyloglucans as well as proteins. This would then involve including a separation process after the extraction stage. The yield and quality of ulvan obtained from any given biomass depend on the conditions of extraction. Temperature, pH, use of extractants and concentration of extractant, particle size of biomass, duration of extraction and pretreatment method are factors which have been identified to affect the yield and quality of ulvan extraction (Kidgell et al. 2019). The yield is defined in terms of the quantity of actual extract compared to the known content of ulvan in the biomass, while the quality is quantified by the degree of polymerization; hence, molecular weight of the ulvan extracted the purity (i.e., absence of other polymers and contaminants) and the level of degradation as a result of depolymerization and desulphurization of the ulvan in the process of extraction.

The key challenge to overcome in the extraction process is that ulvan has relatively low solubility in water, and the cell wall has a rather stable structure with strong intermolecular bonding between the cell wall polymers within which ulvan is entangled (Robic et al. 2009). These challenges are addressed by altering the pH which in turn affects the solubility of the ulvan and the intermolecular interactions between ulvan and the cell wall polymers as these parameters are pH-dependent. In other words, the intermolecular bonds between ulvan and the other molecules making up the green algae cell wall are weakened, while its tendency to be isolated into the solution is increased.

A further challenge in extraction of ulvan is the isolation of ulvan while excluding the other soluble polysaccharides and proteins present in the green algae biomass. The solubility of these other polymers is also pH- and temperature-dependent. Glucuronan, for example, has higher solubility in alkaline condition. This is addressed by selecting the right pH which favors only ulvan, making the extraction process more selective toward ulvan. The presence of salt also affects the extraction process as salts cause aggregation of ulvan. This is addressed by treatment with warm water to dissolve off the salt and increase osmotic pressure prior to extraction (Glasson et al. 2017).

At lower pH, the extraction is more selective toward ulvan as the solubility of other polysaccharides such as glucuronan and proteins is reduced at this condition. For example, protein impurities in ulvan extracted from *U. ohnoi* are significantly lower when hydrochloric acid extraction is used compared to when sodium oxalate extraction is used. The protein content of the extract increased from 4 to 7 μg protein per mg extract to 114–162 μg protein per mg extract. The drying and milling process is also important to minimize the particle size and optimize contact surface between biomass and extractant. Degradation during extraction results in an ulvan extract with shorter chains and less option for depolymerization to obtain more versatile fractions. While lower pH favors more selective extraction with less impurities, under acidic conditions and high temperature the ulvan extract is likely to undergo more degradation during extraction. The processor is therefore faced with the choice of higher yield at the expense of a more degraded product having lower degree of polymerization.

Recommended extraction conditions based on that used in various extractions reported in the literature are the extraction at a pH between 2 and 4.5, temperature between 80 and 90 °C and extraction period of 1–3 h. Some studies carry out repeat extraction on the same biomass to further extract any ulvan still left in the sample. Additional cost of energy and time should be considered for the second extraction which would have lower yield than the first and a lower concentration of ulvan requiring more evaporation per gram of ulvan extracted. Furthermore, since the structure of ulvans varies for different sources and also varies with harvest season, for extract with most uniform chemical structure, biomass from the same source and harvest should be used in the batch.

Unit operations which follow the extraction process are required to isolate and purify the product. These processes vary depending on availability and quality or form of desired final product. The first step after extraction is the separation of

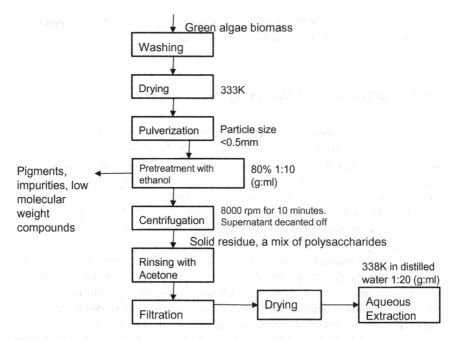

Fig. 8.3 Flowchart summarizing the extraction process of ulvans from green algae

the dissolved crude ulvan from the solid mass. This can be done using filtration or centrifugation. The extracted liquid at his point is likely to have smaller molecules and other polymers, etc. dissolved within it or smaller particles which are too small to be separated. The ulvan is precipitated out by addition of ethanol. Since ulvan will not dissolve in ethanol, this allows the ulvan to precipitate out of the solution. The gel-like precipitate can then be evaporated or concentrated using ultrafiltration. The process of washing, redissolving and reprecipitation can be repeated to improve purity. Impurities of salt and small molecules can be further removed by dialysis and ultrafiltration. For separation into different fractions with a range of molecular weights, methods such as size-exclusion chromatography can be employed.

The extraction process used and the level of purification depend on the target use of the ulvan and standard requirement of the industry. Figure 8.3 summarizes a typical extraction process of ulvan.

8.6 Environmental Implications

While ulvans offer numerous benefits and potential for application in food, nutraceuticals and pharmaceuticals, it is important to understand the environmental impact the production of ulvans. This requires an understanding of the processes, materials and chemicals involved in the whole process from harvesting of the raw material to the final product leaving the production factory gate.

8.6.1 Cultivation and Harvest of Green Algae

In the wild aquatic ecosystem, green algae play a significant role in carbon and nitrogen fixation as well as maintaining oxygen levels. These are essential to other life forms in the aquatic ecosystem. Harvesting of green algae from the wild should therefore give careful consideration to the impact of disrupting the existing balance within the environment where green algae play a key role. The system of harvesting from the wild should therefore implement a method where the total population of algae at any given time is maintained by balancing the rate of harvesting and the rate of replenishing.

Another issue with cultivation of green algae for commercial purpose is the danger of introducing exotic strains into the wild, thereby propagating algae blooms or artificially creating an environment which favors unnatural growth rate of green algae. This is a likely outcome in algae cultivation in aquaculture systems operating within the open sea (mariculture). Recent investigations into the cause of the green tides experienced along the Jiangsu coast of China revealed that the *U. prolifera* which is the predominant species causing the green tides in the region is genetically similar to that growing in the Porphyra mariculture located along the same Jiangsu coast (Zhang et al. 2019).

In China, for example, the economic loss from the green tide algae bloom has been estimated at over 183 million USD (1.3 billion RMB) (Zhang et al. 2019). This loss comes from the loss of income resulting from decline in aquatic organism as a result of hypoxia (oxygen depletion in the water), loss of income from recreational and tourist activities on the shoreline as a result of loss of aesthetics, foul smell and health risk associated with the green tides.

The cultivation of green algae for production of ulvans must therefore be carefully considered to prevent the risk of propagating green tides which result in serious adverse environmental impact.

8.6.2 Water Consumption

The extraction process is an aqueous one such that water serves as the solvent for extraction. Water is also used in the washing of the raw material during cleaning and pretreatment. Following extraction and separation of the gel, the water which contains impurities then needs to be treated before release into the environment. Depending on the level of treatment, this inevitably involves release with a level of contamination into the environment.

The amount of water used in the washing of the biomass was not specified in the said study (Tabarsa et al. 2018). Assuming a 1:10 ratio of dry biomass to tap water, such that 10 mL of water is required per gram for washing stage. For the main extraction, 20 mL of water is used per gram of Ulva. To therefore obtain the crude ulvan extract, a minimum estimated 30 mL of water (10 mL tap water and 20 mL distilled water) is used. Compared to more water-consuming process such as alginate extraction (see chapter on alginate), ulvan extraction consumed a fair amount of water.

8.6.3 Alcohol

In the pretreatment stage, each gram of pulverized ulvan powder required 10 mL of ethanol at 80% v/v concentration. This is used to dissolve off lipids, alcohol-soluble low molecular weight compounds and pigments. More ethanol is then used for precipitation at a concentration of 70% with estimated volume of 20 mL per gram of ulvan. Additional 99% ethanol (unspecified amount) is used for dehydration. Such that for each gram of ulvan produced in the reported study, at least 30 mL of ethanol at 70–80% concentration is required for pretreatment and precipitation with more used in dehydration post-extraction. As a greener alternative, bioethanol from renewable source can be used. However, the sustainability of such extraction process will need further consideration. The alcohol can also be cleaned and recovered through solvent recovery processes such as adsorption, distillation and condensation.

8.6.4 Acetone

Following pretreatment with ethanol to remove lipids, pigments and low molecular compounds, the ulvan is washed with acetone (amount unspecified). Additional acetone is used after the aqueous extraction in the final washing and dehydration prior to drying of the crude extract. Acetone is an organic solvent sourced from non-renewable sources. In the production of biopolymers such as ulvan, the goal is to source polymers from renewable sources, and it is therefore important to seek alternative solvents in the process of extraction.

8.6.5 Energy Consumption

In the extraction of the crude ulvan, energy consumed goes into heating of the system to temperatures between 80 and 90 °C (Tabarsa et al. 2018). Heating 1 kg of water by 1 K consumes 4.18×10^{-3} J of energy (Bird et al. 2005). Further energy is consumed in the drying process. Further energy consumption takes place in the other

Table 8.1 Typical consumption in ulvan extraction process

Component	Amount
Green algae biomass	3–12.5 g per gram algae[a]
Water	30 ml/g ulva
Ammonium oxalate	0.02M
Ethanol	80% 10 ml
	90% 10 ml
Acetone	98%
Energy	
• Grinding	0.5 mm particle size
• Heating	65 °C 2 h
• Drying	60 °C overnight
• Centrifugation	8000 rpm 10 min
	10,000 rpm 10 min

[a]Based on yield of ulvan ranging between 6 and 30%

unit operations which follow extraction such as separation and drying. Separation is done using centrifugation at 10,000 rpm at 10 °C for 10 min. Centrifuges operate from low rpm to medium rpm of about 2000 rpm to high rpm > 6000 rpm with 60,000 rpm being in the ultracentrifugation range. Therefore at 10,000 rpm, the ulvan production separation process is making use of relatively high centrifugal force, thus implying that the energy used up at this stage is considerably high. Manufacturers also need to consider the energy consumed further down in the production process for fractionating into different molecular weight ranges to obtain a less polydisperse sample and for purification which is important in improving the value of the product. As discussed within the chapter, molecular weight and purity have significant effect on the bioactivity of ulvans.

Table 8.1 gives a summary of consumption in a typical ulvan extraction process. These values are based on values reported in research studies (Tabarsa et al. 2018).

8.7 Applications

Bioactive properties of ulvans include antioxidant activity, cholesterol- and triglyceride-lowering effects among others. We know that ulvans from different species of green algae show species related variation in their chemical properties such as molecular weight, sulfation, constituent monosaccharides and level of branching (Kidgell et al. 2019). These variations in chemical properties result in variation in the bioactive properties such that specific species have been identified with some bioactive properties that are not seen in other species. In reviewing the applications of ulvan, we identify some of the species which can be used in the discussed application.

8.7.1 Immune Modulation

The ability of certain substances to influence immune response is referred to as immunomodulation and entities which are able to do these are referred to as immunomodulators. This activity is primarily dependent on the activation of inflammatory process whereby the body releases chemicals to destroy pathogens from the body. Ulvans aids in immunomodulatory process by activating the protein complexes which control production of cytokines which are key to the inflammatory process. Increased production of cytokines, enzymes and the products has been shown by macrophage cells treated with ulvan from species of *U. pertusa* (Li et al. 2018a, b), *U. rigida* (Leiro et al. 2007), *U. intestinalis* (Peasura et al. 2016), *U. Ohnoi* (Fernández-Díaz et al. 2017) *Enteromorpha prolifera* (Kim et al. 2011a) and *U. prolifera* (Cho et al. 2010). Beyond macrophage cell studies, investigation of immunomodulatory activity of ulvans has also been carried out in organisms such as fish (Castro et al. 2006), rats, mice and chickens. Ulvans aid in the bodies' defense against defective or pathogenes by enhancing macrophage functions such as phagocytosis (Karnjanapratum et al. 2012).

The inflammatory cytokine production, hence immunomodulatory activity of ulvan, is comparable to that of lipopolysaccharides albeit requiring a higher concentration of ulvan to achieve the same effect. For instance in one study, a 500 μg mL of ulvan was required to achieve the same immunomodulatory effect as 100 ng/mL of lipopolysaccharides.

Molecular weight and degree of sulfation jointly affect the occurrence and potency of the immunomodulatory activity of ulvans. Purity is also an important factor in the effectiveness of the immunomodulatory activity of ulvans with purer samples showing better immunomodulatory effect. This can be attributed to the reduction in interference with the active functional groups as well as structural chemistry which influences the bioactivity. In one study, ulvans extracted and purified from *U. pertusa* with molecular weight in the range 14,450–1690 kDa showed an increase in immunomodulatory activity quantified as macrophage cell activation when compared with ulvans with lower molecular weight of less than 365 kDa from the same source (Tabarsa et al. 2012). However, this increased immunomodulatory activity resulting from increased molecular weight is not replicated in other ulvan extracts from other species in a different molecular weight range. For examples, ulvan from *U. intestinalis* with a molecular weight of 28.7 kDa showed much higher immunomodulatory activity compared to ulvans from same source with higher molecular weight of 87.2 kDa (Tabarsa et al. 2018), thus indicating that the relationship between molecular weight and immunomodulatory activity is not the same either across all molecular weight range or across species. Degree of sulfation also affects the immunomodulatory activity of ulvan. Most studies show that higher level of immunomodulation is achieved with higher degree of sulfation (Leiro et al. 2007).

The immunomodulatory activity of ulvans makes them potentially applicable as immune-stimulating agents if nutraceutical or pharmaceutical products. Species of green algae not yet confirmed to possess or do not possess immunomodulatory bioactivity include *U. conglobata*, *U. compressa*, *U. flexuosa*, *U. gigantea* and many others (Kidgell et al. 2019).

8.7.2   Antioxidant

Ulvans like other sulfated polysaccharides act as exogenous antioxidants and enhance the activities of the endogenous antioxidants. They do so through their radical scavenging activity and enhancement of the activities of antioxidant enzymes. The antioxidant activity of ulvans makes them applicable in addressing a broad range of human defects which are associated with the presence of reactive oxygen and nitrogen species. Such conditions include cancer, neurodegenerative diseases, cardiovascular diseases and aging. Low molecular weight and high sulfate content have been shown to correlate with the antioxidant activity of ulvans.

Although the mechanism by which ulvans boost endogenous antioxidant activity is not yet fully established, it is thought that they do this by interfering in signaling pathways for expression of antioxidant producing enzymes. This ability of ulvans to stimulate antioxidant enzyme production has been demonstrated by ulvans from species such as *U. amoricana* (Berri et al. 2017). Therefore, ulvan antioxidant activity is both as a result of their chemical structure eliminating oxidative species and also their ability to influence the gene expression for production of enzymes responsible for endogenous antioxidant activity in the body.

8.7.3   Anticancer

Although no clinical trials have been reported yet, studies on animals such as mice and rats and a range of human cancer cell lines such as colon cancer, breast cancer and gastric cancer cell lines have provided evidence that ulvans could potentially have anticancer properties. Ulvan from species such as fasciata, tubulosa, pertusa, latusa, intestinalis and others (Kidgell et al. 2019; Kim et al. 2011b) has shown anticancer properties.

Various approaches are used in the treatment of cancer: preventing the formation of cancerous cells by removing factors such as oxidative processes which could lead to mutation, prevention of growth of the cancer cells or inducing death of the cancer cells. Thus, far evidence points to ulvans potentially play anticancer roles by either preventing the proliferation of cancer cells or promoting apoptosis (programmed cell death) in cancer cells. Example of studies indicates that the ability of ulvans to promote cancer cell apoptosis is by Wang et al. (2014). In the aforementioned study, the increase in pro-apoptotic tumor suppressor expression was observed in cancer cell lines treated with ulvan from *Enteromorpha intestinalis*. The pro-apoptotic activity of ulvan is also demonstrated by the reduction in expression of anti-apoptotic proteins.

The other anticancer approach ulvans is thought to use is the prevention of cancer cell proliferation. This has been investigated by measuring the reduction in the level of DNA replication activity in the cancer cells. This is quantified by the reduction in the level of proliferating cell nuclear antigen (Hussein et al. 2015). The level of DNA replication activity reduced to noticeable levels in rat hepatocytes when treated with ulvan from *Ulva lactuca*. Anticancer activity of ulvans has also been measured as a function or tumor mass and cytotoxicity where a reduction in tumor mass and a high cytotoxicity are indicative of reduced cell proliferation and hence anticancer activity.

Although a number of studies have presented anticancer properties of ulvans, there are also studies which show no cytotoxicity of ulvan against cancer cells. An example of such is studies on *Ulvan intestinalis* which show no cytotoxicity against sarcoma tumor cells at a relatively high concentration of 50–800 μg per ml. However, this study also further showed that ulvans from different species could take different pathways to achieving anticancer activities. While no cytotoxicity was measured in vitro, there was 61–7% (Jiao et al. 2009) reduction in the tumor cell mass in vivo when applied to mice in vivo at a dose of 100–400 mg/kg.

The anticancer activity is very low compared to conventional chemotherapy agents. With such a low level of effectiveness, ulvan can at best serve in adjunct roles in cancer treatment. As a polymeric material with pH-dependent rheological properties, ulvan can serve roles in anticancer therapy delivery system as pH-responsive polysaccharide systems (Yang et al. 2017) and nanoparticle delivery systems (Li et al. 2018a, b). No conclusive data yet exists to relate the chemical structure such as degree of sulfation or molecular weight to the anticancer activity of ulvan.

8.7.4 Anticoagulant Activity

Undesired blood clots could lead to severe health risks and could result from conditions such as operated blood vessels. Anticoagulants have different mechanisms of action with the general goal being to interfere with the pathways leading to blood clotting. Anticoagulant activity of ulvan is affected by degree of sulfation at a particular molecular weight range. Most studies are in agreement that a higher degree of sulfation leads to increased anticoagulant activity of ulvans. However, the effect of molecular weight varies. For ulvans extracted from *Enteromorpha prolifera*, for example, beyond a given molecular weight of 200 kDa, further increase in degree of sulfation did not yield any anticoagulant property (Li et al. 2018a, b). Increasing the degree of sulfation resulted in increased anticoagulant activity for ulvans of *E. prolifera* with molecular weight less than 200 kDa. However, when the molecular weight is increased beyond this value, the ulvan no longer had any anticoagulant effect.

Anticoagulant effect of ulvans is much less than those of currently commercially available anticoagulants such as heparin (Qi et al. 2013). It also varies from species and source since the degree of sulfation and molecular weight is also species and source dependent. Nonetheless, the low level of anticoagulant activity demonstrated in various studies is potentially relevant in developing anticoagulant agent with potential for interfering in the body's intrinsic anticoagulant mechanism.

8.7.5 Antihyperlipidemic

When there is an imbalance in the way the body distributes and eliminates lipids and cholesterol, this leads to metabolic syndrome often associated with hyperlipidemia. This is often a cause of cardiovascular diseases. The components of macroalgae have been linked with having antihyperlipidemic effects and a diet including macroalgae has been shown to have preventive or curative effect on hyperlipidemia. To isolate specific components in order to develop therapies for hyperlipidemia, the sulfated polysaccharides of macroalgae have been investigated for their antihyperlipidemic effect.

Ulvans from species which include *U. pertusa*, *U. lactuca*, *U. prolifera* and *Ulva fasciata* have shown significant hyperlipidemic effect when measured as a function of serum total cholesterol using rat and mice models. Evidence suggests that increased degree of sulfation of ulvan may improve the antihyperlipidemic effect of the ulvan. Oversulfated ulvan with up to 40.6% sulfation achieves a total cholesterol-lowering effect of 44% compared to a 28% reduction measured for crude ulvan with a degree of sulfation of 22.5% (Qi and Sheng 2015). Although it is obvious that the structure of ulvans affects the mechanism and level of antihyperlipidemic effect, the relationship between the chemical structure of ulvan and its antihyperlipidemic effect is unclear. Possible mechanisms presented include gene regulatory effect and interference in metabolism and biosynthesis (Ali and Agha 2009).

8.7.6 Antiviral Property

While ulvans alone show moderate to low antiviral properties, more important is their potential impact on vaccination. Combination of ulvans with vaccination can improve vaccination effectiveness by as much as 100% compared to when vaccination is done alone (Song et al. 2016). This enhancement of vaccination is thought to be associated with ulvan's immunomodulatory activity. The impact of this could be need for less quantities of vaccine in achieving the same level of vaccination, particularly in cases when such vaccines are scarce.

Although some evidence points to improve antiviral activity with higher molecular weight and higher degree of sulfation, the number of studies and species investigated are not enough to make a conclusive relationship between the structural parameters and the antiviral property of ulvan.

Antiviral activities have been investigated in ulvan from species of *U. clathrata*, *U. pertusa* and *U. compressa* (Aguilar-Briseno et al. 2015; Song et al.

2016; Lopes et al. 2017) among others. These antiviral activities have been tested on viruses such as influenza, herpes simplex, Japanese encephalitis, yellow fever, avian influenza, dengue, Newcastle disease, dengue, West Nile and others (Kidgell et al. 2019). Thus far, much of the studies on the antiviral activity of ulvans have been carried out in cell cultures and on mice models.

8.8 Commercial Production

Starting in the early twenty-first century to date, commercial agal farming has gained much attention from researchers, government and venture capitalists. The interest in algae ranges from biofuel production from algae oil, bioethanol production from the algal biomass, water remediation and extraction of the various bioactives for pharmaceutical applications. All with the promise of largely profitable algae are based on commercial venture.

Ulvan, being one of the polysaccharides produced by green algae, has potential commercial application in food and pharmaceutics with the applications as discussed in earlier sections within this chapter. However, commercial production of ulvan is presently still at its infancy due to limitations in the commercially feasible extraction methods to obtain uniform standard extracts.

Commercial ulvan production would benefit from the relatively high productivity of the Ulva green algae, if well managed. One of such fast-growing species that has high nutrient uptake rate is the *U. prolifera* (Zhang et al. 2019). Ulvan is the most valuable polysaccharide of the Ulva green algae, and therefore, the main incentive for commercial cultivation and processing of these green algae is either as food or to extract ulvan. Presently, the use as food seems to gain precedence owing to it being a less complex option in terms of finance and process technology. Consumed as food, the Ulva already is known for improving gastrointestinal health and prevention of certain chronic diseases (Kidgell et al. 2019). However, extraction of the sulfated polysaccharide to which these health benefits are attributed results in improved effectiveness and development of more effective derivatives. As processes to extract purer and more uniform ulvan samples further develop this could encourage more commercialization of ulvan-based products for personal care and clinical applications.

8.9 Conclusion

The ulvan producing green algae species are readily available and highly productive. The cultivation of these species comes with the risk of propagating or contributing to existing algae blooms. On the other hand, well-managed recovery of Ulva from the wild or cultivation in well-managed systems can benefit from the rapid growth rate of these species for ulvan production. Ulvans have the bioactive properties associated with sulfated polysaccharides. In addition to the general sulfated polysaccharide bioactivities, ulvans are particularly promising as immunomodulatory, antioxidant and antihyperlipidemic agents. Much of the studies on ulvans bioactive properties are at an early stage using cellular and animal test models. It is nonetheless important to understand the potential application and the state of technology in the respective applications. Such will be useful in advising further efforts in the use of ulvans.

References

Adrien A, Bonnet A, Dufour D, Baudouin S, Maugard T, Bridiau N (2017) Pilot production of ulvans from *Ulva* sp. and their effects on hyaluronan and collagen production in cultured dermal fibroblasts. Carbohyd Polym 157:1306–1314

Aguilar-Briseno JA, Cruz-Suarez LE, Sassi JF, Ricque-Marie D, Zapata-Benavides P, Gamboa EM, Rodriguez-Padilla C, Trejo-Avila LM (2015) Sulphated polysaccharides from ulva clathrata and cladosiphon okamuranus Seaweeds both inhibit viral attachment/entry and cell-cell fusion, in NDV Infection. Marine Drugs 13(2):697–712

Ali MM, Agha FG (2009) Amelioration of streptozotocin-induced diabetes mellitus, oxidative stress and dyslipidemia in rats by tomato extract lycopene. Scand J Clin Lab Inv 69:371–379

Berri M, Slugocki C, Olivier M, Helloin E, Jacques I, Salmon H, Demais H, Le Goff M, Nyvall P, Collen (2016) Marine-sulfated polysaccharides extract of *Ulva armoricana* green algae exhibits an antimicrobial activity and stimulates cytokine expression by intestinal epithelial cells. J Appl Phycol 28:2999–3008

Berri M, Olivier M, Holbert S, Dupont J, Demais H, Le Goff M, Nyvall Collen P (2017) Ulvan from *Ulva armoricana* (Chlorophyta) activates the P13K/Akt signalling pathway via TLR4 to induce intestinal cytokine production. Algal Res 28:39–47

Bird RB, Stewart WE, Lightfoot EN (2005) Transport phenomena, 2nd edn. Wiley-India, New Delhi, pp 270

Castro R, Piazzon MC, Zarra I, Leiro J, Noya M, Lamas J (2006) Stimulation of turbot phagocytes by ulva rigida *C. agardh* polysaccharides. Aquaculture 254:9–20

Chen D, Wu XZ, Wen ZY (2008) Sulphated polysaccharides and immune response: promoters or inhibitors. Panminerva Med 50:177–183

Cho M, Yang C, Kim SM, You S (2010) Molecular characterization and biological activities of water-soluble sulfated polysaccharides from *Enteromorpha prolifera*. Food Sci Biotechnol 19:525–533

Fernández-Díaz C, Coste O, Malta EJ (2017) Polymer chitosan nanoparticles functionalized with *Ulva ohnoi* extracts boost in vitro ulvan immunostimulant effect in *Solea senegalensis* macrophages. Algal Res 26:135–142

Glasson CRK, Sims IM, Carnachan SM, de Nys R, Magnusson M (2017) A cascading biorefinery process targeting sulfated polysaccharides (ulvan) from *Ulva ohnoi*. Algal Res 27:383–391

Hussein UK, Mahmoud HM, Farrag AG, Bishayee A (2015) Chemoprevention of diethylnitrosamine-initiated and phenobarbital-promoted hepatocarcinogenesis in rats by sulfated polysaccharides and aqueous extract of *Ulva lactuca*. Integr Cancer Ther 14:525–545

Jaswir I, Monsur HA (2011) Anti-inflammatory compounds of macro algae origin: a review. J Med Plant Res 5:7146–7154

Jiao LL, Li X, Li TB, Jiang P, Zhang LX, Wu MJ, Zhang LP (2009) Characterization and anti-tumor activity of alkali-extracted polysaccharide from *Enteromorpha intestinalis*. Int Immunopharmacol 9(3):324–329

Karnjanaprakorn S, Tabarsa M, Chou M, You SG (2012) Characterization and immunomodulatory activities of sulfated polysaccharides from *Capsosiphon fulvescens*. Int J Biol Macromol 51:720–729

Kidgell JT, Magnusson M, de Nys R, Glasson CRK (2019) Ulvan: a systematic review of extraction, composition and function 39(101422):1–20

Kim JK, Cho MI, Karnjanaprakorn S, Shin IS, You SG (2011a) In vitro and in vivo immunomodulatory activity of sulfated polysaccharides from *Enteromorpha prolifera*. Int J Biol Macromol 49:1051–1058

Kim S-K, Thomas NV, Li X (2011b) Anticancer compounds from marine macroalgae and their application as medicinal foods. Adv Food Nutr Res 64:213–224

Lahaye M (1998) NMR spectroscopic characterisation of oligosaccharides from two Ulva rigida ulvan samples (Ulvales, Chlorophyta) degraded by a lyase. Carbohydr Res 314:1–12

Lahaye M, Robic A (2007) Structure and functional properties of Ulvan, a polysaccharide from green seaweeds. Biomacromol 8:1765–1774

Leiro JM, Castro R, Arranz JA, Lamas J (2007) Immunomodulating activities of acidic sulphated polysaccharides obtained from the seaweed *Ulva rigida C. agardh*. Int Immunopharmacol 7:879–888

Li J, Jiang F, Chi Z, Han D, Yu L, Liu C (2018a) Development of *Enteromorpha prolifera* polysaccharide-based nanoparticles for delivery of curcumin to cancer cells. Int J Biol Macromol 112:413–421

Li W, Wang K, Jiang N, Liu X, Wan M, Chang X, Liu D, Qi H, Liu S (2018b) Antioxidant and antihyperlipidemic activities of purified polysaccharides from *Ulva pertusa*. J Appl Phycol 30(4):2619–2627

Lopes N, Ray S, Espada SF, Bomfim WA, Ray B, Faccin-Galhardi LC, Linhares REC, Nozawa C (2017) Green seaweed *Enteromorpha compressa* (Chlorophyta, Ulvaceae) derived sulphated polysaccharides inhibit herpes simplex virus. Int J Biol Macromol 102:605–612

Michel G, Czjzek C (2014) Polysaccharide-degrading enzymes from marine bacteria. Marine enzymes for biocatalysis: Sources, biocatalytic characteristics and bioprocesses of marine enzymes. Woodhead Publishing Ser Biomed 5:429–464

Misurcova L, Skrivankova S, Samek D, Ambrozova J, Machu L (2012) Health benefits of algal polysaccharides in human nutrition. Adv Food Nutr Res 66:75–145

Peasura N, Laohakunjit N, Kerdchoechuen O, Vongsawasdi P, Chao LK (2016) Assessment of biochemical and immunomodulatory activity of sulphated polysaccharides from *Ulva intestinalis*. Int J Biol Macromol 91:269–277

Pengzhan Y, Quanbin Z, Hong Z, Xizhen N, Zhien L (2004) Preparation of polysaccharides in different molecular weights from *Ulva pertusa* Kjellman (Chlorophyta). Chin J Oceanol Limnol 22:381–385

Qi H, Sheng J (2015) The antihyperlipidemic mechanism of high sulfate content ulvan in rats. Mar Drugs 13:3407–3421

Qi H, Zhao T, Zhang Q, Li Z, Zhao Z, Xing R (2005) Antioxidant activity of different molecular weight sulfated polysaccharides from *Ulva pertusa* Kjellman (Chlorophyta). J Appl Phycol 17(6):527–534

Qi H, Huang L, Liu X, Liu D, Zhang Q, Liu S (2012) Antihyperlipidemic activity of high sulfate content derivative of polysaccharide extracted from *Ulva pertusa* (Chlorophyta). Carbohydr Polym 87:1637–1640

Qi XH, Mao WJ, Chen Y, Chen YL, Zhao CQ, Li N, Wang CY (2013) Chemical characteristics and anticoagulant activities of two sulfated polysaccharides from *Enteromorpha linza* (Chlorophyta). J Oceanogr Univ China 12:175–182

Reisky L, Stanetty C, Mihovilovic MD, Schweder T, Hehemann JH, Bornscheuer UT (2018) Biochemical characterization of an ulvan lyase from the marine flavobacterium *Formosa agariphila* KMM 3901T. Appl Microbiol Biotechnol 102(16):6987–6996

Rioux L, Turgeon SL (2015) Seaweed carbohydrates. In: Tiwari BK, Troy DJ (eds) Seaweed sustainability: food and non-food application. Academic press, pp 141–192

Robic A, Rondeau-Mouro C, Sassi JF, Lerat Y, Lahaye M (2009) Structure and interactions of ulvan in the cell wall of the marine green algae *Ulva rotundata* (Ulvales, Chlorophyceae). Carbohyd Polym 77:206–216

Rybak A, Messyasz B, Laska R (2012) Freshwater Ulva (Chlorophyta) as a bioaccumulator of selected heavy metals (Cd, Ni and Pb) and alkaline earth metals (Ca and Mg). Chemosphere 89(9):1066–1076

Scharnweber D, Hubner L, Rother S, Hempel U, Anderegg U, Samsonov SA, Pisabarro MT, Hofbauer L, Schnabelrauch M, Franz S, Simon J, Hintze V (2015) Glycosaminoglycan derivatives: promising candidates for the design of functional biomaterials. J Mater Sci Mater Med 26(9):232–249

Shao Q, He Y, Jiang, S (2011) Molecular dynamics simulation study of ion interactions with zwitterions. J Phys Chem B 115:8358–8363

Song J, Chen X, Liu X, Zhang F, Hu L, Yue Y, Li K, Li P (2016) Characterization and comparison of the structural features, immune-modulatory and anti-avian influenza virus activities conferred by three algal sulfated polysaccharides. Mar Drugs 14(1):4

Tabarsa M, Han JH, Kim CY, You SG (2012) Molecular characteristics and immunomodulatory activities of water-soluble sulfated polysaccharides from *Ulva pertusa*. J Med Food 15:135–144

Tabarsa M, You s, Dabaghian EH, Surayot U (2018) Water-soluble polysaccharides from Ulva intestinalis: molecular properties, structural elucidation and immunomodulatory activities. J Food Drug Anal 26:599–608

Varnier AL, Sanchez L, Vatsa P, Boudesocque L, Garcia-Brugger A, Rabenoelina F, Sorokin A, Renault JH, Kauffmann S, Pugin A, Clement C, Baillieul F, Dorey S (2009) Bacterial rhamnolipids are novel MAMPs conferring resistance to *Botrytis cinerea* in grapevine. Plant Cell Environ 32:178–193

Wang XX, Chen Y, Wang JJ, Liu ZX, Zhao SG (2014) Antitumor activity of sulfated polysaccharide from *Enteromorpha intestinalis* targeted against hepatoma through mitochondrial pathway. Tumor Biol 35:1641–1647

Yabe T, Ishii Y, Amano Y, Koga T, Hayashi S, Nohara S, Tatsumoto H (2009) Green tide formed by free-floating *Ulva* spp. at Yatsu tidal flat, Japan. Limnology 10:239–245

Yang F, Fang X, Jiang W, Chen T (2017) Bioresponsive cancer-targeted polysaccharide nanosystem to inhibit angiogenesis. Int J Nanomed 12:7419–7431

Young CS, Gobler CJ (2016) Ocean acidification accelerates the growth of two bloom-forming, estuarine macroalgae. PLoS ONE 11(5):e0155152

Zhang Y, He P, Li H, Li G, Liu J, Jiao F, Zhang J, Huo Y, Shi X, Su R, Ye N, Liu D, Yu R, Wang Z, Zhou M, Jiao N (2019) *Ulva prolifera* green-tide outbreaks and their environmental impact in the Yellow Sea, China. Nat Sci Rev 0(0):1–14

Chapter 9
Laminarins

Abstract Laminarins are short-chain polysaccharides present in the cell wall of brown algae. The simple structure of short-chain 1-3 linked β glucose unit that makes it attractive in bioethanol production. Laminarin also has some bioactive properties which are of research interests and potential commercial application. Extraction of laminarin from brown algae can be carried out alongside the extraction of the other polysaccharides in the brown algae cell wall, such that it contributes toward more efficient utilization of the algal biomass. This chapter reviews the production process and evaluates the environmental impacts of laminarin production, the commercial potential in terms of availability of raw material as well as existing and potential applications. The occurrence of laminarin in nature and its chemistry are also presented within the chapter.

Keywords Laminarins · Algae · Bioethanol · Aquatic · Biopolymer

9.1 Introduction

Laminarin is a type of sulfated polysaccharide found in brown algae. It serves as a storage polysaccharide. Also referred to as laminarans, they are soluble polysaccharides found in algae. The main repeating unit in laminarin structure is a 1-3 linked β glucose unit. Some branching at the β 1-6 linkage is not uncommon in laminarin chains, and other units such as mannitol and mannose have been reported to be present along some laminarin polymer chains (Yvin et al. 1993). They are usually short-chain polysaccharides with a degree of polymerization between 26 and 31 (Deville et al. 2007) or up to 60 repeating units (Yvin et al. 1993). Algae species such as *Ecklonia kurome*, *Laminaria japonica*, *Laminaria digitata* and *Eisenia bicyclis* contain laminarin. They belong to the group of polysaccharides known as glucans, a diverse group of polymers made up of glucose repeating units. These glucans vary in terms of the orientation and linkages of these repeating units, which can be in forms such as alpha or β, linear or branched, L or D forms (Caipang and Lazado 2015). Glucans also have varying chain lengths from short-chain oligomers in tens of repeating units or less to long-chain polymer structures with thousands of repeating units. Other examples of glucans are cellulose and starch. Glucans exist in a variety of organisms such as

© Springer International Publishing 2020   189
O. Olatunji, *Aquatic Biopolymers*, Springer Series on Polymer and Composite Materials,
https://doi.org/10.1007/978-3-030-34709-3_9

yeast, bacteria and algae. Laminarins can therefore be described as short-chain β glucans with some degree of branching.

There is an increasing need for alternative sources of energy as the conventional fossil fuels are being depleted exponentially as the population rises globally. While crops such as corn and sugarcane are sources of carbohydrates which can be further hydrolyzed and fermented into ethanol used as biofuel, the main issue with such sources of alternative fuel is the competition with food, water and land for human consumption and use for survival. Other crops which are sources of polymeric materials such as gum arabic have these similar issues. Laminarin is sourced from brown algae, and this addresses half of the issues of other alternative fuels and biomaterials. Laminarin is a glucan which can be hydrolyzed to glucose and other simple sugars which can be fermented by microbes which in turn produce ethanol as a by-product.

As the world population grows, there is increasing demand to keep people alive and disease-free. Furthermore, as disease resistance of bacteria to conventional drugs increases, there is a need for more complex compounds for combating diseases and improving immunity. For this reason, exploring naturally occurring polymers such as laminarin for their bioactivities such as immunomodulatory and antimicrobial activity becomes more important.

In this chapter, we explore another interesting biopolymer which can be obtained from an aquatic organism, the brown algae, laminarin. The sources of laminarin, chemical structure, applications and the environmental impacts of laminarin extraction process are all discussed with the reference to various recent studies from different research studies around the world.

9.2   Occurrence in Nature

Glucans, in general, exist in yeast, bacteria, mushrooms, grains of cereals and algae (Caipang and Lazado 2015). In brown algae, laminarin exists in the plastids where it serves as a storage polysaccharide (Rioux and Turgeon 2015). Brown algae contain ~40% polysaccharides by dry weight (Deville et al. 2007), and one of these polysaccharides could include laminarin. Brown algae which are sources of laminations are referred to as Laminariales (Yvin et al. 1993). While alginate serves as a structural polysaccharide in brown algae, laminarin serves as a storage polysaccharide (Konda et al. 2015). The laminarin content could vary from 0 to 18% (Rioux and Turgeon 2015; Konda et al. 2015) in different species of brown algae, other studies report higher laminarin content of up to 35% laminarin per dry mass of brown macroalgae (Motone et al. 2016).

Storage polysaccharides generally have simpler structures such that they can easily be broken down to release energy and carbon when required by the organism. Laminarin content in any given species varies in different growing seasons and growth conditions. For instance, at low nitrate and nitrite content in the water, brown algae proceed to produce laminarin as a stored carbon source; however, where nitrate and nitrite are present, the brown algae directs its energy toward growth (Rioux and

Turgeon 2015). The brown algae species of Laminaria and Saccharina are the more abundant and more common sources of Laminaria; nonetheless, laminarin is also found in other brown algae such as Fucus, Undaria and Ascophyllum species. These species also serve as sources for other algae polysaccharides such as alginates and fucoidan. Laminaria varies in the chain length and level of branching depending on different species and growth factors. These chemical structure variations also have an effect on solubility (Kadam et al. 2015) and bioactivity (Liu et al. 2018) of the specific laminarin.

Laminarin production is thought to be part of the brown algae coping strategy to survive the winter season. It is produced at the end of the peak growth rate in spring, such that laminarin serves as a storage reserve source of carbon during the winter season (Misurcova et al. 2012). Water-soluble carbohydrates are a more accessible source of energy storage compared to insoluble carbon sources such as cellulose (Hildebrand et al. 2017). Laminarin is therefore one of the soluble polysaccharides the brown algae uses as a source of carbon.

9.3 Chemistry of Laminarin

Laminarin is built up of β,1-3 linked glucan with a low level of branching at the 1-6 linkages. Some β,1-6 intrachain linkages may also be present (Kadam et al. 2015). Laminarin has a similar structure to Lichenan found in moss and lentinan found in mushrooms but differs in the degree of branching and nature of glycosidic linkages (Ojima et al. 2018).

9.3.1 Repeating Units

Laminarin is composed of mostly neutral sugars with small amounts of uronic acid (Misurcova et al. 2012). The composition of the sugars varies for different species. For example, while *Saccharina longicruris* contains up to 99% neutral sugars which is the highest composition of neutral sugars thus far observed in laminarins, *Ascophyllum nodosum* has 89.6% neutral sugars and *Fucus vesiculosus* contains 84.1% neutral sugars. Nonetheless, the neutral sugars are always significantly more abundant in laminarin. The arrangements and conformation of the repeating units also vary for different species. Figure 9.1 compares the structure of laminarin extracted from *E. bicyclis* with that from *L. digitata* (Liu et al. 2018).

To understand the difference between the laminarin from the two different species of brown algae, the laminarin forms the two sources that were hydrolyzed using microorganism *Coprinopsis cinerea* which metabolizes the enzyme endo-β-1,3-glucanase which breaks the β-1,3 glycosidic bond in laminarin. By analyzing the residues from the hydrolysis of the different laminarin using high-performance anion exchange, chromatography combined with mass spectrometry revealed the building

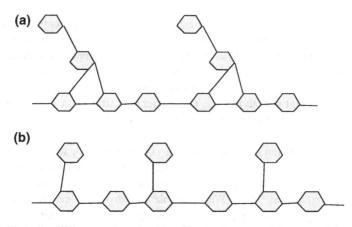

Fig. 9.1 Illustration of the secondary structure of laminarin extracted from two different species of brown algae **a** *E. bicyclis* and **b** *L. digitata*, showing the arrangements of the repeating units

blocks of the laminarins. These building blocks are used to deduce the structure of the different laminarins. This revealed a higher degree of branching in the *E. bicyclis* which had a 25.99% branching while *L. digitata* had a 7.68% branching. This difference in the degree of branching was indicated by the presence of β-1,6 glycosidic bonds in the residue which were not broken down by the glucanase whose activity is specific to the β-1,3 glycosidic bond.

9.3.2 Molecular Weight

Laminarin is generally a low molecular weight polysaccharide with between 26 and 31 repeating units per strand (Deville et al. 2007). Some studies have reported degree of polymerization of laminarin in different ranges of 31–40 while others report a maximum degree of polymerization of 12 (Stone 2009). Laminarin extracted from *L. digitata* is reported to have a degree of polymerization of ~25 (Hrmova and Fincher 2009). The molecular weight of laminarin extracted from a given species could also vary with the stage of growth. For example, laminarin extracted from the mature brown algae *Fucus evanescens* had a higher molecular weight than that extracted when it was at an earlier growth stage (Misurcova et al. 2012). This must therefore be considered when cultivating or harvesting brown algae for laminarin production. Thus, far matured brown algae harvested in winter grown under reduced nitrate and nitrite content seem to show the optimal composition of laminarin.

9.3.3 Solubility Branching

The degree of branching varies in different species; for instance, laminarin extracted from the *L. digitata* brown algae has a degree of branching of about 12.5% while that of *E. bicyclis* has a much higher degree of branching (Ojima et al. 2018). Laminarin is generally referred to as a soluble sulphated polysaccharide. However, the solubility of laminarin varies with structure and medium. Highly branched laminarin is more readily soluble in cold or hot water while less branched laminarin will only dissolve in hot water (Misurcova et al. 2012). This solubility variation also serves as a basis for the extraction of laminarin where the degree of branching of isolated fraction can be controlled by the temperature. The more branched can first be isolated at lower temperature, and the more linear laminarin can be isolated at elevated temperature.

When dissolved in the aqueous solvent, laminarin, like other polymers, will result in an increase in the viscosity of the solution. This increase in viscosity is related to the molecular weight. As a low molecular weight polymer laminarin solutions will not have the thickening effect of other higher molecular weight polysaccharides such as carrageenan. However, the lower viscosity at higher solution concentration will give less flow limitations during extraction and processing of liquid phases.

9.3.4 Methacrylated Laminarin

Modification of laminarin can be achieved by replacing some of the –OH groups with methyl groups. These methacrylated laminarins are applied in tissue engineering for enhancing cell adhesion, for example (Martins et al. 2018; Wang et al. 2018). Polysaccharides form hydrogels; however, they are limited in terms of maintaining their mechanical properties when exposed to the physiological conditions in the body. Although other materials exist for producing hydrogen with desirable mechanical properties at physiological condition, biocompatibility and biodegradability limit the application of such hydrogel-forming materials for biomedical applications. Modifying laminarin with glycidyl methacrylate prior to cross-linking results in a hydrogel with improved mechanical properties compared to unmodified laminarin hydrogels.

An example process for the production of methacrylated laminarin is as follows (Wang et al. 2018): laminarin is dissolved in dimethyl sulfoxide (DMSO) in a mass (g) of laminarin to volume (mL) of DMSO ratio of 1:10. This is carried out under nitrogen atmosphere to prevent reaction with atmospheric air. This takes 1 h to dissolve and is followed by the addition of 167 mg of 4-(N,N Dimethylamino) pyridine and then dropwise addition of glycidyl methacrylate. The amount of glycidyl methacrylate added determines the level of methacrylation. The methacrylation process takes 48 h under room temperature. The reaction is terminated by bringing the pH to neutral by adding hydrochloric acid. On completion, the methacrylated laminarin is then separated from the reaction components to obtain purified methacrylated laminarin. This purification can be achieved using dialysis for a period of 7 days. An example dialysis condition for separation is dialysis against deionized water and a molecular weight cutoff point of 2000 Da. Using this method, laminarin with a degree of methacrylation of 5–10, 10–20 and 20–30% is obtained when 0.75, 1.5 and 3.0 mmol of glycidyl methacrylate were used, respectively (Wang et al. 2018).

The methacrylated laminarin is then cross-linked to obtain hydrogels. The hydrogels obtained from methacrylated laminarin have a tensile strength of up to 0.63 MPa, a compressive strength of 3.2 MPa, and will stretch by 650% of their original length before breakage (Wang et al. 2018). Figure 9.2a shows the structure of laminarin, and Fig. 9.2b shows the structure of methacrylated laminarin.

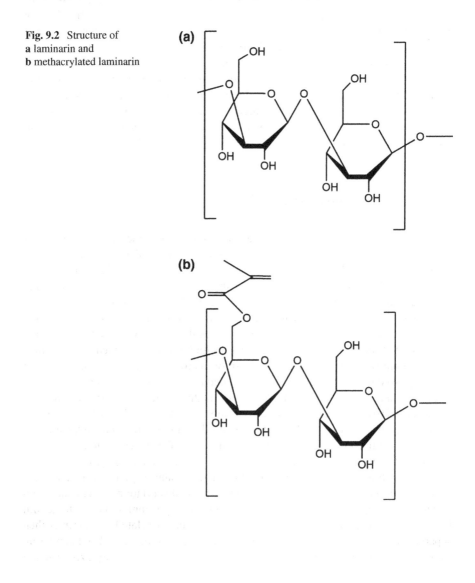

Fig. 9.2 Structure of **a** laminarin and **b** methacrylated laminarin

9.3.5 Biodegradation of Laminarin

Laminarin degrades into its repeating units which are glucose and other reducing sugars such as mannose. Microbes which metabolize these laminarinase include *C. cinerea* and *Microbacterium oxydans* (Kim et al. 2013). At an inoculum volume concentration of 20% (v/v), pH 6.0, and a temperature of 30 °C, a 10 g L^{-1} concentration of laminarin substrate produced 5.11 g L^{-1} of reducing sugars and 2.88 g L^{-1} glucose after a 6-day culture period using *M. oxydans* (Kim et al. 2013). The microbial degradation of laminarin into simple sugars is important both for safe environmental purpose and for commercialization of the sugars.

9.3.6 Chrysolaminarin

It is important to mention here, chrysolaminarin, also referred to as leucosin, another biopolymer found in unicellular algae (Beattie et al. 1961). Diatoms are part of the plankton and benthic algae which serve as a significant part of the aquatic food chain. Diatoms have been used in art and are continuously studied for wide variety of potential applications to advance human life such as use in photo-induced green synthesis of nanoparticles (Chetia et al. 2017). Diatoms produce a polysaccharide which is similar to laminarin known as chrysolaminarin. It is synthesized during the day in the presence of light and used up at night to fuel heterotrophic metabolic activities (Caballero et al. 2016). The similarity in these two polysaccharides lies in the fact that they are both made up of β,1-3 glucose and both water-soluble polysaccharides. Chrysolaminarin contains up to 99.5% glucose repeating units while laminarin has other units, mainly mannitol within the structure. There are no mannitol units present in chrysolaminarin (Beattie et al. 1961). Chrysolaminarin is a linear polymer and has a more crystalline secondary structure while laminarin could have low or high level of branching. While laminarin serves as the storage polysaccharides for the Phaeophyceae, chrysolaminarin serves as the storage polysaccharides for the diatoms. Chrysolaminarin makes up between 12 and 33% dry weight of the diatom. This could be up to 80% when grown at optimal conditions for chrysolaminarin production (Hildebrand et al. 2017). Due to their simpler structure, other than the potential bioactivities, chrysolaminarin has a potential to serve as a source of glucose for ethanol production.

9.4 Availability of Raw Material

Laminarin can be sourced from a variety of brown algae which grow in different marine areas. According to FAO in 2016, 31.2 million tonnes of algae are produced in 2016 (FAO 2018). Of this total, 34,000 tonnes of brown algae live weight was cultivated in aquaculture globally. The live weight contains around 45% moisture prior to natural drying in storage. After drying at room temperature, the algae used as feedstock are generally around 25% moisture content (Konda et al. 2015). This figure excluded *L. japonica* species, and this is one of the main Phaeophyceae which acts as a major source of laminarin. Production of *L. japonica* in 2016 was 8,219,000 tonnes cultivated using aquaculture. China is the lead producers of algae followed by Indonesia producing live weight of 14,387,000 and 11,631,000 tonnes, respectively, in 2016. The figures from 2005 to 2016 show a continuous rise in the production algae from different countries, with some few exceptional cases where a drop in production is experienced as a result of isolated cases. For example, a drop in production of *Undaria pinnatifida* species is reported from 2014 to 2015 to 2016, where the cultivated amount was 2,359,000, 2,297,000 and 22,070,000, respectively. Algae cultivated using aquaculture is given more attention here since 96.5% of the global production of algae is from aquaculture. Although a small amount of algae obtained from wild stock and some countries have implemented policies to conserve the population of certain species of algae in the areas where they grow naturally and contribute significantly to the aquatic ecosystem, aquaculture remains the main method of cultivating algae for large-scale production for either direct consumption as food or for extraction of biocompounds.

Although alginate is the most common biopolymer extracted from brown algae, processes have been presented which allow the extraction of laminarin, fucoidan and alginate from brown algae in a single process. Such developments make it possible to optimize the utilization of the brown algae resource and minimize the component which is discarded as waste. Furthermore, since the species of brown algae required for alginate production are the same which are used for laminarin production, the availability of brown algae for laminarin production is further maintained as it is a raw material for multiple purposes.

9.5   Extraction of Laminarin

As a soluble polysaccharide, laminarin can be extracted by dissolving out in a suitable solvent at specific conditions within which it is soluble. Laminarin is extracted from the thallus of the brown algae. Extraction method commonly used for laminarin is summarized in Fig. 9.3.

The highly branched laminarins are soluble in both hot water and cold water, while the laminarins with less branching are only soluble in hot water. This difference in solubility can be used as a basis for obtaining laminarin of desired level of branching. Extraction is generally achieved by treatment with hot or cold acids or calcium chloride. Acids commonly used are hydrochloric acid or sulfuric acid. Laminarin is then precipitated with alcohol. The extract can then be redispersed in a suitable solvent and then purified. Most common purification presented in many research studies is the ultrafiltration and dialysis.

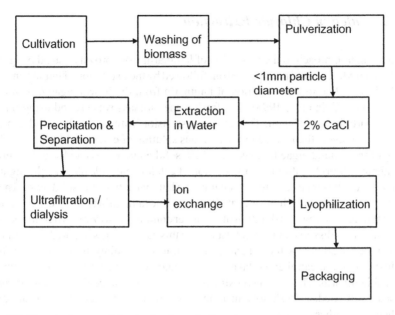

Fig. 9.3 Flowchart summarizing laminarin production stages

9.5.1 Acid Extraction

Acid-based extraction is more commonly used. The process involves the treatment of the brown algae biomass with acidic solution, and this could be hydrochloric acid or sulfuric acid at a 0.1 M concentration at high temperature during which the laminarin is released in the water (Kadam et al. 2015). This usually contains a mixture of laminarin and other soluble polysaccharides. The laminarin is then precipitated out with alcohol. Further, separation is required to obtain purified form of laminarin, and this is done using methods such as ultrafiltration, gel permeation chromatography or dialysis. Laminanrin's lower molecular weight is used as a basis to separate it from higher molecular weight polysaccharides which also dissolve in the same condition.

In a method presented for extraction of laminarin from *Laminaria saccharina* (Yvin et al. 1993), 300 g of the algae biomass was ground in a cryogenic grinder at −40 °C. This ground the biomass into smaller particle sizes with 50–100 μm average diameter. This is then followed by treatment in 900 ml of sulfuric acid at a 0.3% concentration at a temperature of 80 °C for a duration of 1 h under continuous stirring. The mass of algae used here is on wet basis with a solid content of 10–12%. In the completion of the extraction, the system was neutralized by adding an alkali. This is then followed by treatment with polyvinylpyrrolidone at a 1 g/ml for a duration of 2 h. The mixture is separated using vacuum filtration, and further, purification is achieved using ultrafiltration and dialysis. The final purified product is lyophilized to form a dry powder. Using this process, 7 g of dry powder is obtained from 300 g of the feedstock.

9.5.2 Calcium Chloride Extraction

In an another approach to the extraction of laminarin from brown algae, the alginate can first be extracted and precipitated out followed by the extraction of laminarin. This method as applied to the extraction of laminarin from the brown algae *A. nodosum* is as follows (Yvin et al. 1993): The fresh algae biomass was ground using mill to smaller particle size <1 mm. This gave a suspension with a 10–12% solid content. The groundmass is then added to 2% aqueous solution of calcium chloride. For this, 300 g of the ground algae biomass (10–12% solid content) was added to 900 ml of the calcium chloride solution. This addition of calcium chloride allowed the alginate to precipitate. It is then required to separate the laminarin from the solid mass. This is achieved by extraction in hot water at 60 °C since the laminarin is soluble in hot water. The extraction is allowed a duration of 7 h under continuous stirring. After this period, the laminarin is dissolved in the hot water, and this can then be separated by filtration. Further, purification is achieved using ultrafiltration and dialysis. The residue from the first extraction can then be taken through another extraction process to remove residual laminarin. This repeated extraction is necessary since the solvent overtime becomes saturated; a fresh solvent is then required to increase the concentration gradient for further extraction.

9.5.3 Enzyme Extraction

Laminarin can also be extracted from brown algae through the use of enzymes. These enzymes can either be enzymes which digest the non-polysaccharide components of the brown algae biomass leaving behind laminarin in a mixture with other polysaccharides. This mixture of polysaccharides can then be separated by hydrothermal treatment, precipitation and/or molecular separation methods. Another approach involves the partial degradation of the laminarin into even shorter-chain laminarin oligosaccharides using laminarin degrading enzymes (Ojima et al. 2018). This form of extraction is desirable where laminarin oligosaccharides are desired for their bioactive properties. For example, laminarin oligosaccharides extracted from the brown algae *E. bicyclis* showed immunomodulatory effects by promoting the immune response of monocytes (Ojima et al. 2018). There is the β-1,3 glucanase which degrades glucans by breaking some specific forms of β-1,3-linkages. The form of glucanase enzymes which are specific to laminarins is generally called laminarinase. The different glucanases differ in their mode of actions and are bond-specific. The endo-1,3;1,4-β-glucanase cleaves the β-1,3-glycosidic bond as well as the β-1,4-linkage while another type of glucanase the lichenase splits only specific terminal β-1,4-linkage and glucan endo-1,3-β-D-glucosidase would break the β-1,3-linkage but only when at least two adjacent β-1,3-linkages are present (Ojima et al. 2018). These enzymes can be obtained from sources such as barley and baker's yeast (Hrmova and Fincher 2009).

Enzyme extraction often results in the production of oligolaminarins and some sugars. Enzyme extraction is carried out at temperatures of ~30 °C. The system is then heated to 100 °C upon completion in order to deactivate the enzymes (Ojima et al. 2018).

9.5.4 Combined Extraction of Laminarin and Other Brown Algae Polymers

For optimal utilization of the aquatic resources such as brown algae for biopolymer extraction, sustainable extraction methods, whereby extraction of one polymer does not completely destroy or inhibit the extraction of the other useful polymers within the brown algae, are very important.

The conventional approach to the extraction of biopolymers from algae is to extract a single product by either dissolving away all the biopolymers from the fiber or isolating the required biopolymer by methods such as precipitation in alcohol. However, in an approach to optimize the use of the biomass, some research efforts have been directed at combined extraction processes which allow the extraction of multiple biopolymers in a single process from the same algae biomass. Higher combined yield of polysaccharides can be achieved by combined extraction of laminarin, fucoidan and alginate from brown algae; however, the molecular weights and other properties such as viscosity and solubility differ from the conventional single extraction method (Abraham et al. 2019). Combined extraction of laminarin fucoidan and alginate has been carried out of species such as the *Durvillaea potatorum* obtaining a combined yield of 43.57% of polysaccharides compared to a single yield of 38.97% of alginate. The combined extraction process required a modified biorefinery process which could achieve the extraction of all three polysaccharides from a single source where the extraction of one component does not hinder or destroy the other products. The scale-up of such process in a large-scale continuous process requires further modification to the entire production process. The cost of the additional unit operations to extract multiple biopolymers should also be weighed against the value of the biopolymers being extracted.

9.6 Environmental Implications

Here we look at some key environmental issues associated with laminarin extraction process from the cultivation of the brown algae to the final product leaving the factory gate.

9.6.1 Cultivation of Brown Algae

Brown algae have been harvested for centuries in different parts of the world where it has been used in food and agriculture and other uses. Among the brown algae, Laminaria has especially higher ability to extract minerals from polluted water and use these as nutrients (Kim and Bhatnagar 2011). One main advantage of brown algae, and algae in general, as a feedstock in biorefinery for the production of compounds such as laminarin has the advantage of not requiring large land area for cultivation. Algae also have a much higher productivity than land plants, and they can grow up to 83,000,000 tonnes per hectare annually, compared to land plants such as sugarcane whose rate of growth are around 3–30,000,000 tonnes per hectare annually (Konda et al. 2015). Therefore, the cultivation of brown algae is less demanding on the environment than traditional crops such as corn and cotton which also have commercial applications.

9.6.2 Energy Consumption

One of the advantages of laminarin in terms of the extraction process is that highly branched laminarin can be obtained at low temperature in cold water extraction process. Even the hot water extraction process required for the less branched laminarins can be carried out at temperatures as low as 60 °C. Compared with other extraction process where temperatures of ~90 °C are required.

When the algae needs to be cultivated in a closed system to maintain sterility, artificial growth light and temperature and humidity control all required additional energy consumption unlike where the brown algae is harvested from the wild. Where laminarin is required for its bioactive properties such closed growth systems might be required to meet regulatory standards.

9.6.3 Water Consumption

The relatively lower viscosity of laminarin compared to other higher molecular weight polymers such as alginate has the advantage of requiring less water for washing and separation. Unlike in alginate extraction, for example, where a large amount of water is added to reduce the viscosity in order to allow flow in filtration. The process of cultivating brown algae for laminarin production does not use water in the same way as used in the cultivation of land plants. The water used in aquaculture for algae is often recycled, and in some cases, the algae are grown in a symbiotic relationship with fish. Table 9.1 only considers the water used in the extraction process. This water is acidified with the addition of sulfuric acid, and it then needs to be

Table 9.1 Summary of consumptions in laminarin extraction process

Consumption	Amount
Brown algae	42 g per gram of laminarin powder
Energy	>80 °C for 1 h
Water	>170 mL per gram of laminarin powder
Acid	129 mL at 3% w/v
Polyvinylpyrrolidone	1 g/mL

neutralized and treated before discharge or reuse. This water treatment process even when recycled requires additional energy usage.

9.6.4 CO₂ Emission

While the process of cultivation of brown algae is considered as carbon negative, i.e., results in removal of CO_2 for the environment, on the other hand, the process of extraction results in the release of CO_2. These two processes are able to balance each other out, particularly if the process of extraction is carried out in an optimally sustainable manner such that the production of laminarin can be considered a carbon-neutral process.

9.6.5 Acid Consumption

When compared to a process like extraction of chitin from mushrooms requires up to 2 M concentration of hydrochloric acid while extraction of laminarin requires a lower acid concentration of 0.1 M. Therefore, laminarin extraction is a milder process on the environment.

9.6.6 Solid Waste Generation

The brown algae biomass comprises of other compounds mainly alginate (32%), mannitol (18%), little or no proteins, ash and salts (35%) (Konda et al. 2015). In the process of extraction of laminarin, these other components are separated and either go into further processing or are discarded and processed as waste. Brown algae are particularly attractive for extraction of biopolymers as they have a lower lignin content which makes the extraction process milder compared to biomass with higher lignin content where more aggressive delignification processes are required. The solid residue generated can be burnt to produce steam for heating in other processes such as heating and powering turbines for the generation of electricity. This way the waste produced is utilized in powering the process.

Laminarin is considered as the secondary biopolymer of brown algae where alginate is the primary biopolymer with well-established commercial application. Extraction of laminarin alongside alginate and fucoidan in a coextraction process will allow optimal utilization of the brown algae biomass according to reported studies. The very important environmental impact of these polymers lies in their serving as alternatives to the non-renewable sourced polymers in similar applications. However, the limitations lie in the cost and materials for production as well as high-cost additional stages such as ultrafiltration. The unique bioactive properties of laminarin could boost its commercial viability as a high-value product required in lower quantity. Table 9.1 gives a summary of typical consumptions in a laminarin extraction process.

9.7   Applications

Laminarin has been explored for a range of applications which includes the production of bioethanol, animal feed and in food processing. The commercial value of biopolymers is significantly increased when they possess bioactive properties. This way they can be sold at low volume and high price. This is more profitable compared to other non-bioactive applications where they are used as feedstock and required in much larger amounts and at a lower cost. Brown algae as a whole have been consumed for centuries in the Asian countries such as Japan, China and Korea as food or medicinal preparations. The algae in general were believed to have several health benefits. The goal of modern medicine is the isolation and purification of the compounds within the algae which have these bioactive properties for optimal effectiveness. Furthermore, separation of these bioactive compounds from the other component of the algae allows optimal utilization of all the parts of the algae for multiple potentially high-value products. For example, extracts of *E. bicyclis* have anticancer, antiallergic and also have some benefits to the cardiovascular system, in particular prevention of undesirable blood clot in the blood vessels through inhibiting pathways which lead to platelet formation and thrombosis (Irfan et al. 2018). As these extracts contain a mixture of compounds, separation of these combines could identify what compounds have specific effects and contribute to optimal use of this aquatic resource.

9.7.1   Bioethanol Production

The ability to produce ethanol from laminarin is significant in the sense that it provides an alternative non-food source for ethanol production. Compared to other glucose sources such as corn and sugarcane which are used more commonly for commercial ethanol production, laminarin from algae source does not compete with staple foods and does not compete with land. Although several species of algae are used as food in many Asian countries, in many other parts of the world they are not consumed as food and therefore not likely to pose a threat to food security.

The limitation to the use of polysaccharides such as cellulose for ethanol production is that there are limited microbes which can break down cellulose into glucose. Microbes such as *Saccharomyces cerevisiae* can break down sucrose, a much simpler structure into glucose and metabolize this and produce ethanol as a by-product in a single process. Inclusion of an additional stage in the production process for breaking down the feedstock into glucose or simple sugars will significantly add to the cost of production.

There are microbes which are capable of fermenting laminarin to produce ethanol such as *Pichia angophorae* (Horn et al. 2000) yeast *S. cerevisiae* and bacterium *Saccharophagus degradans* (Motone et al. 2016). In an age, where fossil fuel depletion is a major concern, new alternative ways to produce fuel are vital to the survival of the modern world which is heavily dependent on fuel consumption. Production of bioethanol from biomass depends on a carbon source and a microbe which synthesizes the enzyme to break down the carbon source for its metabolism and in the process produces ethanol as a by-product. Laminarinase enzyme extracted from organisms such as the baker's yeast *S. cerevisiae* and bacterium *S. degradans* can degrade laminarin into its sugar units. In a coculture of these two organisms with a medium containing 20 g/L of laminarin, 5.2 g/L of ethanol is obtainable (Motone et al. 2016). In a separate study in a batch fermentation process, a yield of 0.43 g ethanol per g substrate was achieved at a pH of 4.5 and oxygen level of 5.8 mmol L^{-1} h^{-1} (Horn et al. 2000).

One of the limitations of bioethanol production from laminarin is the relatively lower yield and preference of the microbe for batch systems (Horn et al. 2000). For large scale and lower cost of production, scaled up continuous process is required by the biorefineries. For the successful production of bioethanol from laminarin, the cost of extraction of laminarin from brown algae needs to be minimized as bioethanol is a high volume and low-cost commodity and feedstock production needs to be as cheap as possible. Compared to, for example, sugar extraction from sugarcane which mainly requires mechanical processes and thus less cost of chemicals for extraction, extraction of laminarin requires additional cost of chemicals such as sodium hydroxide or calcium carbonate for extraction. Additional unit operations such as ultrafiltration also add to the cost of laminarin as a feedstock for ethanol production. Therefore, although a one-step fermentation process to obtain ethanol from laminarin is possible, the cost implications presently limit the commercial application.

9.7.2 Gut Health

Laminarin has potential to serve as a dietary fiber in human nutrition with some bioactive benefits. The possibility of laminarin acting in the modulation of the composition and function of the gut is one which has been investigated by researchers in a number of reported in vitro and in vivo studies. Although they are indigestible by any of the known digestive enzymes of the human body and cannot be broken down into smaller units for metabolism, their passage through the alimentary canal can potentially pose some health benefits. In an in vitro study which modeled the human digestive system based on an anaerobic batch fermenter system which replicates the microflora and biochemical composition of the human gut system, the variation in the composition with the inclusion of laminarin was investigated. The study revealed that laminarin is digested by the bacteria in the human gut by >90%. In further studies on rats, it was shown that laminarin affected the mucus content which in turn affects bacterial attachment to the walls of the intestine. Laminarin also affects the pH and biochemical composition of the gut, particularly the short-chain fatty acid content (Deville et al. 2007). These parameters, which can be affected by laminarin, contribute to the metabolism in the intestine. Laminarin has a pH lowering effect in the intestine, and this lower pH is beneficial to metabolism, it promotes beneficial bacterial growth, enhances the absorption of minerals and has been associated with cancer prevention.

One of the attractions of dietary fibers is the possibilities of having probiotic effects, that is, enhancing the growth of beneficial bacteria in the human digestive system, mainly Bifidobacterium and Lactobacillus (Deville et al. 2007). There are conflicting reports as to whether laminarin has prebiotic effects or not (Michel et al. 1999; Deville et al. 2007).

The studies on the effect of laminarin on gut health have advanced to in vivo test on mammals such as piglets. For example, studies on piglets following weaning stage of growth showed that laminarin improved their resistance to the pathogenic microbes, *Salmonella typhimurium* (Bouwhuis et al. 2017). This can be achieved either by administering laminarin directly to the piglet or to the show during weaning and then transferred to the piglet in the breast milk. This was achieved with a dosage of 300 ppm laminarin per tonne of feed; therefore, this improvement in gut health can be achieved with a relatively low amount of laminarin. The improvement in gut health is due to a combination of factors; the laminarin either induced the expression of genes which gave the piglet improved defense against the pathogen or the direct antimicrobial effect of laminarin fed in the diet on *S. typhimurium* in the gut or it improved the architecture of the gut, making it more effective at eliminating the pathogenic microbe.

9.7.3 Anticancer Activity

There have been several studies which investigate the anticancer effect of laminarin and the evidences presented show that laminarin has promising use as a functional food with anticancer properties. Cancer prevention and treatment occur by many routes; anticancer drugs and compounds can act by either inducing cell death or enhancing the body's own defense against the cancer cells. The anticancer and anti-tumor bioactivities of laminarin have been demonstrated in human melanoma cells, colon cancer cells and adenocarcinoma cells of rats among others. For example, apoptosis (programmed cell death) of LoVo human metastatic colon cells has been activated by laminarin, and laminarin hydrolysates have shown ability to inhibit growth of some human melanoma and colon cancer cells (Ji and Ji 2014). In other studies, laminarin has been shown to prevent the spreading of cancer cells beyond the point of origin to other tissue (metastasis) by inhibiting the cancer cells' ability to disrupt the extracellular matrix, hence preventing invasion (Deville et al. 2004). It should be noted that the studies mentioned here and much of the existing studies on the anticancer effects of laminarin have only been carried out at the cellular level and on laboratory rats. The findings point to some potential of laminarin as a functional dietary fiber which the intake could yield these anticancer effect. This is a long way from drug development where the route by which the active compound gets to the cell in a more complex organism, such as mammals and humans, introduces several challenges.

9.7.4 Immunomodulatory Effect

Disease resistance is important in animal farming. Spreading of disease could result in devastating financial loss of livestock and could also reduce the growth rate and hence the yield of the animal. Low-cost disease management through boosting the immunity of animals is an effective means of disease management and prevention. Laminarin has a low molecular weight glucan that has been reported to have immune-modulatory effect in grouper fish (Yin et al. 2014). Laminarin immunological effect is also dose-dependent; fish fed with a concentration of laminarin of 0.5, 1 and 1.5% over a period of 48 days. Fish fed with a diet-containing laminarin showed an increased specific growth rate, and they were able to more effectively utilize nutrients. Furthermore, an increase in the number of immune response cells were observed in fish fed with laminarin compared to those with laminarin-free diets. Boosting immune response in fish through the nutrient present in the feed adds economic value to the fish feed and also presents the fish farmer with lower risk of fish developing diseases and improved yield.

Laminarin has been explored extensively for its effect on the growth and develop-ment of pigs at different developmental stages, from development of young piglets to the maternal health of the sow. Laminarin has even been shown to improve the development of porcine embryo (Jiang et al. 2018).

Other species have also shown improved growth rate in response to laminarin administration. These include young chicks and piglets. For example in post-hatch chicks, it is important that they have a moderate growth rate and attain sufficient immunity to survive the early life stage as lots of the livestock could be lost during this fragile stage. It is possible to achieve this required immunity through diet-containing immune modulatory agents. Laminarin tested on such post-hatchery chicks have shown that chicks could grow faster and develop intestine with more efficient nutrient uptake and immune response (Sweeney et al. 2017). However, the immunomodulatory effect of laminarin seems to be limited as it did not improve resistance to *Campylobacter jejuni*, a major pathogen. Therefore, while laminarin might provide some immunomodulatory effect, it does not provide complete immunity against pathogens.

The immunomodulatory bioactivity of laminarin can be partly attributed to the ability of laminarin like other algae polysaccharides to bind with certain proteins called the lipopolysaccharide and beta-1,3-glucan-binding protein (LGBP). These proteins are able to recognize specific polymer chain patterns of these algae polysaccharides such as laminarin. These patterns are referred to as the pathogen-associated molecular patterns (PAMPs). It is deduced that the LGBPs bind with laminarins to form complexes which then aid in the ability to attach to and destroy pathogens. This protein binding ability of laminarin has been investigated in shrimp, resulting in the activation of the prophenoloxidase system which boosts the shrimp's innate immunity (Chen et al. 2016). Most of the studies on laminarin's immunomodulatory bioactivity are toward its use as a functional food for optimizing productivity in farmed animals.

9.7.5 Plant Growth Agent

While the use of fertilizers and manure provides nutrients to support plant growth, the growth of the plant also depends on the ability of the plant to effectively take up and utilize the nutrients, water and sunlight. This is down to the intrinsic functioning of the plant. This intrinsic growth factor can be enhanced by using compounds which can interact with the plant's cellular function and enhance its metabolic and hence growth process as well as defense against pathogenes. Laminarin has been shown to accelerate the process of seed germination through interfering with pathways which lead to the formation of alpha-amylase (Yvin et al. 1993). Laminarin is thought to do this by acting as an elicitor: a compound which induces particular responses in an organism which enables its ability to survive its environment. This occurs by difference means such as by triggering the production of certain enzymes or initiating or inhibiting certain biochemical processes. Laminarin can be used in powdered or liquid form either mixed with additives or dispersed in water and applied to the plant. Concentrations of 0.005–100 g per liter have been presented for plants such as tomatoes, potatoes and carrots in a patented formulation (Yvin et al. 1993)

The use of laminarin to accelerate seed germination and plant growth has significant economic benefits. Plants which germinate faster would reduce the time required to be spent in the nursery thereby reducing the amount of intensive care required for the plants. It also allows for optimal utilization of nutrients and fertilizers. The use of laminarin which is from a natural source could also be used as a means to accelerate growth in organic farming, depending on the regulation of the region.

9.7.6 Food Processing and Preservation

The shelf life of food products plays a major role in their marketing and distribution as well as the use. Fresh meat products are required to retain their organoleptic properties such as taste and texture throughout the period of storage and consumption. The organoleptic properties of these foods are affected by the chemical and microbial composition. It has been observed that the food fed to the animals during rearing significantly affects the organoleptic properties of the meat after slaughter. One of the main factors which lead to loss of organoleptic properties of meat is the lipid content. Before the action of spoilage microbes the loss of quality of the meat occurs due to the deterioration of other food quality index. The oxidation of the lipid within the meat leads to rancidity. Furthermore, as consumers are becoming increasingly health conscious, meat products with lower saturated fats are increasing in demand. In one study, it is reported that pigs fed with a diet including 450 mg of l per kg of feed laminarin for a period of 6 weeks have lower saturated fat content and lower lipid oxidation in storage (Moroney et al. 2015). Although the reduced lipid oxidation from laminarin feed in the pork did not occur in cooked minced pork, cooked minced pork from pigs fed with a diet-including laminarin showed reduced saturated fatty acid content.

9.7.7 Antimicrobial Activity

Laminarin extracted from *E. bicyclis* and *L. digitata* shows some antimicrobial activity against gram-positive and gram-negative bacteria. *E. bicyclis* with a higher degree of branching showed a stronger antimicrobial activity against gram-negative bacteria. There are results indicating that a higher degree of branching could support the antimicrobial activity of laminarins (Liu et al. 2018). However, a weaker activity was observed against gram-negative bacteria. Antimicrobial activity of laminarin is based on early stage studies. It is mentioned here as a potential future application of laminarin.

Other bioactive properties of laminarin include anticoagulant, hypocholesterolemia. However, although studies in the late 50s found that laminarin, when administered intravenously, has anticoagulant properties comparable to that of heparin, it was also extremely toxic to the rabbits' tested on at the pharmacokinetically relevant doses (Thorpe and Adams 1957). Consequently, no further studies have been carried out on the anticoagulant application of laminarin. Oral administration of laminarin for food and other bioactive properties do not pose the toxic effect as when delivered directly to the blood intravenously. The first-pass metabolism through the alimentary canal abates this.

9.8   Commercial Production

Macroalgae cost contributes to about 24% of the total annual running cost of macroalgae-based biochemicals refinery for bioethanol and biochemical production (Konda et al. 2015). The rest of the annual running cost goes into the purchase of consumables, facilities, labor and utilities. A megatone of brown algae is estimated to be priced at 100 USD, if laminarin is processed into sugars, the minimum selling price for the sugars produced from a corefinery of laminarin alongside alginate would be 21–47 cents per pound. If fermented to ethanol, the minimum selling price for the ethanol will be 3.6–8.5 USD per gallon (Konda et al. 2015). This costs much higher than fossil fuels or ethanol from conventional process. The commercial application of laminarin is therefore of more economic relevance in the bioactive applications where the high value can compensate for the relatively higher production cost.

As research interest increases in the numerous benefits of laminarin as a potential bioactive compound, so does demand for analytical grade laminarin which is used to study the structure, properties, develop new applications and better understanding of the structure property relations of laminarin. Companies such as Sigma-Aldrich, Sigma Chimie SARL and Seikagaku Kogyo supply laminarins of different specifications. These have been used in various studies on laminarins (e.g., Ojima et al. 2018; Yvin et al. 1993). These supply companies therefore play a significant role in the advancement of the research in laminarin through ensuring their availability. In studies focused on the structure and applications of laminarin, several hours and resources could be saved in repeated extractions by buying already prepared samples.

9.9   Conclusion

Laminarin, another biopolymer obtained from the brown algae, has several bioactive properties which makes it a valuable biopolymer in the food and animal feed industry. Its lower molecular weight makes separation by ultrafiltration—an useful method in its extraction. It can be extracted either as a sole product or alongside other polysaccharides in the brown algae such as alginate and fucoidan. It seems more attention and has been placed on the application of laminarin in animal feed as a feasible commercial application in the near future. Application for human use includes food preservation and in the production of bioethanol. The environmental impact of the extraction process lies in the consumption of fossil energy to power the processes, the land-use change and CO_2 emissions in the transportation and packaging processes. This is weighed against the CO_2 consumed during the process of cultivation of the algae, the use of algae to clean polluted water and the potential for use in the replacement of fossil-based fuel if the process of bioethanol production from laminarin is commercialized.

References

Abraham RE, Su P, Puri M, Raston CL, Zhang W (2019) Optimization of biorefinery of alginate, fucoidan and laminarin from brown seaweed *Durvillaea potatorum*. Algal Res 38. Article 101389

Beattie A, Hirst EL, Percival E (1961) Studies on the metabolism of the Chrysophyceae. Comparative structural investigations on leucosin (chrysolaminarin) separated from diatoms and laminarin from brown algae. Biochem J 79:531–537

Bouwhuis MA, Sweeney T, Mukhopadhyay A, McDonnell MJ, O'Doherty JV (2017) Maternal laminarin supplementation decreases *Salmonella typhimurium* shedding and intestinal health in piglets following an experimental challenge with *S. typhimurium* post-weaning. Anim Feed Sci Technol 223:156–168

Caballero MA, Jallet D, Shi L, Rithner C, Zhang Y, Peers G (2016) Quantification of chrysolaminarin from the model diatom *Phaeodactylum tricornutum*. Algal Res 20:180–188

Caipang CMA, Lazado CC (2015) Nutritional impacts on fish mucosa: immunostimulants, pre- and probiotics. In: Beck BH, Peatman E (eds) Mucosal health and aquaculture, pp 211–272

Chen Y, Chen J, Kuo Y, Lin Y, Huang C (2016) Lipopolysaccharide and β-1,3-glucan-binding protein (LGBP) bind to seaweed polysaccharides and activate the prophenoloxidase system in white shrimp *Litopenaeus vannamei*. Dev Comp Immunol 55:144–151

Chetia L, Kalita D, Ahmed GA (2017) Synthesis of Ag nanoparticles using diatom cells for ammonia sensing. Sens Bio-Sensing Res 16:55–61

Deville C, Damas J, Forget P, Dandrifosse G, Peulen O (2004) Laminarin in the dietary fibre concept. J Sci Food Agric 84:1030–1038

Deville C, Gharbi M, Dandrifosse G, Peulen O (2007) Study on the effects of laminarin, a polysaccharide from seaweed, on gut characteristics. J Sci Food Agric 87:1717–1725

FAO (2018) The state of world fisheries and aquaculture 2018: meeting the sustainable development goals. Rome. Licence: CC BY-NC-SA 3.0 IGO. ISBN 978-92-5-130562-1

Hildebrand M, Manandhar-Shrestha MK, Abbriano R (2017) Effects of chrysolaminarin synthase knockdown in the diatom *Thalassiosira pseudonana*: implications of reduced carbohydrate storage relative to green algae. Algal Res 23:66–77

Horn SJ, Aasen IM, Ostgaard K (2000) Ethanol production from seaweed extract. J Ind Microbiol Biotechnol 25(5):249–254

Hrmova M, Fincher GB (2009) Plant and microbial enzymes involved in the depolymerization of (1,3)-β-D-glucans and related polysaccharides. In: Bacic A, Fincher GB, Stone BA (eds) Chemistry, biochemistry and biology of 1-3 beta glucans and related polysaccharides. Academic Press, pp 119–170

Irfan M, Kwon TH, Yun BS, Park NH, Rhee MH (2018) *Eisenia bicyclis* (brown alga) modulates platelet function and inhibits thrombus formation via impaired P2Y12 receptor signaling pathway. Phytomedicine 1(4):79–87

Ji CF, Ji YB (2014) Laminarin-induced apoptosis in human colon cancer LoVo cells. Oncol Lett 7(5):1728–1732

Jiang H, Liang S, Yao XR, Jin YX, Kim NH (2018) Laminarin improves developmental competence of porcine early stage embryos by inhibiting oxidative stress. Theriogenology 115:38–44

Kadam SU, Alvarez C, Tiwari BK, O'Donnell CP (2015) Extraction of biomolecules from seaweeds. In: Tiwari BK, Troy DJ (eds) Seaweed sustainability. Academic Press, pp 243–269

Kim SK, Bhatnagar I (2011) Physical, chemical and biological properties of wonder kelp—laminaria. In: Kim SK (ed) Marine medicinal foods. Advances in food and nutrition research, vol 64, pp 85–96

Kim EJ, Fathoni A, Jeong G, Jeong HD, Kim JK (2013) *Microbacterium oxydans*, a novel alginate- and laminarin-degrading bacterium for the reutilization of brown-seaweed waste. J Environ Manage 130:153–159

Konda M, Singh S, Simmons BA, Klein-Marcuschamer D (2015) An investigation on the economic feasibility of macroalgae as a potential feedstock for biorefineries. Bioenergy Res 8:1046–1056

Liu Z, Xiong Y, Yi L, Dai R, Yuan S (2018) Endo-β-1,3-glucanase digestion combined with the HPAEC-PAD-MS/MS analysis reveals the structural differences between two laminarins with different bioactivities. Carbohyd Polym 194:339–349

Martins CR, Custodio CA, Mano JF (2018) Multifunctional laminarin microparticles for cell adhesion and expansion. Carbohyd Polym 202:91–98

Michel C, Bernard C, Lahaye M, Formaglio D, Kaefer B, Quemener B (1999) Algal oligosaccharides as functional foods: in vitro study of their cellular and fermentative effects. Sci Ailm 19:311–332

Misurcova L, Skrivankova S, Samek D, Ambrozova J, Machu L (2012) Health benefits of algal polysaccharides in human nutrition. Adv Food Nutr Res 66:75–145

Moroney NC, O'Grady MN, Robertson RC, Stanton C, Kerry JP (2015) Influence of level and duration of feeding polysaccharide (laminarin and fucoidan) extracts from brown seaweed (*Laminaria digitata*) on quality indices of fresh pork. Meat Sci 99:132–141

Motone K, Takagi T, Sasaki Y, Kuroda K, Ueda M (2016) Direct ethanol fermentation of the algal storage polysaccharide laminarin with an optimized combination of engineered yeasts. J Biotechnol 231:129–135

Ojima T, Rahman MM, Kumagai Y, Nishiyama R, Narciso J, Inoue A (2018) Polysaccharide-degrading enzymes from marine gastropods. Methods Enzymol 605:457–497

Rioux LE, Turgeon SL (2015) Seaweed carbohydrates. In: Tiwari BK, Troy DJ (eds) Seaweed sustainability. Academic Press, pp 141–192

Stone BA (2009) Chemistry of β-glucans. In: Bacic A, Fincher GB, Stone BA (eds) Chemistry, biochemistry and biology of 1-3 beta glucans and related polysaccharides. Academic Press, pp 5–46

Sweeney T, Meredith H, Vigors S, Mcdonnell MJ, Ryan M, Thornton K, O'Doherty JV (2017) Extracts of laminarin and laminarin/fucoidan from the marine macroalgae species *Laminaria digitata* improved growth rate and intestinal structure in young chicks, but does not influence *Campylobacter jejuni* colonisation. Anim Feed Sci Technol 232:71–79

Thorpe HM, Adams SS (1957) The anticoagulant activity and toxicity of laminarin sulphate K. J Pharm Pharmacol 9(1):459–463

Wang H, Xu Z, Wu Y, Li H, Liu W (2018) A high strength semi-degradable polysaccharide-based hybrid hydrogel for promoting cell adhesion and proliferation. J Mater Sci 53:6302–6312

Yin G, Li W, Lin Q, Lin X, Lin J, Zhu Q, Jiang H, Huang Z (2014) Dietary administration of laminarin improves the growth performance and immune responses in *Epinephelus coioides*. Fish Shellfish Immunol 41(2):402–406

Yvin JC, Levasseur F, Amin-Gendy DCP, Tran TK, Patier P, Rochas C, Lienart Y, Cloarec B (1993) Laminarin as a seed germination and plant growth accelerator. EP0649279A1

Chapter 10
Aquatic Plants and Algae Proteins

Abstract Proteins are quite diverse group of polymers. For this reason, this book divides proteins into three chapters—the present chapter on proteins from algae and aquatic plants, a chapter on collagen and a third chapter on enzymes. Proteins from aquatic plants and algae are of particular importance since they serve as a source of plant protein which do not require land space for cultivation. As the demand for protein for nutrition rises alongside global population and arable land spaces decrease due to desertification and demand for living space, an alternative way to grow protein is in the aquatic environment. To this end, this chapter reviews the process of extracting proteins from aquatic plants and algae and discusses the environmental and economic significance. In the process, chemistry and applications are also discussed.

Keywords Proteins · Algae · Aquatic · Phycobiliproteins · Biopolymer

10.1 Introduction

Proteins are rather ubiquitous; they are found in all living things. Described most simply as the building blocks of life, proteins are biopolymers that can be found in a wide range of organisms across the five kingdoms—animals, plants, protista, fungi and prokaryotes. Aquatic-sourced proteins are applicable in various industries from food to cosmetics to biomedical and many others. They have been sourced for their nutritional value in food, skin enhancing effects in cosmetics, productions of scaffolds in tissue engineering, catalyzing chemical processes, in the production of biofuel, in cancer treatment and much more. Within this book, proteins are covered in three chapters—the present chapter which covers proteins from aquatic plants and algae, a chapter on collagen and another chapter covering enzymes.

Aquatic plants such as duckweed and water hyacinth have been investigated for their protein components and how these can be used in applications such as animal and human nutrition as a source of protein is also investigated (Adeyemi and Osubor 2016; Aguilera-Morales et al. 2018). Algae, which include both microalgae and macroalgae, have so far gained more attention as aquatic source of proteins. The microalgae include the diatoms, the blue-green algae, the golden algae and the

© Springer International Publishing 2020   211
O. Olatunji, *Aquatic Biopolymers*, Springer Series on Polymer and Composite Materials,
https://doi.org/10.1007/978-3-030-34709-3_10

microscopic green algae (Bacillariophyta, Cyanophyta, Chrysophyta and Chloro-phyta, respectively). All of these algae are a diverse group with thousands of species. This gives the possibility for a largely diverse range of proteins.

Their ability to grow in the aquatic environment means these aquatic organisms have adapted some features which are distinct from terrestrial plants. Every different feature means these plants and algae have different proteins which are different from those of animals, since these proteins are important in developing these specific features. The aquatic environment therefore serves as a promising source of new proteins and peptides or at the very least a more productive source, considering the higher biomass accumulation rate of aquatic plants and algae compared to those of terrestrial plants.

Aquatic plants and algae continue to play an increasingly important role in the livelihoods of individuals, they can serve as a source of food, means of livelihood and their economic value can be significantly increased through exploring their protein content. These proteins can be further processed into even higher value commodities such as functional foods and bioactive agents in supplements and drugs. This chapter explores some of the aquatic plants and algae, their protein components and existing and potential commercial applications of these proteins. Within this chapter, aquatic plants and algae proteins are reviewed as the total protein content in the biomass. Sub-sequent chapters will then explore specific proteins, namely enzymes and collagen in separate chapters.

It should be noted that while some of these plants and algae could serve as a source of beneficial proteins, their mismanagement or the mismanagement of the aquatic environment within which they grow could lead to adverse commercial, health and environmental impact. A particularly common one of such is the algae bloom which results in release of toxins into the environment which are harmful to human beings (Pulz and Gross 2004). This has become a serious environmental concern with increasing frequency in aquatic habitats such as along the coast of the Yellow Sea in the Jiangsu Province and Shandong Province in China (Zhang et al. 2019).

As the world population continues to increase so does the demand for food. Protein forms a large part of the daily nutritional requirement, and much of these come from terrestrial plants and animals. Aquatic source of protein is mainly fish and aquatic animals. Algae and aquatic plants can provide an alternative or additional protein source which do not require land for cultivation and grow at a much faster rate as well as offer some additional health impacts.

10.2 Occurrence in Nature

Aquatic plants and algae contain considerable amounts of protein which make them commercially relevant sources of protein. For example, the blue-green cyanobac-teria Spirulina contain between 60 and 70% protein by dry weight of its biomass (Yucetepe et al. 2018). Many aquatic organisms produce more than one biopolymer

Table 10.1 Protein yields from some aquatic plants and algae

Organism	Protein yield (%)	References
Kappaphycus alvarezii	12.69–23.61	Kumar et al. (2015)
Undaria pinnatifida	15	Suetsuna and Nakano (2000)
Azolla (water fern)	28	Brouwer et al. (2019)
Lemna gibba duckweed	21.5	Aguilera-Morales et al. (2018)
Ulva lactuca	17.2	Aguilera-Morales et al. (2018)
P. tenera (red algae)	47	Fleurence (1999)
Nannochloropsis gaditana (microalgae)	17.4 and 24.8	Safi et al. (2017)
Spirulina platensis (blue-green algae)	29.05	Yucetepe et al. (2018)

of commercial value. The rate of accumulation of each polymer varies; furthermore, the optimal conditions for maximum accumulation of one polymer are most likely not the optimal condition for the others. For example, the red marine alga *Porphyridium marinum* contains up 46% protein by dry weight when grown in medium with 17.6 mM potassium nitrate. However, when grown with 5.9 mM potassium nitrate, the protein content is 35% and when grown with 3.5% mM potassium nitrate, the protein content is even lower at 25% dry weight (Li et al. 2019). This red microalgae accumulates maximum starch at light intensity of 100 μmol photons m^{-2} s^{-1}, sodium nitrate ($NaNO_3$) concentration of 1 g/L and a sodium chloride concentration of 20 g/L (Hlima et al. 2019). However, this condition is not the optimal condition for protein accumulation. Furthermore, while the starch content increases the light intensity and salinity of growth environment increases the starch content in these algae, it does the opposite for the protein content. This is explained to be as a result of the conditions favoring biochemical processes which lead to starch formation that does not favor those biochemical reactions leading to protein formation. Table 10.1 gives yields of protein obtained from different species based on values reported from different studies in the literature.

10.3 Chemistry of Aquatic Proteins

Proteins are polymeric structures made up of polypeptides which are chains made up of several amino acids which are repeating units joined together by amide links that links the amine group with a carboxylic acid group. They are heteropolymers whose repeating units can be trillions of variations of arrangements of 20 different amino units. Unlike other aquatic polymers that have been explored in this book such as fucoidans and carrageenans, proteins are monodisperse; they have uniform molecular weight, hence number of repeating units within their polymer chains.

Proteins differ from one another by the sequences of amino acids in their primary structure, the interaction of the polypeptide chain in the secondary structure and the folding of these chains in the tertiary structure. Some proteins further have a quaternary structure defined by the combination of different folded polypeptide structures by hydrogen bonds. One such protein with a quaternary structure is hemoglobin. Some researchers have explored how to develop other monodisperse polyamides from proteins as a route to achieving monodispersity in polymerization reactions (Yang et al. 2003).

Some proteins from aquatic organisms share the same general chemistry as proteins from terrestrial organisms. Proteins such as collagen, luciferase and amylase are examples of proteins found in land as well as aquatic organisms. However, some of these proteins have some distinct features in terms of their chemical structure that differs from one another. The rest of this section will highlight some of the features specific to aquatic proteins.

Despite some of the adverse effects associated with the consumption of high quantity of meat, animal protein has the highest amount of all essential amino acids required for healthy nutrition for humans. To counter the risks associated with meat consumption such as high fat content leading to cardiovascular disease, it is recommended to consume a diet consisting of mainly plant-based proteins and achieve the required essential amino acids through a combination of a range of plant protein sources such as legumes, grains, fruits and vegetables (FAO, WHO 1991).

Despite being considered as highly nutritious consisting of proteins, vitamins and minerals, amino acid composition of algae is rather limited. Some of the essential proteins, which are not synthesized by humans and required by the body, are not present in algae. This is the general case with plant protein, whereby some essential amino acids are missing in plants of both terrestrial and aquatic. These essential amino acids missing in plants include leucine, histidine, tryptophan, lysine, valine, threonine, methionine and phenylalanine (Young and Pellett 1994). The amino acid missing in particular plant or algae varies from species to species. For example, in red algae, leucine and isoleucine are usually present in low amounts in brown algae cysteine, while lysine and methionine are the missing or limited amino acids in brown algae. Tryptophan and lysine are generally in limited amounts in all algae (Bleakley and Hayes 2017). Amino acids which are more abundant in algae include aspartic acid and glutamic acid which could be up to 22–44% in species Fucus sp. and up to 32% in Ulva sp. (Fleurence 1999).

While algae and aquatic plants cannot serve as a main source of proteins due to the absence of some essential amino acids, they nonetheless can serve as an additional more efficient protein source. Polypeptides from plant sources tend to be cyclic peptides, and these are known to have more bioactive properties than the linear structures. The cyclic structures tend to be more stable than the linear structures (Tan and Zhou 2006). This stability could ensure they retain their secondary structure until they reach the bloodstream, thus allowing them a better chance of eliciting their bioactivities.

Algae and aquatic plants, therefore, can contribute to a more diversified protein source for nutrition and other applications. Due to the increased rate of productivity of aquatic plants and algae, they could contribute a more reliable source of plant type proteins which can be used for nutrition and as bioactive agents.

10.4 Availability of Raw Materials

In 2016, 89,000 tonnes of microalgae was farmed across 11 countries of the world—88,600 tonnes of this was from China. These include species such as *Haematococcus pluvialis*, Nannochloropsis spp., Chlorella spp. and Spirulina spp. All are being farmed in large, medium and small scales (FAO et al. 2018). While macroalgae get a larger revenue from their food sales, microalgae are mostly sold as high-value functional products. Therefore, despite the lower annual tonnes produced, microalgae are valued at around a billion USD annually, compared to that of macroalgae at 6 billion USD (Bleakley and Hayes 2017).

The protein content in macroalgae is comparable to those found in animal-based proteins and is higher than those found in land plants such as soybean, wheat and legumes. Algae yield around 2.5–7.5 tonnes per hectare annually, while microalgae yield 4–15 tonnes per hectare annually. These yields are rather high compared to the conventional plant-based proteins such as wheat, soybeans and legumes which yield 1.1, 0.6–1.2 and 1–2 tonnes per hectare annually (Van Krimpen et al. 2013). Macroalgae and microalgae could contain similar or even more protein that terrestrial plants typically used as protein source. Spirulina, a microalgae which have gained much popularity as a nutrient source, could contain up to 63% protein per dry weight (Tokuşoglu and Üunal 2003). The red algae species *Porphyra tenera* contains up to 47% protein per dry weight (Fleurence 1999).

Protein content in any particular algae varies with factors such as growth season, temperature, harvest period, region and nutrient content of water. The types of proteins present also vary. How different species react to particular changes in growth condition in turn affects the types of proteins they metabolize. The algae make use of these proteins to survive and function within its environment; therefore, the stimulus it gets from these environments determines what protein it is prompted to produce. These factors can be used to manipulate particular algae to produce desired type of protein by controlling the growth environment. This requires an in-depth understanding of the correlation between environmental factors and the metabolism of the specific species. For example, highest protein yield is obtained from the algae *Kappaphycus alvarezii* in August, November and February when studied over 12 months from September 2004 to April 2006 in Northwestern India (Kumar et al. 2015).

Aquatic plants such as water fern, duckweed and water hyacinth contain relatively moderate-to-high amounts of proteins. 28% protein content by dry weight has been reported for water fern (Brouwer et al. 2019) and 21.5% for duckweed (Aguilera-Morales et al. 2018). While there are up to hundreds of thousands of species of algae, some of which belong to the plant kingdom, there are just about a hundred species of non-algae aquatic plants. The recent rise in algae bloom incidents in different parts of the world further eludes to the fast rate of growth of algae. Water hyacinth, ferns and duckweed are also fast-growing plants which can be explored as abundant sources of aquatic proteins. Despite the fact that aquatic plants and algae could serve as an abundant source of proteins, well managed and controlled systems for the cultivation of aquatic plants and algae is important in order to avoid undesirable population blooms which could result in devastating environmental impacts.

10.5 Extraction of Proteins from Algae and Aquatic Plants

Protein extraction from aquatic plant could be with different goals. It could be carried out with the aim of processing the proteins from algae into more digestible or more appealing forms. Example of such is incorporation of algae proteins into snacks (Lucas et al. 2018). The proteins could be extracted for their bioactive properties for use in non-food applications such as cosmetics, pharmaceutical or biotechnology applications. The latter section of this book will explore some of the applications of such proteins from aquatic plants and algae.

Conventional extraction processes can be carried out using water, acids or alkali. Other methods also exist which make use of more advanced techniques, some in combination with the conventional techniques, to improve ease and effectiveness of extraction. One of the factors which limit full commercial exploitation of proteins from algae for processed food is the cost and environmental implications of the extraction process to prepare them into edible forms. However, when considering high-value applications such as for use as therapeutic agents, these added costs could then be compensated.

In algae, proteins are usually present alongside polysaccharides such as cellulose, alginate, ulvans and carrageenans. The extraction strategy could therefore be to either break down these polysaccharides in order to free up the proteins, or to break down the protein into a more soluble form which is then dissolved out of the fibrous structure. This is then followed by other unit operations such as separation and drying.

10.5.1 Physical Extraction

This process involves the application of mechanical force to the biomass which then frees up the water-soluble protein from the fiber. This can be achieved by applying shear through grinding, homogenizing or pressing (Barbarino and Lourenço 2005). It can also be achieved by applying osmotic pressure through immersion in water for an extended period of time (Marrion et al. 2003). Ultrapure water is used to have minimal concentration of minerals in the water and optimize concentration gradient and consequently osmotic pressure. Using this method to extract both water-soluble and insoluble proteins from *Palmaria palmata*, one study reports a yields of 0.0677% for the osmotic method and a yield of 0.0692% (mass protein per dry weight of algae) using shear force (Harnedy and FitzGerald 2013). Extraction yield of up to 40% has been reported using the physical method (Harnedy and FitzGerald 2013).

10.5.2 Chemical Extraction

This method requires the use of an alkali or acidic solution for extraction of the protein. The protein becomes soluble at altered pH and in some cases elevated temperatures. This is likely to obtain the protein in a denatured form. Yields of up to 59% can be obtained using this method (Barbarino and Lourenço 2005). The alkali or acid could also be used to disrupt the cell wall structure, allowing the protein to be released more easily. Sodium hydroxide and hydrochloric acid are commonly used; others include polyethylene glycol, potassium carbonate and N-acetyl-cysteine (Bleakley and Hayes 2017).

In some cases, the protein can be extracted alongside other components which can be used alongside the protein, for example, as animal feed. In this case, the processor can eliminate the cost of separating the protein from these components and simply needs to obtain the protein from the biomass. An example of this is extraction of leaf protein concentrate from the water hyacinth. Here the extraction of the leaf protein concentrate was carried out by first blanching the washed water hyacinth leaves for 5 min with acetic acid at 5% concentration. The blanched leaves where then rinsed with deionized water to get rid of the acids on the surface after which the biomass was dried. The next step is to extract the fats, and this was done by soaking in 95% ethanol within which the fat is soluble. This process was allowed 6 h to allow the alcohol penetrate into the walls and dissolve the fats and other ethanol-soluble compounds. This leaves behind the carbohydrates, proteins, minerals and other compounds such as alkaloids and phenolic compounds. The residual fat in the leap protein concentrate depends on the yield of extraction. This concentrate can then be used as a protein source in applications such as animal feed.

10.5.3 Enzyme Extraction

Another way to extract proteins from the algal biomass is through the use of enzymes. The role of the enzyme is to breakdown the polysaccharides to sugar units which are soluble and separating the protein. The water-soluble proteins can be separated from the water-soluble sugars by separation method such as precipitation of the protein or crystallization of sugars. The enzymes used depend on the type of algae and its components. Enzymes such as carrageenase, xylanase and cellulase are used for extraction of protein from red algae (Joubert and Fleurence 2008; Fleurence et al. 1995). One limitation of the enzyme extraction method is the high concentration of enzymes required to achieve commercially feasible yields. This could be around (48.0 $\times$ 10^3 units/100 g) (Bleakley and Hayes 2017). Extraction yields of the enzymatic method are higher compared to the other conventional methods. Extraction yields of up to 87% (weight of protein extracted per weight of protein present in the biomass) have been reported (Harnedy and FitzGerald 2013).

10.5.4 Ultrafiltration and Diafiltration

These processes can be used in combination to optimize the yield and/or quality of protein extract. For example, in the extraction of water-soluble protein from *Nannochloropsis gaditana* microalgae, the cell disruption in order to free the proteins prior to ultrafiltration or diafiltration was carried out with either a combination of high-pressure homogenization and ultrafiltration/diafiltration or a combination of enzyme treatment with protease enzyme and ultrafiltration/diafiltration (Safi et al. 2017). This method is presented as a mild biorefinery process which has less environmental impact in terms of minimal use of solvents and acids or alkali. High-pressure homogenization at 150 GPa permeating at 9 L h^{-1} was sufficient to break the cell walls of the microalgae and release the proteins in the aqueous system. The ultracentrifugation followed by diafiltration was then required to separate the polysaccharides from the proteins. Although the system was maintained at room temperature with no additional heat input, due to the frictional forces resulting in heat generation during homogenization, a cooling system was necessary to maintain the temperature at 30 °C. When cell disruption was achieved using enzymatic treatment, this was carried out at a temperature of 50 °C and a pH of 8. The enzyme treatment was allowed a period of 4 h using alcalase, a protease enzyme.

The combination of enzyme treatment with ultrafiltration and diafiltration resulted in a higher yield than when high-pressure homogenization was combined with ultrafiltration/diafiltration in the particular study where a yield of 24.8% was obtained for the former and 17.4% for the latter. This does not necessarily mean that the enzyme method would always result in a higher yield for all algae; this needs to be carried out for a range of species and a correlation drawn between yield and species for both methods before such conclusions can be made. Figure 10.1 shows a flowchart of the extraction process involving cell disruption, ultrafiltration and diafiltration.

10.5.5 Assisted Extraction

The conventional extraction methods can be modified by using additional techniques to assist the extraction process. An example is the use of ultrasound. This could be used to create tiny bubbles within the extraction medium, weather neutral, alkaline

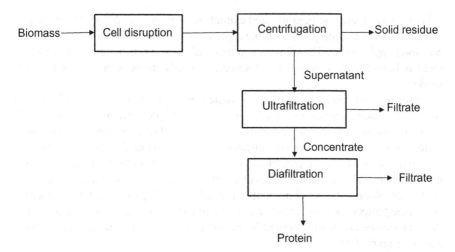

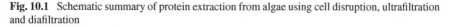

Fig. 10.1 Schematic summary of protein extraction from algae using cell disruption, ultrafiltration and diafiltration

or acidic. These acoustic cavities exert pressure and generate heat within the cell wall causing disruption of the cell wall (Mason et al. 1996). This makes it easier for the water, alkali, acid or enzyme to penetrate the cell wall and act toward releasing the protein. In one example, when applied prior to acid extraction of protein from the alga *Ascophyllum nodosum*, ultrasound increased the protein yield by 540%. The required extraction time was also reduced by 50 min (Kadam et al. 2017). Such reduction in processing time and increase in yield are significant toward a more feasible commercial-scale extraction of algal protein. An integrated extraction process combined the ultrasound with other methods to extract protein from microalgae. The ultrasound method combined with the sugaring out method and liquid biphasic flotation achieved a yield of up to 93.33% (protein extracted as a percentage of protein present in the biomass) at optimum conditions (Sankaran et al. 2018). The process at large scale had an efficiency of 85.25%. While costly equipments and technical expertise are required to achieve a less fossil-sourced energy-demanding process, it is important to have high efficiency to compensate for the cost incurred in equipment and technical manpower for production.

Another way to improve the conventional extraction method is through the use of pulsed electric field. This method is used for other applications such as drug delivery where the electric field is used to open up the skin or other tissue, to enhance the transport of drugs which will otherwise not permeate the tissue or cell membrane and to permeate faster. This involves application of high voltage to the biomass for a short period of time, usually a fraction of a second. When pulsed electric field is applied to a membrane, this opens up temporary channels through which larger molecules can permeate the membrane (Silva and Sulaiman 2019). This principle has also been applied for extraction of protein from algal biomass. For example, protein yield from Spirulina was increased by 13%, while that from chlorella was increased by 27%

using the pulsed electric field-assisted extraction (Garcia et al. 2018). These are two of the common microalgae which have been commercially explored for food and non-food applications. Therefore, a commercially feasible method which increases yield is bound to have significant commercial implications with a ready-to-adapt market.

Other methods include microwave-assisted method (Barba et al. 2015), use of subcritical water or supercritical CO_2 (Herrero et al. 2006; Dumay and Morancais 2016) and membrane technology where a semipermeable membrane is used to separate a component against a concentration gradient. This method is often preceded by cell disruption by methods such as homogenization or enzyme treatment to make the protein available in solution (Safi et al. 2017). These processes have the common goal of developing means of disrupting the cell wall using extreme fluid properties at specific temperature and pressure in order to aid extraction of the protein. All of these processes are mostly limited by financial requirement of necessary facilities and scale up of the process.

10.6   Environmental Implications

Extraction using more novel methods such as electrophoresis without additional chemicals or heat input can be considered since it significantly results in significantly higher yields and the electric current can be generated using green energy sources. As the technology for extraction of protein from algae and plant biomass advances, particularly in an age where more attention is being paid to environmental impact, the research in this area is bound to move toward developing more environmentally sustainable extraction processes which are financially feasible.

10.6.1 CO_2 Emission

Use of supercritical CO_2 could be a way of not only minimizing emissions but also removal of CO_2. This will, however, incur more cost in production which is not commercially feasible. A tonne of microalgae requires to produce an estimated 1.8 tonnes of CO_2 (Kliphuis et al. 2010). Since the process of growing the biomass removes CO_2 from the environment, it can be generally regarded as a CO_2-negative process. This is then weighed against the carbon emitted in the other processes involved in the extraction of protein and going further down the value chain to the packaging and transportation to final consumer.

10.6.2   Solid Waste Generation

Compared to lignocellulosic land plants such as corn and sugarcane, algae biomass has simpler composition. Brown algae, for example, comprise 15% laminarin, 18% mannitol, 32% alginate and 35% ash and salts. Lignocellulosic plants, for example, corn stover, contain 35% cellulose, 19% hemicellulose, 16% lignin and other compounds up to 27% (Konda et al. 2015). This means the solid residue generated from the extraction of protein from algae results in less diverse range of compounds which make the utilization of this waste less complex compared to residue from lignocellulosic biomass.

10.6.3   Water Consumption

In extraction of proteins, water is used as the main extraction medium. Water is also consumed in the washing of the biomass prior to extraction. In the cultivation process, aquatic plants and algae have one major advantage over terrestrial plants of not requiring freshwater. They can, in fact, be used to clean up water by extracting the nutrients in the water for their metabolism. However, careful control of the levels of the different nutrients is required for optimal growth and control of the yield of the desired biopolymer from the aquatic plant or algae. In the extraction of protein from *Spirulina platensis*, for example, 1 g of biomass required 15 ml of distilled water as extraction medium (Yucetepe et al. 2018). Much of the water used in the extraction process is then evaporated in the drying process, and this requires some energy input. The water used in washing is either sent to external water treatment plants or recycled internally. Where acids or alkalies are used, the water needs to be neutralized and this requires the use of either more acid or alkali in the water treatment.

10.6.4   Acids and Alkali

Although the enzyme extraction method relies on enzymes to break down the cell walls with the aim of eliminating or at least minimizing the need for additional chemicals, in some cases it is necessary to use acids or base to achieve the required pH for enzymes to function optimally. Enzymes are proteins, and the activity of these proteins is determined by their configuration as well as conformation (O'Brien et al. 2012). The conformation is dependent on the pH of the environment within which they are acting, therefore for specific enzymes, a specific pH is required. For example, while alcalase is used, a pH 8 was required (Safi et al. 2017). This will require the addition of an alkali to achieve this pH. However, this is relatively mild as it is close to a neutral pH.

10.6.5   Energy Consumption

Energy consumed depends on the method being used and the efficiency of the operating system. While ultrasound is being used, one study reports 0.4 kWh required per liter of the biomass suspension being processed during the disruption of microalgal cell (Keris-Sen et al. 2014). While physical method such as high-pressure homogenization is being used, energy to generate the amount of shear to disrupt the cell is required.

Amount of heat required for extraction varies with the method being used. For example, in extraction of water-soluble proteins from microalga *N. gaditana* where cell disruption is carried out by high-pressure homogenization, the process of homogenization results in generation of heat from the friction forces and the exothermic processes of cell disruption. This process thereby requires cooling to prevent overheating of the system (Safi et al. 2017). Proteins are prone to denaturation at elevated temperature; therefore, the temperature should be kept as close to ambient as possible. To achieve optimal cell disruption, a pressure of 150 GPa was sufficient. This resulted in over 90% of the cells being disrupted in a single pass. This process required an energy input of 0.44 kWh per kg of biomass. Repeated passes and increasing homogenization pressure did not result in higher percentage of cell disruption. Such studies are necessary in terms of saving cost and time. Indeed, two passages would have required an energy input of 7.5 kWh per kg resulting in only 3% increase in yield.

Energy also used in the additional unit operations is required to separate and purify. Centrifugation of the suspension following cell disruption by physical, chemical or enzyme treatment requires around 10,000 g for 10 min (Safi et al. 2017). Use of enzymes reduces the heat energy required in the energy-demanding methods.

10.7   Applications

A variety of proteins can be found in algae, and some of these have distinct functions. Some edible algae and aquatic plants are consumed in their whole form as food, skin care, aquaculture or animal feed. Here we look at applications which involve extracting protein from aquatic plants and algae for protein specific applications which range from food to clinical applications. Some of the proteins described here are not only limited to aquatic plants and algae and can be found in other terrestrial plants and animals.

10.7.1   HIV Microbicides

A type of protein which is currently receiving attention in HIV prevention research is lectins. Lectins are present in animals, microorganisms and terrestrial. They are also present in algae such as cyanobacterium (Singh et al. 2017). They are a diverse group of proteins which include scytovirin, griffithsin and cyanovirin. Lectins have the ability to bind to polysaccharides, glycans and glycolipids in a reversible manner (Harnedy and FitzGerald 2011; Singh et al. 2015). This binding ability allows them to cause cell agglutination and precipitation of glycoconjugates. This makes them applicable in diverse applications such as antiviral, antitumor and antiinflammatory agents and well as development of protein expression systems and nutraceuticals. The use of lectins as microbicides to prevent the transmission of HIV has been explored in several research studies (Janahi et al. 2018; Hopper et al. 2017; Alexandre et al. 2010). The lectins can selectively bind to the HIV cells by binding to the glycans and polysaccharides on the surface of the virus which shields it from attack by the antimicrobial agents or from being recognized by the body's immune system. Lectins also act by preventing the virus cells from binding to uninfected cells. This has been demonstrated in in vitro studies on vaginal mucosa using scytovirin an algal lectin (Janahi et al. 2018).

10.7.2 Phycobiliproteins

Phycobiliproteins are another form of proteins present in algae. These are water-soluble proteins that serve a role in capturing of light in photosynthesizing organisms. They are common in red algae and cyanobacteria (Dumay and Morancais 2016). Phycobiliproteins find commercial applications as natural dyes for food and cosmetics (Spolaore et al. 2006). They also have been explored for other applications such as fluorescent imaging and flow cytometry (Aneiros and Garateix 2004). More recently, phycobiliproteins have been associated with some bioactive properties such as antiviral, antiinflammatory and antioxidant activities (Sekar and Chandramohan 2008). Algae are also a source of certain bioactive polypeptides (Fan et al. 2014). These have 2–30 repeating units of amino acids which have some bioactive properties depending on the types and sequence of amino acids within these short chains. These bioactive proteins are usually derived from the long-chain proteins and polypeptides through hydrolysis or fermentation to break down the chains into shorter ones.

10.7.3 Food

Aquatic plant and algae play an increasingly important role in the future food security. The world population is expected to reach 9 million by 2050, and the rate of food production is being threatened due to several factors such as dangerous weather conditions due to climate change and social problems leading to violence which has resulted in reduced farming activities in these conflict regions (FAO et al. 2018). There is an urgent need to develop alternative ways to produce food and produce it much faster than conventional farming methods. With a much faster rate of nutrient accumulation and growth, algae and aquatic plants are much more productive than land plants. Furthermore, they do not require land for growth which means they do not compete with space for living and can be grown away from disputed land areas. Proteins are an important food content and are necessary in maintaining a balanced diet. FAO estimates that 10.9% of people in the world are undernourished as of 2017 and that figure is said to have been growing in the past two years. The number of malnourished people in the world rose from 804 million as of 2016 to 821 million in 2017 (FAO et al. 2018). Alongside carbohydrates and fats, protein forms one of the macronutrients required by humans for energy and growth.

In many parts of the world, fish serves as a major source of protein. With recent algal blooms which result in the release of toxins into the water, fish in some parts is at threat of being unsafe for human consumption. Aquatic plants and algae play two roles in this aspect. The possible use to clean up water and as alternative sources of proteins; either directly consumed as food, nutraceuticals or used as ingredients in food products (Sinha et al. 2019).

An aquatic plant can be regarded as edible if it contains some nutrients which can be broken down by the enzymes in the body and if it does not contain any toxins. Recent studies present a protein digestibility of 99.36% for protein extracted from the blue-green algae Spirulina. This is relatively high compared to that of soybean which is 85% (Yucetepe et al. 2018). Although this value was derived from in vitro studies, it indicates that such protein is suitable for food applications and indeed some Spirulina-based food products are available in the market today. Thousands of species of macroalgae and microalgae exists, such that the different proteins within them also largely vary. The digestibility and types of amino acids which make up the proteins vary for different algae (Boisen and Eggum 1991; Bleakley and Hayes 2017).

When consumed directly as food, algae have relatively low bioavailability. This is thought to be due to the fact that proteins are usually present alongside other polysaccharides within the cell wall of the algae. These fibers with which the proteins are attached are indigestible; for the protein to be available as a nutrient, it must be released from the fiber and then broken down into forms which can be absorbed into the body along the alimentary canal. The digestibility therefore depends on the enzymes present within the body which can break down the protein and free it from the fibrous structure. Most often, it is therefore necessary to extract these proteins and then make them into edible preparations as processed foods before they can have relevant digestibility or bioavailability.

While egg and casein protein have digestibility coefficients of 94.2 and 95.1%, respectively, microalgae Chlorella sp., *Scenedesmus obliquus* and Spirulina sp. have digestibility coefficients of 76.6, 88.0 and 77.6%, respectively (Becker 2007). The digestibility varies widely among algae species. For example, *Undaria pinnatifida* has a bioaccessibility of 87%, *Ulva lactuca* has a digestibility of about 85.7%, while *P. tenera* has a bioavailability of 78%. Some studies report red algae have to be more digestible than brown algae, digestibility of red algae studied ranged between 83 and 87% while that of the brown algae ranged between 78.7 and 82% (Tibbetts et al. 2016). Much of the digestibilities referred to herein are based on in vitro studies.

Algae have been made into plant protein-based snacks of various forms. These have been shown to have similar sensory appeal to conventional snacks. Lucas et al. (2018) reported a sensory acceptability index of 82% for snacks which have been enriched with Spirulina as a protein and nutrient enrichment. Algae has been part of human diet for centuries; therefore, this could serve as a basis for using proteins from algae as protein enrichment in foods. This is important in, for example, producing protein supplements where a high protein content is required. Isolation of specific component such as protein also allows for standardization and quality control of products through identification of the specific compounds present and their quantity.

People of different regions have used freshwater plants for different applications for centuries. Example of such freshwater plants includes wild rice (Zizania spp.), Chinese water chestnut (*Eleocharis dulcis*), Indian lotus (*Nelumbo nucifera*), watercress (*Rorippa nasturtium-aquaticum*), water spinach (*Ipomoea aquatica*), water caltrop (*Trapa natans*), water mimosa (*Neptunia oleracea*), wild taro (*Colocasia esculenta*) and cattails (Typha spp.). Some have been consumed as food and used as animal feed or in construction of structures, while some are simply valued for their aesthetic appeal and revered as traditional/religious symbols—an example of this is the Indian lotus. There are some interests in extracting proteins from some of the aquatic plants which are popularly known to cause some environmental nuisance which often also lead to economic losses through the destruction of aquatic wildlife. For example, aqueous extract of water hyacinth was shown to contain over 56.38% protein and had 17 out of the 20 essential amino acids for humans. The water hyacinth leaf protein concentrate was also reported to have metal composition within acceptable limits which could make it a potential source of protein for human use (Adeyemi and Osubor 2016). The protein content in the water hyacinth leaf protein concentrate was much higher than the carbohydrate (33.61%), fat (4.11), ash (4.88%) and crude fiber (1.02%). Such that, over half of the leaf protein concentrate of the water hyacinth is protein; this makes it potentially suitable as a high-protein food or supplement.

10.7.4 Animal Feed

The protein in algae is used in animal, poultry and fish feed with the aim of improving yield and quality since protein forms an important part of the diet. However, much of the work in this area has looked at using whole or semi-processed macro- and microalgae. Few studies have reported on exclusively extracting the proteins from algae and incorporating in animal feed. This is due to the fact that animal feed is required to be sufficiently low cost; therefore, expensive protein isolation prior to incorporation in the feed would not be commercially feasible. A number of commercially available animal feeds incorporate algae as a rich source of protein alongside the other nutrients present in algae. Ruminants are particularly good candidates for algae-derived protein-based feed as they can better digest the fiber and have a better bioavailability of the protein within. This biological process is quite similar to the extraction of algal protein using the enzymes which break down the polysaccharides within the fiber, thereby making the protein available for extraction.

Animals fed with diet which included algae as a protein source have shown improvement in immune system, increased productivity and improved quality of meat and milk. This has been observed in a variety of commercially bred animals such as steers used for beef and pigs used for pork (Allen et al. 2001; Saker et al. 2001; Montgomery et al. 2001; Braden et al. 2004; Gatrell et al. 2014). In poultry, chicken broilers fed with microalgae Spirulina show improved fertility and a more aesthetically pleasing egg yolk color (Zahroojian et al. 2013). However, these attributes are not directly linked with the protein content. The improvement in egg yolk color is thought to be due to the presence of beta carotene (Anderson et al. 1991). In aquaculture, both microalgae and macroalgae are used as feed for aquaculture animals such as mollusks, sea bream and shrimps (Müller-Feuga 2013). They also act to regulate the nutrient and mineral content in the water (Chuntapa et al. 2003).

Aquatic plants such as water fern, duckweed and American pondweed have been used as animal feed as a protein source as well as other nutrients. Some aquatic plants such as duckweed species, *Lemna gibba*, have relatively high protein content. For example, a crude protein content of 21.5% was measured for *L. gibba*, while green algae species, *U. lactuca*, was 17.2% in the same study (Aguilera-Morales et al. 2018). Both the duckweed and *U. lactuca* contained eight of the essential amino acids necessary for fish feed.

10.7.5 Antihypertensive

The antihypertensive properties of compounds can be measured by their ability to inhibit the activity of the angiotensin-converting enzyme (ACE I). This enzyme converts the inactive angiotensin into an active ACE I which is then converted to ACE II which caused constriction of the cardiac blood vessels leading to hypertension. Bioactive peptides act as antihypertensive agents by preventing the conversion of inactive ACE to the active ACE I, which in turn prevents the formation of ACE II (Seca and Pinto 2018; Daskaya-Dikmen et al. 2017).

Another therapeutic approach to hypertension is the enhancement of vasodilation to reduce blood pressure. Like flow in a channel, as the vessel is widened, the pressure in the vessel is reduced. Vasodilation is facilitated by the enzyme bradykinin through a series of processes. The body's natural control system releases the appropriate enzyme in response to the pressure in the blood vessels. A normal blood pressure is therefore dependent on this control system, where the vasoconstricting enzymes and substrates are working in conjunction with the vasodilating enzymes to maintain homeostasis. Angiotensin I is also thought to be responsible for degradation of the bradykinin, thus preventing vasodilation (Daskaya-Dikmen et al. 2017). Therefore, by inhibiting the angiotensin enzyme, vasodilation is also promoted.

The antihypertensive properties shown by these peptides are absent in the parent proteins. These parent proteins need to be broken down into the peptide forms before they can manifest the antihypertensive bioactivity. This is usually achieved artificially through enzymatic hydrolysis, heat treatment or fermentation. This could possibly happen naturally in the body and perhaps why the seaweeds have been associated with such properties when consumed as foods (Sabirin et al. 2016).

Proteins and peptides are susceptible to enzymatic breakdown in the gastrointestinal tract. Orally consumed peptides may therefore not have sufficient bioactivity to elicit their antihypertensive effect when taken into the body through this route. The peptides' bioactivity is highly dependent on the sequence of amino acid and the primary and secondary structures of the short chain (usually 2–20 amino acid residues). When taken orally, these are susceptible to being digested as other food proteins and broken down and absorbed into the body as amino acids. Therefore, for these peptides to be able to elicit antihypertensive effect, they need to either have resistance to the enzymes within the gastrointestinal tract, completely bypass the gastrointestinal tract by using other routes of delivery into the bloodstream or be prepared in forms which protects the active ingredient (in this case the polypeptide) until it gets to the target.

On the other hand, the intestinal enzyme activity could act on these peptides and convert them to more active antihypertensive short-chain peptides. This has been demonstrated in peptides extracted from *Ulva rigida*, where the peptides containing three amino acid residues produced from the hydrolysis of the parent protein were further broken down by the intestinal mucosa peptidase into a smaller peptide containing two amino acid residues (Paiva et al. 2016). In this case, the intestinal enzyme activity was beneficial in improving the antihypertensive effect of the peptides by partial digestion.

Indeed, the studies on the antihypertensive effect of these algae-derived peptides are relatively advanced and have been carried out on animal and human test subjects (Saito et al. 2002; Saito and Hagino 2005). Algae species from which peptides have been shown to have antihypertensive activity have been extracted that include *U. pinnatifida*, *Porphyra yezoensis* and *U. rigida* (Seca and Pinto 2018).

The presence of particular groups such as D-amino acids thiophene oxazole and some alpha and beta amino acids within the polypeptide structures of some marine plants gives them some excellent bioactive properties. Some products also exist in the market as functional foods with antihypertensive effects. Examples are Evolus® and Ameal-S 120® (Seca and Pinto 2018). The potency of the different peptides varies, and the hypertensive effect is also dose dependent. Example prescribed dosage is 1.8 g daily of *Pyropia yezoensis*-derived peptides to human patients (Saito et al. 2002) and 10 mg per day per kg of body weight to hypertensive rats through oral administration (Suetsuna and Nakano 2000).

10.7.6 Antioxidant

Compounds which inhibit the activities of reactive oxygen species or remove them for the body are referred to as antioxidants (Yucetepe et al. 2018). Algae and other aquatic plants have been investigated widely for their antioxidant properties. The antioxidant properties of proteins from algae can be attributed to their bioactive proteins. Some of the antioxidant properties of some polysaccharides such as fucoidans which show antioxidant activity have been thought to be due to the presence of proteins attached to them.

Proteins extracted from aquatic plants and algae have been explored for a range of other applications which include antiinflammatory, anticancer, antiviral, lipid and glucose-lowering effect, anticancer, immunomodulatory effect, appetite suppression, antimicrobial, cytomodulatory and hypocholesterolemic.

10.8   Commercial Production

Protein is an important component of the human diet. It is one of the three main macronutrients which are vital for energy production, the others being fats and carbohydrates. For much of the world, sea-sourced food is the primary source of protein, largely fish. With the growing demand for increased food production rate and protein being a very important part of this, algae have a large potential present and future market to serve as a rich protein source. Extraction of the protein from the algae and aquatic plants improved digestibility and also reduces the cost of transportation of the whole algae or aquatic plant biomass, which usually has high water content.

A non-essential amino acid, glutamic acid, has a sodium salt, monosodium glutamate, which naturally occurs in the red sea weed *Saccharina japonica*, commonly known as MSG, which is commercially used as a flavor enhancer. Although not a protein, it is an amino acid. It is more commercially produced from fermentation of glutamic acid by corynebacterium species (Kakogawa et al. 1972) and made one of its earliest commercial appearance in the form of the flavor enhancer brand named "Aji-no-moto" (Choudhury and Sarkar 2017). One of the particular appeals of an algae-sourced amino-acid-based flavor is that it serves as an option to meat- or fish-based seasoning to vegans and vegetarians.

The extraction of proteins can be carried out alongside extraction of polysaccharides and lipids as a way to make optimal use of all the commercially valuable compounds within the algae. Presently, companies and researchers who are focused on extraction of particular compound from algae tend to discard the other parts of the algae which include the proteins. This is particularly due to the additional costs required for further processing and the complexity of developing a multiproduct business model. Companies who are presently involved in production of algae protein-based products include Nova Scotia in Canada, Calpis Co Ltd in Japan and Valio Ltd in Finland.

10.9 Conclusion

Aquatic plants and algae are a rich source of proteins. While they are not a total source of proteins, they can be used to augment dietary protein requirement and serve as a relatively cheap and abundant source of protein for direct and indirect human consumption. The higher biomass accumulation rate and higher quality of protein of aquatic plants and algae compared to terrestrial crops give them an advantage. Beyond food, these aquatic polymers also have important bioactive properties such antiviral and antihypertensive activities which make them applicable as nutraceuticals and in the potential future clinical applications such as HIV prevention.

References

Adeyemi O, Osubor CC (2016) Assessment of nutritional quality of water hyacinth protein concentrate. Eqypt J Aquat Res 42:269–272

Aguilera-Morales ME, Canales-Martinez MM, Avila-Gonzalez E, Flores-Ortiz CM (2018) Nutrients and bioactive compounds of *Lemna gibba* and *Ulva lactuca* as possible ingredients to functional foods. Latin Am J Aquat Res 46(4):709–716

Alexandre KB, Gray ES, Lambson BE, Moore PL, Choge IA, Mlisana K, Abdool Karim SS, McMahon J, O'Keefe B, Chikwamba R, Morris L (2010) Mannose-rich glycosylation patterns on HIV-1 subtype C gp120 and sensitivity to the lectins, griffithsin, cyanovirin-N and scytovirin. Virology 402(1):187–196

Allen V, Pond K, Saker K, Fontenot J, Bagley C, Ivy R, Evans R, Brown C, Miller M, Montgomery J (2001) Tasco-Forage: III. Influence of a seaweed extract on performance, monocyte immune cell response, and carcass characteristics in feedlot-finished steers. J Anim Sci 79:1032–1040

Anderson DW, Tang CS, Ross E (1991) The xanthophylls of Spirulina and their effect on egg-yolk pigmentation. Poult Sci 70:115–119

Aneiros A, Garateix A (2004) Bioactive peptides from marine sources: pharmacological properties and isolation procedures. J Chromatogr B 803:41–53

Barba FJ, Grimi N, Vorobiev E (2015) New approaches for the use of non-conventional cell disruption technologies to extract potential food additives and nutraceuticals from microalgae. Food Eng Rev 7:45–62

Barbarino E, Lourenço SO (2005) An evaluation of methods for extraction and quantification of protein from marine macro- and microalgae. J Appl Phycol 17:447–460

Becker E (2007) Micro-algae as a source of protein. Biotechnol Adv 25:207–210

Bleakley S, Hayes M (2017) Algal proteins: extraction, application, and challenges concerning production. Foods 6:33 (1–34). https://doi.org/10.3390/foods6050033

Boisen S, Eggum B (1991) Critical evaluation of in vitro methods for estimating digestibility in simple-stomach animals. Nutr Res Rev 4:141–162

Braden K, Blanton J, Allen V, Pond K, Miller M (2004) *Ascophyllum nodosum* supplementation: a preharvest intervention for reducing *Escherichia coli* O157:H7 and Salmonella spp. in feedlot steers. J Food Proteins 67:1824–1828

Brouwer P, Nierop KGJ, Huijgen WJJ, Schluepmann H (2019) Aquatic weeds as novel protein sources: alkaline extraction of tannin-rich *Azolla*. Biotechnol Rep E00368 (in press)

Choudhury S, Sarkar NS (2017) Algae as a source of natural flavour enhancers—a mini review. Plant Sci Today 4(40):172–176

Chuntapa B, Powtongsook S, Menasveta P (2003) Water quality control using *Spirulina platensis* in shrimp culture tanks. Aquaculture 220:355–366

Daskaya-Dikmen C, Yucetepe A, Karbancioglu-Guler F, Daskaya H, Ozcelik B (2017) Angiotensin-converting enzyme (ACE) inhibitory peptides from plants. Nutrients 9(316):1–19

Dumay J, Morancais A (2016) Proteins and pigments. In: Seaweed in health and disease prevention, pp 275–318

Fan X, Bai L, Zhu L, Yang L, Zhang X (2014) Marine algae-derived bioactive peptides for human nutrition and health. J Agric Food Chem 62:9211–9222

FAO, WHO (1991) Protein quality evaluation—report of joint FAO/WHO expert consultation. FAO, Rome, Italy

FAO, IFAD, UNICEF, WFP, WHO (2018) The state of food security and nutrition in the world 2018. Building climate resilience for food security and nutrition. Rome, FAO. Licence: CC BY-NC-SA 3.0 IGO. ISBN 978-92-5-130571-3

Fleurence J (1999) Seaweed proteins: biochemical, nutritional aspects and potential uses. Trends Food Sci Technol 10:25–28

Fleurence J, Massiani L, Guyader O, Mabeau S (1995) Use of enzymatic cell wall degradation for improvement of protein extraction from *Chondrus crispus*, *Gracilaria verrucosa* and *Palmaria palmata*. J Appl Phycol 7:393–397

Garcia SE, van Leeuwen J, Safi C, Sijtsma L, Eppink MHM, Wijffels RH, van den Berg C (2018) Selective and energy efficient extraction of functional proteins from microalgae for food applications. Bioresour Technol 268:197–203

Gatrell S, Lum K, Kim J, Lei X (2014) Nonruminant nutrition symposium: potential of defatted microalgae from the biofuel industry as an ingredient to replace corn and soybean meal in swine and poultry diets. J Anim Sci 92:1306–1314

Harnedy PA, FitzGerald RJ (2011) Bioactive proteins, peptides, and amino acids from macroalgae. J Phycol 47:218–232

Harnedy PA, FitzGerald RJ (2013) Extraction of protein from the macroalga *Palmaria palmata*. LWT Food Sci Technol 51:375–382

Herrero M, Cifuentes A, Ibanez E (2006) Sub- and supercritical fluid extraction of functional ingredients from different natural sources: plants, food-by-products, algae and microalgae: a review. Food Chem 98:136–148

Hlima HB, Dammak M, Karkouch N, Hentati F, Laroche C, Michaud P, Fendri I, Abdelkafi S (2019) Optimal cultivation towards enhanced biomass and floridean starch production by *Porphyridium marinum*. Int J Biol Macromol 15(129):152–161

Hopper TSJ, Ambrose S, Grant CO, Krumm SA, Allison TM, Degiacomi MT, Tully MD, Pritchard LK, Ozorowski G, Ward AB, Crispin M, Doores KJ, Woods RJ, Benesch JLP, Robinson CV, Struwe WB (2017) The tetrameric plant lectin BanLec neutralizes HIV through bidentate binding to specific viral glycans. Structure 25(5):773–782

Janahi EMA, Haque S, Akhter N, Wahid M, Jawed A, Mandal RK, Lohani M, Areeshi MY, Almaki S, Das S, Dar SA (2018) Bioengineered intravaginal isolate of *Lactobacillus plantarum* expresses algal lectin scytovirin demonstrating anti-HIV-1 activity. Microb Pathog 122:1–60

Joubert Y, Fleurence J (2008) Simultaneous extraction of proteins and DNA by an enzymatic treatment of the cell wall of *Palmaria palmata* (Rhodophyta). J Appl Phycol 20:55–61

Kadam SU, Álvarez C, Tiwari BK, O'Donnell CP (2017) Extraction and characterization of protein from Irish brown seaweed *Ascophyllum nodosum*. Food Res Int 99(3):1021–1027

Kakogawa TS, Takasago UJ, Kakogawa TK, Takasago YT (1972) Process for producing monosodium glutamate. Patent 3655746

Keris-Sen UD, Sen U, Soydemir G, Gurol MD (2014) An investigation of ultrasound effect on microalgal cell integrity and lipid extraction efficiency. Bioresour Technol 152:407–413

Kliphuis AM, de Winter L, Vejrazka C, Martens DE, Janssen M, Wijiffels RH (2010) Photosynthetic efficiency of *Chlorella sorokiniana* in a turbulently mixed short light-path photobioreactor. Biotechnol Prog 26:687–696

Konda M, Singh S, Simmons BA, Klein-Marcuschamer D (2015) An investigation on the economic feasibility of macroalgae as a potential feedstock for biorefineries. BioEnergy Res 8(3):1–11

Kumar SK, Ganesan K, Rao SVP (2015) Seasonal variation in nutritional composition of *Kappaphycus alvarezii* (Doty) Doty—an edible seaweed. J Food Sci Technol 52(5):2751–2760

Li T, Xu J, Wu H, Jiang P, Chen Z, Xiang W (2019) Growth and biochemical composition of *Porphyridium purpureum* SCS-02 under different nitrogen concentrations. Mar Drugs 17(124):1–16

Lucas BF, de Morais MG, Santos TD, Costa JAV (2018) Spirulina for snack enrichment: nutritional, physical and sensory evaluations. LWT Food Sci Technol 90:270–276

Marrion O, Schwertz A, Fleurence J, Gueant JL, Villaume C (2003) Improvement of the digestibility of the proteins of the red alga *Palmaria palmata* by physical processes and fermentation. Mol Nutr Food Res 47:339–344

Mason T, Paniwnyk L, Lorimer J (1996) The uses of ultrasound in food technology. Ultrason Sonochem 3:S253–S260

Montgomery J, Allen V, Pond K, Miller M, Wester D, Brown C, Evans R, Bagley C, Ivy R, Fontenot J (2001) Tasco-Forage: IV. Influence of a seaweed extract applied to tall fescue pastures on sensory characteristics, shelf-life, and vitamin E status in feedlot-finished steers. J Anim Sci 79:884–894

Müller-Feuga A (2013) Microalgae for aquaculture: the current global situation and future trends. In: Richmond A, Hu Q (eds) Handbook of microalgal culture: applied phycology and biotechnology, 2nd edn. Wiley, Oxford, UK, pp 352–364

O'Brien E, Brooks BR, Thirumalai D (2012) Effects of pH on proteins: predictions for ensemble and single molecule pulling experiments. J Am Chem Soc 134(2):979–987

Paiva LS, Lima EMC, Neto AI, Baptista J (2016) Isolation and characterization of angiotensin I-converting enzyme (ACE) inhibitory peptides from *Ulva rigida* C. Agardh protein hydrolysate. J Funct Foods 26:65–76

Pulz O, Gross W (2004) Valuable products from biotechnology of microalgae. Appl Microbiol Biotechnol 65:635–648

Sabirin F, Soo KK, Ziau HS, Kuen LS (2016) Antihypertensive effects of edible brown seaweeds in rats. Int J Adv Appl Sci 3:103–109

Safi C, Olivieri G, Campos RP, Engelen-Smit N, Mulder WJ, van den Broek LAM, Sijtsma L (2017) Biorefinery of microalgal soluble proteins by sequential processing and membrane filtration. Bioresour Technol 225:151–158

Saito M, Hagino H (2005) Antihypertensive effect of oligopeptides derived from nori (*Porphyra yezoensis*) and Ala-Lys-Tyr-Ser-Tyr in rats. J Jpn Soc Nutr Food Sci 58:177–184

Saito M, Kawai M, Hagino H, Okada J, Yamamoto K, Hayashida M, Ikeda T (2002) Antihypertensive effect of Nori-peptides derived from red alga *Porphyra yezoensis* in hypertensive patients. Am J Hypertens 15:210A

Saker K, Allen V, Fontenot J, Bagley C, Ivy R, Evans R, Wester D (2001) Tasco-Forage: II. Monocyte immune cell response and performance of beef steers grazing tall fescue treated with a seaweed extract. J Anim Sci 79:1022–1031

Sankaran R, Manickam S, Yap YJ, Ling TC, Chang JS, Show PL (2018) Extraction of proteins from microalgae using integrated method of sugaring-out assisted liquid biphasic flotation (LBF) and ultrasound. Ultrason Sonochem 48:231–239

Seca AML, Pinto DCGA (2018) Overview on the antihypertensive and anti obesity effects of secondary metabolites from seaweeds. Mar Drugs 16(237):1–18

Sekar S, Chandramohan M (2008) Phycobiliproteins as a commodity: trends in applied research, patents and commercialization. J Appl Phycol 20:113–136

Silva FVM, Sulaiman A (2019) Polyphenoloxidase in fruit and vegetables: inactivation by thermal and non-thermal processes. In: Encyclopedia of food chemistry. Reference module in food science, pp 287–301. https://doi.org/10.1016/B978-0-08-100596-5.21636-3

Singh RS, Thakur SR, Bansal P (2015) Algal lectins as promising biomolecules for biomedical research. Crit Rev Microbiol 1(1):77–88

Singh RS, Walia AK, Khattar JS, Singh DP, Kennedy JF (2017) Cyanobacterial lectins characteristics and their role as antiviral agents. Int J Biol Macromol 102:475–496

Sinha S, Patro N, Patro IK (2019) Amelioration of neurobehavioral and cognitive abilities of F1 progeny following dietary supplementation with Spirulina to protein malnourished mothers. Brain Behav immun (in press)

Spolaore P, Joannis-Cassan C, Duran E, Isambert A (2006) Commercial applications of microalgae. J Biosci Bioeng 101:87–96

Suetsuna K, Nakano T (2000) Identification of an antihypertensive peptide from peptic digest of wakame Undaria pinnatifida. J Nutr Biochem 11:450–454. https://doi.org/10.1016/S0955-2863(00)00110-8

Tan NH, Zhou J (2006) Plant cyclopeptides. Chem Rev 6:840–895

Tibbetts SM, Milley JE, Lall SP (2016) Nutritional quality of some wild and cultivated seaweeds: nutrient composition, total phenolic content and in vitro digestibility. J Appl Phycol 28:3575–3585

Tokuşoglu Ö, Üunal M (2003) Biomass nutrient profiles of three microalgae: *Spirulina platensis, Chlorella vulgaris*, and *Isochrisis galbana*. J Food Sci 68:1144–1148

Van Krimpen M, Bikker P, van der Meer I, van der Peet-Schwering C, Vereijken J (2013) Cultivation, processing and nutritional aspects for pigs and poultry of European protein sources as alternatives for imported soybean products. Wageningen UR Livestock Research, Lelystad, The Netherlands, p 48

Yang J, Gittlin I, Krishnamurthy VM, Vazquez JA, Castello CE, Whitesides GM (2003) Synthesis of monodisperse polymers from proteins. J Am Chem Soc 125(41):12392–12393

Young VR, Pellett PL (1994) Plant proteins in relation to human protein and amino acid nutrition. Am J Clin Nutr 59:1203S–1212S

Yucetepe A, Saroglu O, Daskaya-Dikmen C, Bildik F, Ozcelik B (2018) Optimisation of ultrasound-assisted extraction of protein from *Spirulina platensis* using RSM. Food Technol Econ Czech J Food Sci 36(1):98–108

Zahroojian N, Moravej H, Shivazad M (2013) Effects of dietary marine algae (*Spirulina platensis*) on egg quality and production performance of laying hens. J Agric Sci Technol 15:1353–1360

Zhang Y, He P, Li H, Li G, Liu J, Jiao F, Zhang J, Huo Y, Shi X, Su R, Ye N, Liu D, Yu R, Wang Z, Zhou M, Jiao N (2019) *Ulva prolifera* green-tide outbreaks and their environmental impact in the Yellow Sea, China. Natl Sci Rev 1–14

Chapter 11
Enzymes

Abstract Enzymes play important roles in catalyzing biological processes such as breaking down complex compounds in food into simpler forms which can be utilized to produce energy and aid metabolism and growth. These enzymes are also used commercially in products such as biological cleaning agents, biofuel, food and medicine. Aquatic sources of enzymes include internal organs of fish, carnivorous plants and microorganisms. Organisms which inhabit extreme aquatic environments are also sources of enzymes which perform better than other enzymes in industrial processes. Some aquatic enzymes also provide the option of more environmentally friendly processes for biofuel production.

Keywords Enzymes · Aquatic · Proteins · Extremozymes · Biopolymers

11.1 Introduction

In other chapters of this book, different forms of proteins such as collagen, aquatic plants and algae proteins have been covered. Enzymes are discussed in a separate chapter as they are indeed a diverse group of compounds, and their source, processing and commercialization can be differentiated from other proteins. While the majority of enzymes are sourced from microorganisms, plants and terrestrial animals, the aquatic environment is a great source of a wide variety of enzymes. Since 70% of the earth is made up of water and life itself is thought to have begun in water, the aquatic environment is therefore expected to be an even more abundant source of enzymes compared to land.

Enzymes play numerous roles in a range of food, pharmaceutical, chemicals and cleaning products. Their use also extends to biofuel production where enzymatic transesterification of fats can be used in biodiesel production as an alternative to the alkali-catalyzed transesterification. A number of processes would either occur too slowly to be commercially feasible or not at all without the use of enzymes. In food processing, enzymes are used in the production of products such as fish sauce, they are also employed in more effective alternative processes for product preparation such as fish descaling. Within the chapter, some of the different industrial applications of different enzymes are discussed. The origins of these enzymes, the

233

O. Olatunji, *Aquatic Biopolymers*, Springer Series on Polymer and Composite Materials,
https://doi.org/10.1007/978-3-030-34709-3_11

environmental implications of their production process, biochemistry, as well as the demand and commercial production are also discussed with the aim of giving the reader an understanding of the integral role of these classes of aquatic polymers within the natural world and in industry.

The diversity of organisms present in the aquatic environment and the intricate food chain makes it a promising source of enzyme-producing organisms. The large variety of organisms present in the aquatic environment, some of which feed on other organisms and non-living matter means these organisms in the aquatic environment produce the enzymes necessary to catalyze a wide variety of biological processes. Organisms including fish which produce proteolytic enzymes for breaking down proteins in the other sea life they feed on, marine bacteria which produce chitinolytic enzymes for breaking down chitins in the shells of crustaceans, and amphipods inhabiting the Challenger Deep of the Mariana Trench, the deepest part of the ocean, produce cellulases and amylases that allow them to digest wood from shipwreck and fallen plants as a means of surviving in the deep dark part of the sea where food is extremely scarce at extreme temperature and pressure (Kobayashi et al. 2012). Other examples of enzymes obtained from aquatic sources include trypsin, pepsin, lipase, collagenase and chymotrypsin.

Fish, particularly the internal organs of fish, fish viscera, is one of the most widely explored aquatic sources of enzymes as an alternative to the conventional sources (microorganisms, terrestrial plants and animals). Enzymes found in fish include hydrolases, proteases and carbohydrases, listed in order of their abundance (Kim et al. 2002). Algae are also a promising aquatic-sourced alternative for enzyme production. Algae have the ability to break down a wide range of compounds in the water and use for energy and carbon production. They can take up nutrients from water and break down carbon compounds. Photosynthesizing algae also make use of light to power their growth and energy production. They also have biochemical processes in place to protect from pathogens and environment. To do all these, they make use of enzymes which facilitate their synthesis. It is therefore expected that when they are harvested some of these enzymes remain within their biomass. Enzymes present in algae include carboxylase, oxygenase, dehydrogenase, amylase, cellulase, lipase, sucrose, phosphate, sulfatase, glycogen synthase, phosphatase, starch synthase, glycolate oxidase and peroxidases. All of these serve specific roles in the algae (Mogharabi and Faramarzi 2016). For example, superperoxide dismutase which is found in red algae as well as blue-green algae catalyzes the process which defends the algae from oxygen toxicity. The amount and composition of the enzymes present vary for different species of algae.

A wide variety of microorganisms exist in different zones of the aquatic environment. These microorganisms have adapted to survive in a range of conditions. Examples are microorganisms which exist in the underwater volcanoes of the deep sea where temperatures could be over 100 °C, while other bacteria are found in higher depths in waters where the temperatures can be as low as -2 °C in, for example, the Antarctic waters. The relatively less complex structure of the microorganisms makes them an attractive option for commercial enzyme production. This also allows the possibility of genetically engineering microorganisms to produce desired enzymes, where the gene coding for enzyme production from an aquatic organism can be placed in the nucleus of that of another organism which grows more effectively at the industrial scale. Microorganisms can be cultured in bioreactors which mimic the conditions in the part of the aquatic environment they inhabit for large-scale enzyme production. Within this chapter, we discuss an example process for production of enzymes from microorganisms obtained from the aquatic habitat.

Thus far within this text, we have discussed the main polymers extracted from algae for commercial applications. These include the common ones such as alginates, carrageenan and agar and those produced in lesser amounts like fucoidan, laminarin and ulvan. In the process of extraction of these biopolymers, the enzymes are mainly discarded or destroyed alongside other residues. However, algae are a source of a variety of enzymes of commercial value. It is therefore worth exploring the potential for commercial enzyme production from algae.

This chapter therefore explores the diversity of enzymes which are obtained from the aquatic environment, and how the aquatic-sourced enzymes differ from those of the terrestrial origin and in doing so the economic and environmental impact of aquatic-derived enzymes.

11.2 Occurrence in Nature

The aquatic environment is host to a diverse range of organisms. These organisms have adapted the means to survive a range of conditions which range from mild to extreme temperature, light, salinity and pressure. The complexity of life in the aquatic environment has resulted in the organisms inhabiting the aquatic environment evolving a range of biochemical processes for survival. These biochemical processes are catalyzed by the diverse range of enzymes, thus making the aquatic environment a formidable source of enzymes. Microorganisms, plants and animals within the aquatic environment metabolize enzymes which have several industrial applications. Some of these such as lipase and proteases are well established commercially, while some are still at infancy in terms of commercial application.

11.2.1 Microorganisms

The ease of genetic manipulation and mass culture for commercial production make microorganisms attractive source of commercial enzymes. Moraxella from Antarctic seawater and marine streptomycete are examples of microorganisms which produce the enzyme lipase (Zhang and Kim 2010). *Vibrio fluvialis*, *Listonella anguillarum* and *Vibrio mimicus* are examples of marine bacteria which produce the enzyme chitinase which hydrolyzes chitin. Microorganisms play a significant role in the breakdown of the dead tissue, and organisms in the aquatic environment, for example, the shed scales of marine zooplankton, are converted to carbon and energy by microorganisms

Table 11.1 Examples of some enzymes produced by different organisms

Enzyme	Example source	Substrate	References
Cellulase	*Cladosporium sphaerospermum* (isolated from Ulva algae)	Cellulose	Trivedi et al. (2015)
Chitinase	*Pseudoalteromonas piscicida*	Chitin	Paulsen et al. (2016)
Lipase	*Pseudomonas putida*	Triglyceride	Sivasubramani et al. (2013)
Protease	*B. mojavensis* (from seawater)	Protein	Haddar et al. (2009)
Agarase	Acinetobacter sp. PS12B (marine bacteria)	Agar	Leema and Sachindra (2018)
Alginate lyase	*Vibrio splendidus*	Alginate	Zhuang et al. (2018)
Alginate lyase	Mollusk (Lambis sp.)	Alginate	Sil'chenko et al. (2013)
Carrageenase	*Cellulophaga lytica*	Carrageenan	Yao et al. (2013)
Amylase	*Hirondellea gigas* (giant amphipod)	Amylose	Kobayashi et al. (2012)
Fucoidanase	Bacterium (Formosa algae)	Fucoidan	Silchenko et al. (2013)
Cutinase	*Thermobifida fusca*	Cutin	Hong et al. (2019)

which degrade chitin, and this serves as a source of food for these microorganisms. Humans have for long developed methods to breakdown polymers for different purposes such as production of fuel and processing of food, using chemical catalysts. Due to limitations and some adverse effects of the use of such chemicals, there is a shift toward the use of biological catalysts to replace the chemical catalysts. Table 11.1 gives some examples of catalysts and their sources.

The enzymes obtained from microorganisms have shown superior stability and activity compared to enzymes from plants and animals. However, the need to optimize the aquatic resource and the existing need to minimize the waste from the processing of resources such as seafood make the production of enzymes from aquatic plants and animals just as important.

11.2.2 Symbiotic Microorganisms

Aquatic organisms often have symbiotic relationships which aid their individual survival. Some of the microorganisms present in larger organisms produce enzymes which have industrial applications. For example, symbiotic bacteria present in the marine shipworm produce protease (Greene et al. 1996). These microorganisms harbored by the fish can then be removed from the respective organ, while the rest of the fish is used for food and other products. Microorganisms existing in the seawater are also a source of enzymes. These microorganisms can be collected and harvested in bioreactors for the production of desired enzymes. One such enzyme-producing microorganism found in seawater is the *Bacillus mojavensis* A21 (Haddar et al. 2009) which produces protease.

11.2.2.1   Biofilms

Some microorganisms form biofilms which protect them from harsh conditions and substances. The biofilm also supports their adhesion to wet surfaces in the natural aquatic environment and in wet surfaces in everyday life. These biofilms are composed of polymeric materials which are present in the exoskeletal matrix of microorganisms such as polysaccharides. Alginate forms a large amount of the biofilm material. At different points in the developmental stage, the biofilms need to be broken down to allow spreading of the film and/or the bacteria. This breakdown is achieved by alginolytic enzymes produced by the microorganisms. Recently, the possibility of using the alginate lyase enzyme produced by these biofilm-forming microorganisms to destroy biofilm-forming bacteria is being explored (Daboor et al. 2019). Biofilms aid resistance of bacteria to antibiotics in humans and antiseptic chemical in cleaning of surfaces. Extraction of alginolytic enzymes from biofilms to then attack the biofilms of these resistant bacteria provides a promising way to address antibiotic resistance.

11.2.3   Fish Viscera

Some of the common commercially available enzymes are sourced from the internal organs (viscera) and muscles of fish (Shahidi and Kamil 2001). These parts are generally not eaten; therefore, their production does not compete with food. It is also in line with the sustainable development goal of optimizing the utilization of fisheries resources since these parts are otherwise discarded. Examples of enzymes from aquatic source are the pepsin and collagenase from crab hepatopancreas (Kim et al. 2002). Pepsin is also obtained from fish. Extracts from squid offal contain collagen-degrading enzymes. The occurrence of enzymes in an aquatic animal can be affected by the type of diet the animal consumed. For example, juvenile green abalone showed a variation in the digestive enzymes they produced when fed with a diet of either algae or sea grasses (Garcia-Carreno et al. 2003). This can be attributed to the fact that the body will produce the type of enzyme required to digest the food it consumes. The activity of the enzyme produced by an organism at certain conditions is dependent on the habitat of the organisms.

11.2.4 Carnivorous Plants

While some plants depend on the nutrient supply from the soil or water they are grown in for their growth and survival, there are a special group of plants which have adapted features that allow then to capture prey, digest them and use their biomass as nutrients. Examples of species of aquatic carnivorous plants are *Aldrovanda vesiculosa, Utricularia vulgaris, U. reflexa* Oliver, *U. stygia* Thor and *U. intermedia* Hayne (Adamec 2010). However, there are over 700 species of known carnivorous plants.

These plants mainly feed on insects which they lure into the trap using sweet nectar they secrete. Once the insect is trapped, the enzymes within the plants digest the insect to produce the nutrients needed by the plants. The enzymes produced by the plants must therefore include all the essential enzymes to break down protein, lipids, chitin, glycogen and other compounds which make up the insect's body that can be converted to produce the carbon, nitrogen, sulfate and phosphate needed for plant growth. These carnivorous plants have been confirmed to have different enzyme compositions for digesting prey (Schulze et al. 2012). The minerals present in the insect body are also taken up by the plant. These enzymes are expected to have quite significantly high activity as some of these plants have been known to digest relatively large rodents.

Other than the enzymes they produce, carnivorous plants also make use of digestive enzymes of the symbiotic bacteria present within them to digest prey. An example of a carnivorous plant which has been confirmed to do this is the *Utricularia breviscapa* (Lima et al. 2018). This carnivorous plant grows in the floodplains in Brazil and can therefore be classified as aquatic. It floats on water by means of its parenchyma which is filled with air and traps insects with the aid of its segmented leaf structure with utricles which capture prey by creating a hydrostatic pressure. Bacteria species which are present in these carnivorous aquatic plants include Bacillus, Aquitalea, Sphingomonas, Azospirillum, Chromobacterium, Novosphingobium and Acidobacteria (Lima et al. 2018). Therefore, a carnivorous aquatic plant can have a host of bacterium in its trap which aids in the digestion of the prey. The types of bacteria and population distribution vary depending on the habitat.

Understanding the mechanisms of action of such plants could contribute toward the development of plant-based biological pest control and enzyme-based pesticides for aquaculture.

11.2.5 Extremophiles

Some organisms have adapted to survive in extreme conditions which require them to produce enzymes not commonly found in organisms of their kind. An example of such is the deep-sea giant amphipod, *Hirondellea gigas*, which has developed enzymes for digesting wood (Kobayashi et al. 2012). These organisms exist in the Mariana Trench, which is the deepest part of the ocean where the most extreme conditions exist and little life or organic matter exists to feed on. Other wood-digesting organisms which live in the higher zones of the water like shipworms and piddocks also exist. The activity of the enzymes and optimal conditions for activity vary as a result of the difference in the conditions under which these organisms exist.

This adaptation to extreme conditions gives enzymes sourced from these extremophiles the particular advantage of having relatively higher activities at extreme conditions (Fernandes 2016). Since the aquatic organisms which are the source of these enzymes are adapted to processing food and metabolic activities using these enzymes at such extreme conditions, the enzymes they produce for these activities are therefore conformed to retain their activity at such conditions. Such property becomes commercially useful where products such as food and pharmaceutics need to be processed at extreme conditions while retaining the activity of the enzymes.

Enzymes are therefore abundant in the aquatic environment and are present in a diverse range of organisms. Their occurrence in a diverse range of conditions such as temperature, salinity, pH, pressure and light intensity gives rise to the availability of enzymes that are optimally active in more varied conditions compared to enzymes sourced from terrestrial organisms. The fact that the aquatic environment makes up a larger portion of the earth and hosts more diversity of organism also results in more naturally occurring enzymes in the aquatic environment.

11.3 Chemistry of Some Aquatic Enzymes

Enzymes are mostly proteins (some enzymes exist that are not proteins), chains of amino acids linked together by peptide bonds. Amino acids being the repeating units of these proteins are a diverse class of compounds which are characterized by the presence of an amine and a carboxylic acid group attached to an alkyl. Enzymes can have molecular weights in the thousands to tens of thousands range. Proteinase A, for example, has a molecular mass of 50 kDa, while proteinase B has a molecular mass of 95 kDa (Komori and Nikai 2013).

The functioning of an enzyme depends on the primary, secondary and tertiary structure of the enzyme. Most enzymes are globular proteins which fold up in specific patterns in their tertiary structure. This unique tertiary structure allows them to combine with specific substrates in a unique manner and by so doing catalyzes specific processes. The activity of enzymes varies at different conditions. For example, alkaline proteinase will act at pH above 7, while acid proteinase will act at the low end of pH and lipases will only act at the interface between water and oil since lipase is hydrophilic and the triglyceride it breaks down is hydrophobic (Bele et al. 2014a, b). Therefore, unlike nutritional protein which is required to be broken down to produce amino acids which are then utilized in this form, enzymes are required to retain their tertiary structures in order to serve their purpose.

Some enzymes are specific to particular bonds, while some can catalyze a wide range of processes. For example, lipases catalyze the breakdown of fats to produce fatty acids, glycerol and monoglycerides, while proteases cleave amide bonds to produce shorter chain peptides. Figure 11.1 illustrates the cleavage of amide bond to form a more stable cyclic peptide obtained from *Ulva rigida* (Seca and Pinto 2018), and Fig. 11.2 illustrates the breakdown of triglyceride into fatty acids and glycerol. These are examples of two different processes which are catalyzed by proteolytic enzymes and lipase enzymes, respectively.

Lipases also catalyze a wide range of reactions which include hydrolysis, acidolysis, esterification, alcoholysis, aminolysis and interesterification (Bele et al. 2014a, b). In catalyzing the range of reactions which break down fat, lipases can also be used to catalyze the synthesis of other lipids. Examples of such are the acidolysis to release the hydroxyl from the COOH group from the fatty acid and alcoholysis which releases the alcohol group of the glycerol allowing the formation of triglyceride. Therefore, lipases are not just able to catalyze the breakdown of triglycerides but also formation of other similar compounds. An example of lipase-catalyzed lipid production is the formation of ergosterol ester (He et al. 2019) from ergosterol, a cholesterol equivalent produced in fungi (Bell 2007). Lipase-catalyzed formation of

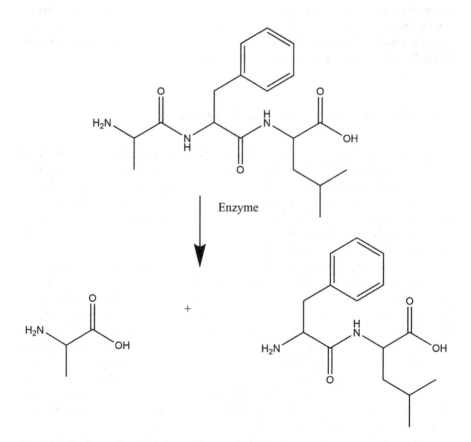

Fig. 11.1 Cleavage of amide linkage to form amino acid

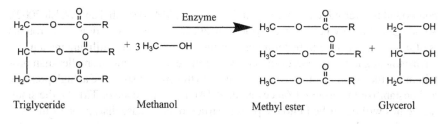

Fig. 11.2 Transesterification reaction

ergosterol ester from ergosterol is an important process because the ergosterol ester has more diverse application than ergosterol. The ester form has a lower melting point, higher solubility in oil and improved stability.

The presence of enzymes in a sample can be confirmed by their level of activity. For example, to confirm the presence and activity of proteolytic enzymes, the amount of protein degraded into amino acid can be measured using the Bradford method and the amount of lipase can be measured by introducing into a lipid sample such as castor oil and quantifying the amount of fats converted to fatty acids in a given time by titrating with an alkali and using an indicator to identify the quantity of alkali required to neutralize the fatty acids formed. The formation of clear white crystals upon introduction of a lipase containing sample onto agar plates containing 1% (v/v) Tween 20 or Tween 80 is also used as an indication of lipase activity (Bele et al. 2014a, b).

Understanding the chemical structure of different enzymes is important in identifying their activities and hence applications. This provides a better understanding of the aquatic world. For example, the ability of salmon sharks to maintain relatively high body temperature in the cold Alaskan waters which could be as low as $-2\,°C$ (Bernal et al. 2005) is partly attributed to the difference in the enzyme activity in red and white muscles and the arrangement of these muscles within the organism's internal organs (Glancy and Balaban 2011).

11.3.1 Enzyme Stability in the Deep Sea

The tertiary protein structure of enzymes needs to be retained in order to ensure they remain active and serve their intended function. The tertiary structure can be destroyed by factors such as temperature, pH, salinity and pressure. The functioning of these enzymes is crucial to the survival of the organism; therefore, for organisms living in extreme conditions, certain mechanisms are put in place to retain the enzyme activity. Understanding the way in which the enzyme structure is retained at such extreme conditions can be adapted to commercial applications of enzymes, for example, in developing enzyme-catalyzed reaction at high pressure.

At high pressures in the sea, marine organisms have developed a special compound which allows them to retain their protein structure at such pressures. This compound is referred to as trimethylamine oxide (TMAO) (Seibel and Walsh 2002). Although not a polymer or an enzyme in itself, its unique interaction with proteins is of much significance to better understanding, conserving and utilizing the aquatic environment and the resources therein. Water molecules are much smaller than protein molecules such that they can penetrate into the protein structure at high pressure and destabilize the protein (Yancey et al. 2014). In the absence of TMAO, the small molecules will be pushed into the protein structure and cause disruption of the protein tertiary and secondary structure thus preventing it from functioning, eventually resulting in the death of the organism.

Although the precise mechanism of action is yet to be uncovered, it is seen in higher concentration in the organisms which live in these deep waters compared to those in the lower depths, and hence, it is attributed to their ability to survive at the high pressures which exist in the deep sea. TMAO is also attributed with the resistance of the enzymes in the fish to urea and also preventing the freezing of the fish's bodily fluid at the low temperatures in the deep sea (Seibel and Walsh 2002).

The compound which helps retain the enzyme structure and hence activity in fish against the high pressure in the deep sea may, however, pose a health risk for humans. TMAO has been associated with adverse cardiovascular events in humans (Velasquez et al. 2016). However, more studies are required to confirm the mechanism by which this occurs and other associated factors.

11.4 Availability of Raw Materials

In 2016, 89,000 tonnes of microalgae was farmed across 11 countries of the world, 88,600 tonnes of which was from China. These include species such as *Haematococcus pluvialis*, Nannochloropsis spp., Chlorella spp. and Spirulina spp., all being farmed in large, medium and small scales. While macroalgae get a larger revenue from their food sales, microalgae are mostly sold as high-value functional products. Therefore, despite the lower annual tonnes produced, microalgae are valued at around a billion USD annually, compared to that of macroalgae at 6 billion USD.

The protein content in macroalgae is comparable to those found in animal-based proteins and is higher than those found in land plants such as soybean, wheat and legumes. Algae yield around 2.5–7.5 tonnes per hectare annually, while microalgae yield 4–15 tonnes per hectare annually. These yields are rather high compared to the conventional plant-based proteins such as wheat, soybeans and legumes which yield 1.1, 0.6–1.2 and 1–2 tonnes per hectare annually (van Krimpen et al. 2013). Macroalgae and microalgae could contain similar or even more protein than terrestrial plants typically used as protein source. Spirulina, microalga which have gained much popularity as a nutrient source, could contain up to 63% protein per dry weight (Tokusoglu and Unal 2003). The red algae species *Porphyra tenera* contains up to 47% protein per dry weight (Fleurence 1999).

Protein content in any particular algae varies with factors such as growth season, temperature, harvest period, region and nutrient content of water. The types of proteins present also vary. How different species react to particular changes in growth condition in turn affects the types of proteins they metabolize. The algae make use of these proteins to survive and function within its environment; therefore, the stimulus it gets from these environments determines what protein it is prompted to produce. These factors can be used to manipulate a particular alga to produce desired type of protein by controlling the growth environment. This requires an in-depth understanding of the correlation between environmental factors and the metabolism of the specific species. For example, highest protein yield is obtained from the alga *K. alvarezii* in August, November and February when studied over 12 months from September 2004 to April 2006 in northwestern India (Kumar et al. 2014).

Aquatic plants such as water fern, duckweed and water hyacinth contain relatively moderate to high amount of proteins, 28% protein content by dry weight has been reported for water fern (Brouwer et al. 2019) and 21.5% for duckweed (Aguilera-Morales et al. 2018). While there are up to hundreds of thousands of species of algae, some of which belong to the plant kingdom, there are just about a hundred species of non-algae aquatic plants. The recent rise in algae bloom incidents in different parts of the world further eludes to the fast rate of growth of algae. Water hyacinth, ferns and duckweed are also fast-growing plants which can be explored as abundant sources of aquatic proteins. While they could serve as an abundant source of proteins, well managed and controlled systems for the cultivation of aquatic plants and algae are important in order to avoid undesirable population blooms which could result in devastating environmental impacts.

11.5 Extraction of Enzymes from Aquatic Organisms

The processes involved in the isolation of enzymes depend on the source from which it is being extracted. Here we take examples of methods reported for the extraction of enzymes from fish viscera, aquatic microorganisms and algae.

11.5.1 Enzyme Extraction from Fish Viscera

The fish viscera can be obtained as a by-product of fish processing factories, for example, in the production of sushi, canned fish, dried fish and packed fish fillets. In all the aforementioned processes, the internal organs of the fish are removed as waste. The internal organs can also be collected from the fishmongers in markets. Here the method used by Jayapriya et al. (2014) is presented as an example for the extraction of digestive enzyme from fish viscera.

The collected organs are washed with distilled water followed by grinding. The pH is then adjusted to 8 by adding Tris-HCl (0.02 M). Crude extract is obtained as the supernatant after centrifugation at 6000 rpm for 15 min. This crude extract contains the enzymes mixed with other compounds. Ammonium sulfate at 80% saturation is then added to precipitate the crude enzyme from the liquid supernatant. The precipitate is then separated from the rest of the liquid by centrifugation at 10,000 rpm for 15 min. To further purify the enzyme, the solid precipitate is dissolved in 0.2 M Tris-HCl buffer at pH 8. This solution is then purified by dialysis. This method can be applied for the extraction of enzymes from waste of fish such as seer fish sardines, great barracuda, red snapper and milk shark. The protease extracted using this method has optimum enzyme activity at a pH of 10.

11.5.2   Extraction of Enzyme from Algae

The process of isolating enzymes from algae requires first freeing up the enzymes from the solid biomass. Like most enzymes, the enzymes present in algae are water-soluble globular proteins. Homogenization of the algae biomass followed by separation of the liquid and solid by centrifugation will yield a liquid supernatant which comprises a mixture of some soluble polysaccharides and other proteins. The enzymes can then be precipitated out of the solution. This is then redissolved and further purified using methods like gel filtration (Mogharabi and Faramarzi 2016).

In one example extraction process reported by Bele et al. (2014a, b), the algae biomass is collected and washed with water to remove debris. It is then dried and ground into powdered form. The powdered algae biomass is mixed with distilled water to allow all water-soluble components to dissolve. The liquid is then separated from the solid mass. The protein is precipitated out of the aqueous phase using ammonium sulfate added at a mass-to-volume ratio of 0.0663 g:1 ml. The solid protein precipitate is then recovered by centrifugation at 10,000 g for 15 min at 4 °C. The solid fraction contains the enzyme. This is then redissolved in trichloroacetic acid (TCA) buffer and further purified by gel filtration using Sephadex grade 100 (Bele et al. 2014a, b). This method was used to extract enzymes from *Ulva lactuca*, *Ulva fasciata*, *Chaetomorpha antenna*, *Gelidium pusillum* and *Enteromorpha compressa*.

Magnesium sulfate has been conventionally used for precipitation of globular proteins for several decades (Howe 1921). Other salts such as sodium sulfate and magnesium sulfate can also be used for similar purposes. The choice of salt depends on the requirements of the processor. For example, protein precipitates formed using magnesium sulfate are gelatinous in nature and this results in relatively slow filtration; sodium sulfate is preferred when working at temperatures above 34 °C, and ammonium sulfate can be used where the nitrogen element in the salt does not pose a challenge in processing or analysis.

Enzyme extraction from marine algae can also be carried out using two-phase extraction with PEG-4000 (Bele et al. 2014a). In this method, sodium sulfate is added to the supernatant of the centrifuged sample 0.75 g in 10 ml mass of sample per volume of salt solution. To this, ~3 ml of PEG-4000 is added at a 50% concentration.

This results in the solution separating into two phases (indicating the presence of globular proteins). The phases can be separated by pipetting out each phase.

In a third method, the protein can be separated using acetone–ether precipitation. An oil–water emulsion is first formed using castor oil and sunflower oil in a bile salt solution at a neutral pH and gum acacia as emulsifier. The ground algae biomass is homogenized with acetone at ice-cold temperature. This is followed by filtration with the solid residue retained and washed with acetone, an equal volume of acetone and ether and then with ether. This process removes the lipids, polysaccharides and other compounds which dissolve in acetone and ether from the algae leaving behind the protein as the solid mass. The water-soluble protein is then obtained from the dry mass by dissolving in cold water followed by centrifugation at 15,000 rpm for 10 min. The supernatant is obtained as the enzyme containing fraction (Bele et al. 2014a, b).

The salts used in precipitation can be removed by methods such as ion-exchange resin or gel filtration using Sephadex G-25. The method of precipitation significantly affects the yield of protein and the enzyme activity. For example, using the ammonium sulfate precipitation method, protein concentration in the extract from the alga *E. intestinalis* was 117 μg/ml and the enzyme activity was 0.123 meq/min/g using castor oil as the fat substrate. Protein concentration from *U. lactuca* was lowest of all the algae tested in the said study. It had a protein concentration of 87 μg/ml and an enzyme activity of 0.109 meq/min/g (Bele et al. 2014a, b).

Upon purification using methods like gel filtration, the purity of the enzyme is confirmed by SDS-PAGE (Jayapriya et al. 2014). This confirms the molecular weight of the polypeptide extracted from the sample. A single band confirms that the protein extracted has a single molecular weight value. For example, purified protease extracted from great barracuda viscera shows a single band at 34 kDa (Jayapriya et al. 2014), while crude lipase extracted from algae shows a band ranging between 45 and 60 kDa (Bele et al. 2014a, b), indicating a less pure sample.

Since in both cases the enzyme is purified mainly of the bases of molecular weight, the extract contains all proteins within a specified molecular weight range. Some of these might not be enzymes and will also contain a combination of different enzymes. The distinction is made by introducing the enzyme into a reaction it catalyzes following incubation at the appropriate conditions. For example, the extract from algae is mixed with castor oil formulation following incubation at 37 °C and shaking at 200 rpm (Bele et al. 2014a, b). The amount of fatty acid produced is then determined by titration against alkali (sodium hydroxide) to obtain a measure for the lipase activity present within the sample.

11.5.3   Enzyme from Marine Microorganisms

Several marine organisms isolated from the waters and from aquatic organisms have proven to be reliable sources of enzymes such as chitinase, cellulase, lipase and agarase. Here we look at the process of extraction of lipase from marine microorganisms. The general approach to commercial production of enzymes from microorganisms is to cultivate the microorganism in a medium which promotes the metabolism of the desired enzyme. For this purpose, it is important to study the growth kinetics of the microorganism in order to identify the growth phase during which it produces the desired enzyme at optimal rate. Once the desired amount of enzyme is produced by the microorganism, the rest of the process is then directed at recovering and purifying the enzyme free of the microorganism. One of the challenges of sourcing enzymes from microorganisms is the risk of contamination of the final product with the microorganism which could be pathogenic, especially where the enzyme is used in food or pharmaceutical products.

Enzyme production from microorganisms is illustrated here using the method presented in the literature for the extraction of lipase from marine bacteria as a case study. In the study by Sivasubramani et al. (2013), marine bacterium *Pseudomonas putida* isolated from the Vellar estuary in the Cuddalore district, Tamil Nadu, Southern India, is used as the source of the lipase enzyme. Other lipase-producing marine bacteria strain includes Bacillus spp. (Valsa et al. 2003) and marine Streptomyces sp. (Yuan et al. 2016).

The temperature, pH, salinity and nutrient composition are critical parameters to be controlled when culturing microorganisms for enzyme production. In this particular example, the optimal values for these parameters were a neutral pH, 35 °C and a sodium hydroxide concentration of 0.6%. The nutrient in the medium consisted of 1% dextrose and 0.2% yeast extract. Companies such as Sigma-Aldrich supply nutrient media which are optimal for required microorganism culture. The culture was incubated for a duration of 48 h.

First, the microorganisms are isolated from the water of the estuary. The sample is then incubated in Tween 80 agar plates to isolate the lipase-producing microbes. The optimal growth conditions of the most productive strain are then identified by growing at different growth conditions and finding the optimum pH, temperature, salinity, nutrient composition and duration. The bacteria strain is then grown in a shake flask using the optimal parameters determined. For a particular microorganism and desired enzyme, the optimal growth conditions will vary. The optimal conditions will also vary for different strains; therefore, the processor must determine the optimal growth conditions for a specific strain and target enzyme.

The inoculum is then introduced into a bioreactor in a ratio of 1:100 (volume of inoculum to volume of bioreactor medium). The culture was grown over a duration of 48 h at the optimal conditions. At the end of the process, the medium was filtered. The liquid filtrate was precipitated with ammonium sulfate followed by centrifugation at 3000 rpm for 30 min at 4 °C. The solid precipitate is then dissolved in 0.05 M Tris-HCl and then purified by dialysis. The enzyme activity of the purified protein was 9.474 U/mg at the highest when tributyrin is used as the triglyceride substrate.

SDS-PAGE analysis showed that the partially purified protein sample containing the enzyme derived using the method described had molecular weights of 34, 45 and 52 kDa, which are within the range of molecular weight of lipase. The actual molecular weight of lipase has been reported to be lower, and others have reported higher molecular weights of lipase from SDS-PAGE analysis. For example, Yuan et al. (2016) reported lipase extracted from marine bacteria with a molecular weight of 29 kDa.

The enzyme activity varies depending on the organism it is sourced from. Table 11.2 compares the enzyme activity of some enzymes from different species of fish, algae, microorganisms and the giant deep-sea amphipod *H. gigas*. The different units are given as presented in the different studies to measure enzyme activities. In the study carried out of *H. gigas* (Kobayashi et al. 2012), the enzyme activity of the different enzymes present in the deep-sea residing amphipod varied significantly for each enzyme. The amylase activity 112.1 mU/μg is much higher than the cellulase activity. The activity of the other enzymes, protease, xylanase and mannanase is even much lower. This is indicative of the types of nutrients available to the organisms for energy and growth.

Table 11.2 Enzyme activity of protease and lipase from different sources

Enzyme	Source	Activity	References
Lipase	*Ulva lactuca* (Puducherry)	0.057 meq/min/g	Bele et al. (2014a, b)
Lipase	*Enteromorpha compressa* (Puducherry)	0.062 meq/min/g	Bele et al. (2014a, b)
Lipase	*Gelidium pusillum* (Kovalam)	0.086 meq/min/g	Bele et al. (2014a, b)
Lipase	*Chaetomorpha antenna* (Puducherry)	0.068	Bele et al. (2014a, b)
Lipase	Marine bacteria Pseudomonas	1.763 U/mg	Sivasubramani et al. (2013)
Lipase	Bacteria from Antarctic water *Arthrobacter gangotriensis*	7.894 U/mg	Rashidah et al. (2006)
Protease	Red snapper	Crude: 3.79 U/mg Purified: 58.80 U/mg	Jayapriya et al. (2014)
Protease	Great barracuda	Crude: 1.51 U/mg Purified: 35.17 U/mg	Jayapriya et al. (2014)
Protease	Sardines	Crude: 2.32 U/mg Purified: 36.04 U/mg	Jayapriya et al. (2014)
Protease	Milk shark	Crude: 1.53 U/mg Purified: 26.43 U/mg	Jayapriya et al. (2014)
Protease	*Aureobasidium pullulans* (marine yeast)	Purified: 623.1 U/mg	Zhang and Kim (2010)
Amylase	*H. gigas*	112.1 mU/μg	Kobayashi et al. (2012)
Mannanase	*H. gigas*	8.22 μU/μg	Kobayashi et al. (2012)
Cellulase	*H. gigas*	3.42 mU/μg	Kobayashi et al. (2012)
Protease	*H. gigas*	0.26 mU/μg	Kobayashi et al. (2012)

It is noted that the process of recovery of the enzymes from the solid mass following cultivation of the microbes or homogenization of plant and animal biomass is quite similar. The common goal is therefore to free up the proteins and then isolate the desired enzymes.

11.6 Immobilization of Enzymes

Since enzymes are biological catalysts, they do not take part in the actual reaction or biochemical process, they serve as surfaces for substrates to attach, and once the process is complete, the enzymes are released and become available again. In industrial processes, the enzymes need to be recovered and stored in a suitable form for reuse. This is achieved by immobilizing the enzyme. Enzymes are mostly proteins which are soluble in water. This makes their recovery quite challenging after a process; evaporation or crystallization at high temperature is not an option since most enzymes are denatured or completely destroyed at temperatures about ~45 °C. Immobilization of enzymes therefore entails converting them into insoluble forms which can be more effectively collected, stored and reused. This can be achieved by chemical or physical immobilization process (Dutta 2008). Other than being a source of enzymes, aquatic organisms are also used in the immobilization of enzymes and these are also discussed within the section.

Whole cells of enzyme-producing organisms can also be immobilized with the activity of the enzyme and organisms remaining active. Although the use of immobilized enzymes and biological catalysts is at industrial scale which is still at infancy (Moreno-Garcia et al. 2018), immobilized enzymes offer benefits such as avoiding the live organism or enzyme from getting into the final product and cost savings. Whole algae cells of *N. muscorum*, for example, can be immobilized in sodium alginate. When immobilized in 2% alginate at 30 °C, a concentration of 0.5 g/L and a stirring rate of 100 rpm, the cells retained their activity even after five cycles of use. The algal cells also showed higher yield in the immobilized form compared to when used in their free form to catalyze the bioconversion of androst-4-ene-3,17-dione to testosterone (Arabi et al. 2010). The rest of this section discusses the different enzyme immobilization techniques, and in the process, some biopolymer used in enzyme immobilization is also discussed.

11.6.1 Chemical Immobilization

The chemical immobilization process can be done by covalently bonding the vacant functional groups on the enzymes to insoluble supports. These supports could be monomers which are then copolymerized with the enzymes. Chemical immobilization can also be achieved through cross-linking the enzyme with multifunctional reagents such as glutaraldehyde. In any form of chemical immobilization, it is important that the active site of the enzymes is not altered; otherwise, the enzyme is destroyed. Therefore, the functional groups involved are those which do not affect the enzyme's activity.

Natural polymers are also used in chemical immobilization of enzymes. Agarose is commonly used for such application. Polymers generally have minimal reactivity with enzymes; therefore, when used for immobilization of enzyme, polymers are first activated with reagents prior to contacting with the enzymes. This pretreatment is targeted at freeing up specific functional groups on the polymer to allow covalent reaction with the inactive sites on the enzyme.

11.6.2 Physical Immobilization

In the physical method, immobilization can be achieved by entrapment, microencapsulation or adsorption of the enzyme within or unto a support. The adsorption method makes use of surface-active compounds unto which the enzymes are physically adsorbed. Cellulose and collagen are examples of natural polymers used as adsorbents for physical enzyme immobilization. Other adsorbents are calcium carbonate, alumina, clays and glass plates. Hydroxyapatite, although not a polymer, is also an interesting material used as physical enzyme immobilization adsorbent which occurs naturally in the aquatic environment. Hydroxyapatite is present in the scales of fish where it makes up part of the extracellular matrix. Adsorption method is particularly preferred as a method of enzyme immobilization as it is relatively simple process, it is reversible and there is little chance of the enzyme being deactivated in the process of immobilization. It, however, has the disadvantage of the weak bond between the enzyme and the support which are mainly due to secondary molecular interactions unlike the chemical method which are formed with covalent bonds.

The entrapment method makes use of cross-linked polymers within which the enzymes are trapped. The polymer is mixed with the enzyme prior to cross-linking. The enzyme dispersed within the polymer is then trapped within the network upon cross-linking. In the microencapsulated method, the enzymes are placed in the semipermeable membrane microcapsules. The microcapsules are formed using interfacial polymerization where the enzymes are encapsulated within the droplets of polymers formed. This method yields an immobilized enzyme system with high surface area.

11.7 Environmental Implications

Table 11.3 gives a summary of consumption in the process of a typical enzyme extraction process using the method presented by Bele et al. (2014a, b) as an example case study. The reader should note that the processes vary for different feedstocks.

Table 11.3 Estimated consumption in enzyme extraction from algae

Consumption	Amount
Water	~20 ml per gram algae
Energy	Drying (atmospheric) Grinding Cooling (4 °C) Centrifugation: 10,000 g at 4 °C Filtration and purification
Biomass	~3.7 g algae biomass required per gram protein
Salts and buffers	Ammonium sulfate ~0.663 g per gram of algae TCA 10 ml per gram enzyme extract
Solid waste generated	82.8%

11.7.1 Cultivation of Algae

A wide variety of enzymes can be obtained from different algae species. Currently, thousands of species of algae are not cultivated commercially (Arabi et al. 2010). Production of enzymes from algae could increase the diversity of algae species being cultivated for different purposes with enzyme production included.

11.7.2 Replacement of Alkali or Acid Catalyst in Biodiesel Production

Production of biodiesel can be made even more environmentally friendly by eliminating the use of acids and alkali and replacing with lipase, a biological catalyst. The current limitation with this approach to biodiesel production is that the enzyme-catalyzed transesterification reaction is slower compared to the chemically catalyzed one. The issue of reusability of the enzyme can be addressed by immobilization of lipase on a support surface such that the enzyme can be used to catalyze several cycles (Narwal and Gupta 2013).

11.7.3 Water Consumption

The consumption of water begins with the first washing of the algae biomass to get rid of debris. More water is then used to dissolve the proteins prior to centrifugation. The exact amount of water used is not specified within the test referenced for the extraction of enzyme from algae. However, taking a common mass-to-volume ratio of 1:10 as a minimal used in similar extraction processes, we can estimate that a minimum of 20 ml of water is used in the extraction of enzyme from 1 g of algae.

11.7.4 Use of Salts and Buffers

Where ammonium sulfate is used, this eventually breaks down into nitrogen, hydrogen and sulfate which can be taken up by plants and other organisms if adequately disposed of. Unlike other more intensive processes such as the extraction of chitin from crustacean shells or extraction of collagen from animal tissue which makes use of strong acids and alkali, enzyme extraction is relatively milder and requires use of milder chemicals.

11.7.5 Energy Consumption

The homogenization of the algae biomass into powdered form achieves improved release of the proteins into water as a result of reduced particle size and mechanical disruption, and this process requires significant amounts of energy. Drying of the samples prior to homogenizing reduces the amount of energy consumed in this process. The quantity of energy consumed at this stage varies depending on the machinery used and the efficiency.

The drying process is done at ambient conditions; high-temperature oven-drying is not an option here as this will result in the destruction of the tertiary structure of the enzyme which is a key to their activity as catalysts.

Energy is also consumed in the centrifugation at different stages at ~10,000 to 15,000 g for periods of 10 and 15 min at low temperatures ~4 °C. Therefore, energy is required for cooling of the system. Cooling is usually achieved using electricity in refrigerated centrifuges. The environmental impact of both centrifugation and cooling energy consumption depends on the source of electricity. This could be hydroelectric, nuclear, gas, solar, etc.

11.7.6 Solid Waste Generated

While up to 60% of fish is eaten as food, the rest of it either goes to waste or used in lower value applications such as fish meal, pet food, fertilizers and fish oil. Production of enzymes from the solid waste generated from fish significantly boosts the value chain. Enzymes have a much broader application in industry compared to the former, and adding this to the value chain of the fish food industry can contribute revenue which could compensate for the cost of production of other products from the fish waste such as production of chitin from the fish scales (chitin is discussed in a separate chapter in this book). Proteins make up around 17.2% of *U. lactuca* (Aguilera-Morales et al. 2018) for example. This means that approximately 82.8% of the rest of the algae form the solid waste generated. This can then be further processed to extract other biopolymers such as ulva and carrageenan.

11.8   Applications

Enzymes from aquatic sources find applications ranging from food processing to cleaning agents. In various chapters of this book, reference has been made to some extraction processes which make use of enzymes such as proteases to extract biopolymers. Therefore, enzymes play a role in the process of production of other valuable products.

11.8.1   Biological Cleaning Agents

Much of the stains in everyday life are food stains, blood stains, ink stains or other stains which are likely to have polymeric components such as protein in blood and food. Enzymes which can break down these compounds are therefore ideal cleaning agents as they can break down the large molecules of the stain into smaller molecules which wash off more easily. An example of enzymes used for such application is protease. Protease has been long known for its cleaning properties. Protease extracted from the marine shipworm's gland of Deshayes has superior cleaning power when compared to the conventional phosphate detergents (Greene et al. 1996). The said protease remains active at temperatures up to 50 °C and is completely inactivated at 70 °C. This is ideal as this is the general temperature range for eco-friendly washing at room temperature. Protease also has potential application as a cleaning agent for contact lenses as it retained stability and cleansing activity in the chemicals such as hydrogen peroxide and sodium hypochlorite used in the cleaning and storage of such product (Greene et al. 1996). Another example of protease from microorganism which shows good cleansing properties is the protease obtained from a culture of the bacterium *B. mojavensis* which is obtained from seawater (Haddar et al. 2009).

Protease from the viscera of fish has been shown to effectively remove blood stain from cloth (Jayapriya et al. 2014). The effectiveness of the enzyme depends on the species it is extracted from. For example, enzyme from great barracuda showed a stronger stain removal activity compared to that from sardines, milk shark, seer fish and red snapper. Other enzymes such as lipase which breaks down triglycerides into fatty acids and glycerol can also be used to remove fat-based stains. Lipase is obtained from marine algae (Bele et al. 2014a, b).

The enzymes are combined with regular detergents, solutions and chemicals to serve as biological cleaning agents. It is therefore important that these enzymes are stable at the pH and temperatures and maintain their activity and stability when in contact with the different chemicals which are used in the cleaning products.

11.8.2 Biodiesel Production

The production of biodiesel is a transesterification reaction between fatty acid and methanol in the presence of a catalyst. This yields glycerol and soap as a by-product alongside the biodiesel. Lipase produced from algae has potential application in biofuel production. As mentioned earlier, one of the reactions catalyzed by lipase is the transesterification reaction. In fact, lipase-catalyzed transesterification has up to sevenfold higher yield than sodium hydroxide-catalyzed transesterification. This has been observed when the transesterification of lipids extracted from green microalga Tetraselmis sp. using both catalysts was compared (Teo et al. 2014). The aforementioned study made use of immobilized lipase.

11.8.3 Antifungal Agents and Pesticides

The key activity of enzymes is in catalyzing the breakdown of compounds such as chitin, cellulose and lipids. This breakdown of compounds can be associated with antimicrobial and pesticide activity. Fungi and insects, for example, are made up of chitin which provides them with a level of resistance to environment or chemical damage. Therefore, enzymes which can break down chitin present in the fungal cell walls and insect exoskeletons can act as antifungal and antimicrobial agents. The marine bacterium *Pseudoalteromonas piscicida*, for example, shows antifungal activity against seven different fungi strains which includes *P. galatheae*, *V. neptunius*, *P. piscicida*, *P. rubra*, *P. fuliginea* and *V. fluvialis* (Paulsen et al. 2016). This form of antifungal and pesticides provides an alternative to the chemical-based ones which are often less selective and could have adverse effects on whole crop and product.

11.8.4 Bioactive Oligomers

Some biopolymers have improved bioactivity when they are partially hydrolyzed into smaller molecular weight oligomers. These are short-chain polymers with a degree of polymerization less than 100. The bioactivity of some polymers such as alginate and fucoidan can be tuned by breaking specific bonds on the polymer chain. For example, alginate which is made up of a combination of block and alternating chains of mannuronic and guluronic acid can be broken down into smaller chains of either purely mannuronic or guluronic acid oligomers or alternating chains of mannuronic and guluronic acid units using enzymes which are specific to the respective sites.

For example, a poly-1 $\rightarrow$ 4-alpha-L-guluronate lyase enzyme cleaves the 1 $\rightarrow$ 4 glycosidic bond of alpha-L-guluronic acids (Sil'chenko et al. 2013). This enzyme was extracted from the mollusk Lambis sp.

11.8.5   Fish Processing: Skin and Scale Removal

Proteolytic enzymes are used in removal of skin and scales in processing of fish food. The use of enzymes for skin removal in fish has been known for a few decades (Amano 1962; Kim et al. 2002). Manually descaling and deskinning fish are often time-consuming and require more man power. This process could be made faster by treating the fish with enzymes where a batch of fish can be processed at once.

The goal of enzyme-assisted skin removal is to remove the fish skin without damaging the muscles which are the part required for the fish food product. The fish skin is often removed as it contains fat and in particular food preparations such as fish fillets, the texture is undesirable. In other seafood such as squid, the removal of the skin is necessary to improve the organoleptic properties and also improve storage. In removal of fish skin, a combination of pepsin and carbohydrase is often used to aid efficient removal. Lipase can be applied in the defatting of fish during processing (Zhang and Kim 2010). This results in a less physically demanding process compared to manual fat expression and is also milder than use of alkaline for defatting.

Proteolytic enzymes from fish are also used in the process of extracting other compounds from aquatic waste. For example, in the extraction of glycosaminoglycans from the viscera of Nile tilapia, proteolytic enzymes are used in the deproteinization of the tissue (Nogueira et al. 2019). GAG is also a polymer which is valued for its anticoagulant properties and other pharmacological applications. The enzymes from aquatic waste, either endogenous or exogenous, are also useful in obtaining other valuable by-products from aquatic waste.

11.8.6   Fish Sauce and Fish Protein Hydrolysate Production

Fish sauce is produced from the fermentation of fish such as sardines and anchovy; generally, fish growing in the pelagic zone of the water are less expensive catch. The process occurs at room temperature, pH between 5.5 and 6.5 and high salinity which is achieved by adding 20–30 weight % of salt. A mixture of endogenous enzymes is allowed to ferment the fish and produce the sauce as the fermented by-product. The high salinity eliminates much of the bacteria which could also ferment the fish to produce undesirable by-products; therefore, the salinity ensures the right conditions for the endogenous enzymes and microorganisms which will result in the production of fish sauce. The process of fermentation takes place between 6 and 12 months at the end of which an ambered colored solution of fish sauce and salt is obtained from the bottom drain.

Although much of the enzymes which act on the fish protein to produce fish sauce are the endogenous enzymes, some bacterial activity, particularly in the first 3 weeks, does result in the production of fatty acids and other metabolites which contribute to the distinct flavor of fish sauce (Kim et al. 2002). These are likely the halophiles which can survive the high salinity. The activity of trypsin and chymotrypsin is significantly reduced during the process due to the high salinity; however, the activity of these enzymes is still important to the production of fish sauce. Proteolytic enzymes remain at maximal activity throughout the process. The level of activity of the microorganisms or enzymes is also dependent on the species of fish and the conditions.

The activities of these enzymes can be accelerated by either increasing the pH to alkaline conditions, reducing the pH to higher acidity which accelerates hydrolysis or increasing the temperature to 45 °C which is high enough to accelerate the process but without denaturing the enzyme. At the varied pH, certain enzymes which promote the breakdown of the fish proteins will be enhanced. For example, at higher pH and lower salinity trypsin, chymotrypsin and alkaline proteinase are more active. These methods could reduce the fermentation time by about 80%; however, taste of the resulting sauce is often inferior to that at longer period.

Fish protein hydrolysate is produced by treating the fish with only proteolytic enzymes. The proteolytic enzymes specifically act on the peptide bonds. In the process of production of fish protein hydrolysate, other enzymes are inactivated in order to ensure only the proteolytic enzymes act on the fish protein. This results in the production of soluble protein which is then separated, concentrated and dried.

Enzymes can also be used to process the wastewater from cooking or processing of fish and marine invertebrates into fish products such as sauces and protein hydrolysate. This further aids the optimal utilization of the aquatic resources.

11.8.7 Caviar Production

Caviar is one of the high-value food products from fish. It is produced from the roe of primarily sturgeon and other fishes such as herring, salmon, cod, trout and catfish. The key challenge in caviar production is the separation of the roe from the connective tissue which links it to the parent body. The process of removing the roe from the connective tissue is referred to as riddling (Kim et al. 2002). This leads to higher cost of production in terms of time and man power. The time and complexity of this process can be reduced by using proteolytic enzymes. The enzymes break down the connective tissue leaving behind the eggs used for caviar. Using this method compared to conventional riddling method can achieve up to 20% improvement in yield (Kim et al. 2002). The yield is a measure of how much of the roe is successfully riddled without damage to the egg. The process is quite delicate; if not carefully done, the egg is damaged; and typically, over half of the eggs are damaged in the riddling process. The enzyme method minimizes the mechanical stress on the egg, resulting in hire successful riddling. This has significant economic impact on the producer as it reduces loss of product and wasted labor. Collagen-degrading enzymes and other protein-degrading enzymes such as collagenase and pepsin are the enzymes used in the riddling process for caviar production.

11.8.8 Biomarkers for Environmental Pollution Measurement

Enzymes from fish can be used as a biomarker to measure exposure to pollutants in the environment. An example of such an enzyme is cholinesterase which is being investigated for use as a biomarker to measure exposure of fish to pesticides (Mena et al. 2014). Pesticides from agricultural activities on land often get into the aquatic environment and are absorbed by the fish. These fish are then consumed by humans and could pose severe health threats.

11.8.9 Pharmaceutical Application

The applications of enzymes in pharmaceutical industry are centered around the ability of these enzymes to break down tissue materials, serve as sites for processes and induce the destruction of pathogenic organisms. Proteases, for example, due to their activity in catalyzing the breakdown of the proteins which make up tissue such as collagen, are used as a mild means of cleaning and removal of dead skin tissue from wounds (Yaakobi et al. 2007) compared to manually cleaning and removal of dead tissue from wounds, whereby the mechanical stress on the wound would cause severe pain and discomfort to the patient. The applications of enzymes in the pharmaceutical industry extend to other areas such as treatment of cardiovascular diseases, inflammations and cancer (Preeti et al. 2011). Producing such enzymes from aquatic sources provides the potential of developing more robust treatment to address a broader range of diseases and disease-causing organisms. The diversity of activity and stability of aquatic-sourced enzyme are particularly promising for developing new ways to counter bacterial resistance to drugs.

11.9 Commercial Production

One of the challenges facing commercial production of enzymes from fish waste is the level of purity required in enzyme production. This results in a product with high production cost and where most of the mass is not used. Compared to the part of the fish which is consumed as food, the crude extracts from fish viscera without the purification stage have some proteolytic activities which could potentially make them useful in applications such as stain removal additives in detergents and dehairing of hides for leather production (Jayapriya et al. 2014). The cost of production could be reduced by using the crude extract without the expensive purification stage, although this would require some compromise in terms of effectiveness and limited applications. Another alternative to the use of purified enzymes is the use of immobilized algal cells with the enzyme present within the cells. These immobilized algae cells could have even higher activity than the cells in the free form (Arabi et al. 2010).

Of all the enzymes obtainable from the aquatic environment, proteases are the most widely used enzymes in industry. Their applications range from production of detergents, food, pharmaceutics and silver recovery in X-ray to production of leather (Jayapriya et al. 2014). Around 60% of enzymes produced in the industry are proteases (Zhang and Kim 2010), thus making them very valuable resources. Within this chapter, the several applications of enzymes from aquatic sources have been discussed and much of these are indeed proteases. The use of these enzymes cuts across several industries which depend on them; hence, they pose significance to the economy.

Southeast Asia records around 300,000 tonnes of fish sauce production annually since as far back as the early 1960s. The enzyme-assisted method of riddling in caviar production has been commercialized for caviar production in some countries like Canada, the USA and Australia (Kim et al. 2002). Companies that have commercialized enzyme production from marine sources include Biotec ASA in Norway, Carnitech in Denmark, Biotec Maximal in Norway and Isnard-Lyraz in France.

The commercialization of enzyme production from aquatic wastes and by-products results in less of these wastes being released into the environment and causing adverse effects. The main limitations that have been highlighted in the commercialization of aquatic-sourced enzymes are the cost of production, the variability of the enzymes from different aquatic organisms which limits large-scale production and the availability of alternatives from terrestrial sources. While the other alternatives remain readily available, producers are less inclined to pursue extraction of enzymes from the aquatic wastes. Therefore, further research is needed in developing improved methods for commercial extraction of enzymes from aquatic sources.

In various industries, enzymes are used for a wide range of applications. They catalyze biochemical processes such as the breakdown of starch into sugars for production of alcohol and breakdown of proteins to produce peptides for cosmetics. Many biochemical processes of commercial importance will either not occur at all or occur too slowly to be commercially feasible.

The diverse nature of algae as polyphyletic organisms made up of thousands of different species means there is potentially a wide spectrum of enzymes which can be obtained from algae. Algae-sourced enzyme offers the advantage of being able to grow the feedstock for enzyme production with increased productivity and without the use of land space since algae grow at a much faster rate than terrestrial plants and do not require land space as do terrestrial plants and animals. There is, however, more demand for better enzymes in terms of stability, low cost and efficiency.

Microalgae in particular are attractive for enzyme production as they can be genetically modified toward a higher yield of specific enzymes. Having an alternative source of enzyme for the catalysis of industrial processes could remove certain limitations of some commercial enzymes. For example, pH limitations of some protease are either active at alkaline or acidic at pH.

11.10 Conclusion

Enzymes play diverse roles in industry. The aquatic environment due to the diversity of organisms living in a wide range of conditions serves as a source of diverse group of enzymes. Aquatic-sourced enzymes have unique characteristics of being adapted to particularly challenging environment and are therefore likely to produce enzymes which retain stability at extreme processing conditions which is beneficial in industry. Applications of enzymes which have been discussed within the chapter include fish processing, biodiesel and cleaning agent production. Extraction processed from different sources has similar recovery and purification stages, and this could potentially be a more robust production facility which could allow production of the same enzyme from multiple sources. Presently, enzymes from aquatic sources are less commonly used in industry than the enzymes from terrestrial plants, microorganisms and animals. Extraction of enzymes from aquatic waste and by-products could contribute significantly to optimal use of aquatic resources and reduction in environmental pollution.

References

Adamec L (2010) Mineral cost of carnivory in aquatic carnivorous plants. Flora 205:618–621

Aguilera-Morales ME, Canales-Martinez MM, Avila-Gronzalez R, Flores-Ortiz CM (2018) Nutrients and bioactive compounds of the *Lemna gibba* and *Ulva lactuca* as possible ingredients to functional foods. Lat Am J Aquat Res 46(4):709–716

Amano K (1962) The influence of fermentation on the nutritive value of fish with special reference to fermented fish products of South-East Asia. In: Heen E, Kreuzer R (eds) Fish in nutrition. Fishing News Books, London, pp 180–197

Arabi H, Yazdi Tabatabaei M, Faramarzi MA (2010) Influence of whole microalgal cell immobilization and organic solvent on the bioconversion of androst-4-en-3,17-dione to testosterone by *Nostoc muscorum*. J Mol Catal B Enzym 62(3–4):213–217

Bele SD, Sharmila S, Rebecca JL (2014a) Isolation and characterization of lipase from marine algae. Int J Pharm Sci Rev Res 27(1):191–195

Bele SD, Sharmila SS, Rebecca JL (2014b) Comparative study of different methods of extraction of lipase from seaweeds. Res J Pharm Biol Chem Sci 5(3):1741–1748

Bell AS (2007) Major antifungal drugs. Ref Modul Chem Mol Sci Chem Eng Compr Med Chem II 7:445–468

Bernal D, Donley JM, Shadwick RE, Syme DA (2005) Mammal-like muscles power swimming in a cold-water shark. Nature 437(27):1349–1352

Bleakley S, Hayes M (2018) Algal proteins: extraction, application and challenges concerning production (2017): 6(33):1–34

Brouwer P, Nierop KGJ, Huijgen WJJ, Schluepmann H (2019) Aquatic weeds as novel protein sources: alkaline extraction of tannin-rich Azolla. Biotechnol Rep 24:e00368. https://doi.org/10.1016/j.btre.2019.e00368

Daboor SM, Raudonis R, Cohen A, Rohde JR, Cheng Z (2019) Marine bacteria, a source for alginolytic enzyme to disrupt *Pseudomonas aeruginosa* biofilms. Mar Drugs 17(5):307–319

Dutta R (2008) Fundamentals of biochemical engineering. Springer, Berlin, Heidelberg, New York, pp 50–52. ISBN 978-3-540-779fYJ-1

Fernandes P (2016) Enzymes in fish and seafood processing. Front Biotechnol 4(59):1–14

Fleurence J (1999) Seaweed proteins: biochemical, nutritional aspects and potential uses. Trends Food Sci Technol 10:25–28

Garcia-Carreno FL, Navarrete del Toro MA, Serviere-Zaragoza E (2003) Digestive enzymes in juvenile green abalone, Haliotisfulgens, fed natural food. Comp Biochem Physiol B Biochem Mol Biol 134(1):143–150

Glancy B, Balaban RS (2011) Protein composition and function of red and white skeletal muscle mitochondria. Am J Cell Physiol 300(6):C1280–C1290

Greene RV, Griffin HL, Cotta MA (1996) Utility of alkaline protease from marine shipworm bacterium in industrial cleansing applications. Biotechnol Lett 18:759–764

Haddar A, Agrebi R, Bougatef A, Amidet N, Sellami-Kamoun A, Nasri M (2009) Two detergent stable alkaline serine-proteases from *Bacillus mojavensis* A21: purification, characterization and potential application as a laundry detergent additive. Bioresour Technol 100:3366–3373

He W, Li L, Zhao J, Xu H, Rui J, Cui D, Li H, Zhang H, Liu X (2019) Candida sp. 99-125 lipase-catalyzed synthesis of ergosterol linolenate and its characterization. Food Chem 280:286–293

Hong R, Su L, Wu J (2019) Cutinases catalyze polyacrylate hydrolysis and prevent their aggregation. Polym Degrad Stab 159:23–30

Howe PE (1921) The use of sodium sulfate as the globulin precipitant in the determination of proteins in blood. J Biol Chem 49 93–107

Jayapriya J, Sabtecha B, Tamilselvi A (2014) Extraction and characterization of proteolytic enzymes from fish visceral waste: potential applications as destainer and dehairing agent. Int J ChemTech Res 6(10):4504–4510

Kim H, Seo HJ, Byun DS, Heu MS, Pyeun JH (2002) Proteolytic enzymes from fish and their utilization. Fish Sci 68:1557–1562

Kobayashi H, Hatada Y, Tsubouchi T, Nagahama T, Takami H (2012) The hadal amphipod *Hirondellea gigas* possessing a unique cellulase for digesting wooden debris buried in the deepest seafloor. PLoS One 7(8):e42727

Komori Y, Nikai T (2013) Gloydius halys venom metalloproteinases. In: Handbook of proteolytic enzymes, vol 1, 3rd edn., pp 965–967

Kumar KS, Ganesan K, Rao PV S (2014) Seasonal variation in nutritional composition of Kappaphycus alvarezii (Doty) Doty- an edible seaweed. J Food Sci Technol. https://doi.org/10.1007/s13197-014-1372-0

Leema RT, Sachindra NM (2018) Purification and characterization of agarase from marine bacteria Acinetobacter sp. PS12B and its use for preparing bioactive hydrolysate from agarophyte red seaweed *Gracilaria verrucosa*. Appl Biochem Biotechnol 186(1):66–84

Lima FR, Ferreira AJ, Menezes G, Miranda VFO, Dourado MN, Araujo WL (2018) Cultivated bacterial diversity associated with the carnivorous plant *Utricularia breviscapa* (Lentibulariaceae) from floodplains in Brazil. Braz J Microbiol 49:714–722

Mena F, Azzopardi M, Pfennig S, Ruepert C, Tedengren M, Castillo LE, Gunnarsson JS (2014) Use of cholinesterase activity as a biomarker of pesticide exposure used on Costa Rican banana plantations in the native tropical fish *Astyanax aeneus* (Gunther, 1860). J Environ Biol 35(1):35–42

Mogharabi M, Faramarzi MA (2016) Are algae the future source of enzymes? Trends Pept Protein Sci 1(1):1–6

Moreno-Garcia J, Garcia-Martinez T, Mauricio JC, Moreno J (2018) Yeast immobilization systems for alcoholic wine fermentations: actual trends and future perspectives. Front Microbiol 9(241):1–13

Narwal SK, Gupta R (2013) Biodiesel production by transesterification using immobilized lipase. Biotechnol Lett 35(4):479–490

Nogueira AV, Rossi GR, Iacomini M, Sassaki GL, Cipriani TR (2019) Viscera of fish as raw material for extraction of glycosaminoglycans of pharmacological interest. Int J Biol Macromol 121:239–248

Paulsen SS, Andersen B, Gram L, Machado H (2016) Biological potential of chitinolytic marine bacteria. Mar Drugs 14(230):1–17

Preeti C, Dimpi G, Drukshakshi J, Jasbir S (2011) Applications of microbial proteases in pharmaceutical industry: an overview. Rev Med Microbiol 2(4):96–101

Rashidah AR, Nazalan N, Razip S, Koay PC (2006) Lipase producing psychrophilic microorganism isolated from Antarctica. J Bacteriol 182:125–132

Schulze WX, Sanggaard KW, Kreuzer I, Knudsen AD, Bemm F, Thorgersen IB, Brautigam A, Thomsen RL, Schliesky S, Dyrlund TF, Escalante-Perez M, Becker D (2012) The protein composition of the digestive fluid from the venus flytrap sheds light on prey digestion mechanisms. Mol Cell Proteomics 11(11):1306–1319

Seca AML, Pinto DCGA (2018) Overview on the antihypertensive and anti-obesity effects of secondary metabolites from seaweeds. Mar Drugs 16(237):1–18

Seibel BA, Walsh PJ (2002) Trimethylamine oxide accumulation in marine animals: relationship to acylglycerol storage. J Exp Biol 205:297–306

Shahidi F, Kamil YVAJ (2001) Enzymes from fish and aquatic invertebrates and their application in the food industry. Trends Food Sci Technol 12(12):435–464

Sil'chenko AS, Kusaikin MI, Zakharenko AM, Zvyagintseva TN (2013) Isolation from the marine mollusk Lambis sp. and catalytic properties of an alginate lyase with rare substrate specificity. Chem Nat Compd 49(2):215–218

Silchenko AS, Kusaykin MI, Kurilenko VV, Zakharenko MA, Isakov VV, Zaporozhets TS, Gazha AK, Zvyagintseva TN (2013) Hydrolysis of fucoidan by fucoidanase isolated from the marine bacterium, Formosa algae. Mar Drugs 11:2413–2430

Sivasubramani K, Singh JR, Jayalakshmi S, Kumar SS, Selvi C (2013) Production and optimization of lipase from marine derived bacteria. Int J Curr Microbiol Appl Sci 2(4):126–135

Teo CL, Jamaluddin H, Zain NAM, Idris A (2014) Biodiesel production via lipase catalysed transesterification of microalgae lipids from Tetraselmis sp. Renew Energy 68:1–5

Tokusoglu O, Unal MK (2003) Fat replacers in meat products. Pak J Nutr 2(3):196–203

Trivedi N, Reddy CRK, Radulovich R, Jha B (2015) Solid state fermentation (SSF)-derived cellulase for saccharification of the green seaweed Ulva for bioethanol production. Algal Res 9:48–54

Valsa AK, Thomas A, Mathew M, Mohan S, Manjula R (2003) Optimization of growth conditions for the production of extracellular lipase of vacillus my colds. Ind J Microbiol 43 67–69

Van Krimpen M, Bikker P, van der Meer I, van der Peet-Schwering C, Vereijken J (2013) Cultivation, processing and nutritional aspects for pigs and poultry of European protein sources as alternatives for imported soybean products. Wageningen UR Livestock Research, Lelystad, The Netherlands, p 48

Velasquez MT, Ramezani A, Manal A, Raj DS (2016) Trimethylamine N-oxide: the good, the bad and the unknown. Toxins (Basel) 8(11):326–337

Yaakobi T, Cohen-Hadar N, Yaron H, Hirszowicz E, Simantov Y, Bass S, Freeman A (2007) Wound debridement by continuous streaming of proteolytic enzyme solutions: effects on experimental chronic wound model in porcin. Wounds 19(7):192–200

Yancey PH, Gerringer ME, Drazen JC, Rowden AA, Jamieson AJ (2014) Marine fish may be biochemically constrained from inhabiting the deepest ocean depths. Proc Natl Acad Sci USA 111:4461–4465

Yao Z, Wang F, Gao Z, Jin L, Wu H (2013) Characterization of a κ-carrageenase from marine Cellulophaga lytica strain N5-2 and analysis of its degradation products. Int J Mol Sci 14(12):24592–24602

Yuan D, Lan D, Xin R, Yang B, Wang Y (2016) Screening and characterization of a thermostable lipase from marine Streptomyces sp. strain W007. Biotechnol Appl Biochem 63(1):41–50

Zhang C, Kim S (2010) Research and application of marine microbial enzymes: status and prospects. Mar Drugs 8:1920–1934

Zhuang J, Zhang K, Liu X, Liu W, Ji A (2018) Characterization of a novel polyM-preferred alginate lyase from marine Vibrio splendidus OU02. Mar Drugs 16(9):295

Chapter 12
Collagen

Abstract Collagen is obtained from the bones, skins and scales of aquatic animals. It is made up of three polypeptide chains linked together to form a tertiary structure. It serves a structural role within the tissue where it provides flexibility within the tissue matrix. Collagen can be processed into various forms, and it finds applications in high-value products such as scaffolds in tissue engineering. Aquatic-derived collagen serves as an alternative to porcine or bovine sourced collagen, thus eliminating associated health risk or ethical concerns associated with the collagen sourced from these land animals. Since collagen can be obtained from the parts of the aquatic animals that are generally not consumed as food, the production of collagen contributes toward management of waste resources as well as optimal utilization of aquatic resources.

Keywords Collagen · Proteins · Biopolymers · Fish · Squids

12.1 Introduction

Collagen is abundant in nature as it is present in the tissue of every animal on land and in the sea. It is limited to animals; therefore, aquatic plants and algae do not serve as a source of collagen. Within the aquatic environment, main commercial sources of collagen are from fish fin, skins and scales (Olatunji and Denloye 2017). Other aquatic sources of collagen are invertebrates such as squids, starfish and jellyfish (Benedetto et al. 2014; Tan et al. 2013). Collagen competes with the fish oil, and fishmeal market as the fish parts is also used in this application. One of the most common uses of collagen in the industry is in foods where it is mostly used in its hydrolyzed form. Collagen from fish has been proven to be interchangeable with bovine and porcine hydrolyzed collagen (Benedetto et al. 2014), with the added advantage of being free from the risk of diseases such as mad cow diseases which are associated with the collagen products obtained from these sources. It also serves as an alternative where ethical or religious concerns exist with hydrolyzed collagen sourced from bovine or porcine.

Like other proteins, collagen is made up of amino acid repeating units which form into polypeptide chains. They however distinguish from other proteins by their triple

O. Olatunji, *Aquatic Biopolymers*, Springer Series on Polymer and Composite Materials,
https://doi.org/10.1007/978-3-030-34709-3_12

helix structure. There are various types of collagen; however, the type I collagen is most common in vertebrates where it could make up to 78% of the collagen. In the intact form, collagen plays mainly structural role; however, when hydrolyzed into its polypeptide forms, it has more diverse applications which range from food to biomedical.

The production of collagen serves some advantage to the environment as it provides a way to optimize the utilization of the global aquatic resources. Further to this, by adding value to the aquatic resource from which it is sourced, it provides an additional source of income to the fishermen and fish traders who can then sell the by-products to other companies who can then convert to collagen and these can then be converted to other products of much higher value. Some adverse environmental impacts arise from water utilization, use of acids, salts and alkali which might be released into the environment and cause some adverse effects as well as carbon emissions. All of which can be addressed through sustainable practices such as recycling and process optimization.

Collagen is fairly well explored aquatic biopolymer resource. Global collagen market is estimated to have an annual growth rate of 5.2% (Research and Markets 2019). Much of this comes from its use in food and other applications are also seeing increased interest. Companies such as Nitta Gelatin existing across the world are involved in the production of gelatin from fish. In addition to this, researchers across the world are also continually exploring new sources and new ways of obtaining and making use of gelatin. Such innovations include 3D printed scaffolds for bone regeneration and corneal replacement and these are discussed in this chapter. With such high-end application from fish by-products which are generally considered as waste, collagen has significant impact on present and future economies.

The rest of the chapter explores collagen as another important aquatic sourced biopolymer, beginning with where it is sourced, the chemistry of aquatic sourced collagen and how this differs or is similar to non-aquatic sourced collagen, the production processes and how these processes impact the environment and the current state of the commercial collagen production.

12.2 Occurrence in Nature

Collagen is relatively abundant in nature, and it is the most abundant protein in vertebrates. It is present in humans and other primates, mammals, vertebrates and invertebrates. It makes up the bones, skin and other connective tissue where it is produced by fibroblast cells. Collagen makes up approximately 30% of protein in animals, in humans it makes up 75% of the dry weight of the skin (Shoulders and Raines 2009). Within the tissue, proteins serve structural roles alongside other tissue components such as elastin and hydroxyapatite, depending on the particular tissue composition. The bone tissue for example comprises a composite with collagen and other non-collagenous proteins forming the matrix filled with minerals and forming the disperse phase of the biological tissue matrix (Aerssens et al. 1994; Robinson

1979). Without collagen, bone will be completely brittle and without the bone minerals, bone would have no rigidity. Therefore, the combination of collagen and minerals gives bone the ideal mechanical property to perform its biological function.

Collagen plays a structural role in the extracellular matrix where it provides thermal and mechanical stability as well as interaction with the surrounding cells and components. Collagen can be of several types mainly type I and type II collagen, although there are up to 28 different types of collagen in vertebrates on land and in water (Shoulders and Raines 2009). Type I collagen consists of two alpha-1-chains and one alpha-2-chains in the triple helix structure (Felician et al. 2019). Type I collagen is more abundant in marine invertebrates such as sea urchins, jellyfish, starfish and squids (Benedetto et al. 2014; Tan et al. 2013; Delphi et al. 2016). Collagens from fish parts such as scales, fins, skin and bones have been mostly identified as type 1 collagen. Since human collagen is about 90% type 1 collagen fish sourced collagen has the potential to have good compatibility with the human system when used externally or internally. Whale sharks collagen is type II collagen (Jeevithan et al. 2015). Collagen is distinct from other proteins by their right-handed triple helix structure as well as the occurrence of the amino acid glycine at every third residue along the polypeptide chain.

12.3 Chemistry of Collagens

The general structure of collagen is a right-handed triple helix structure. A fibrous protein does not have a tertiary structure as this helix structure does not fold up. Collagen has high molecular weight. For example, collagen extracted from the jellyfish *Rhopilema esculentum* had a molecular weight of between 100 and 150 kDa when measured using sodium dodecyl sulfate-polyacrylamide gel electrophoresis (SDS-PAGE) and Fourier transform infrared (FTIR) (Felician et al. 2019). This was then broken down with enzymes to polypeptides with 10–15 kDa and then further reduced to 25 kDa with further enzyme treatment. Different types of enzyme break different types of bonds such that the molecular weight of the proteins can be varied by varying the enzyme used. The molecular weight has an impact on the bioactivity of collagen (Chi et al. 2014); therefore, the ability to modify the molecular weight of collagen increases the diversity of their applications (Fig. 12.1).

In fish collagen which could be obtained from the skin, fins or scales, alanine and glycine were most abundant in the stated order, while tryptophan was not present in the collagen structure (Mahboob 2015). The other more prominent amino acids that occur on the collagen polypeptide chains are proline and hydroxyproline. There are cases where the polypeptides making up the collagen triple helix are identical polymeric chains made up of the same amino acid repeating units, and this is referred to as homotrimeric triple helix. It is however more common to have the case where the polypeptides are not identical polymer chains; this is referred to as heterotrimeric.

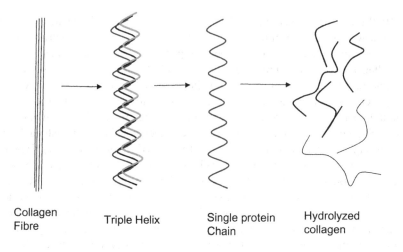

| Collagen Fibre | Triple Helix | Single protein Chain | Hydrolyzed collagen |

Fig. 12.1 Collagen and hydrolyzed forms

Normal human collagen type 1 is heterotrimeric, and it comprises of two alpha-1-chains and one alpha-2-chain (Chang et al. 2012). Homotrimeric collagen type 1 occurs only in fetal tissue or in cases of cancer or fibrosis.

The polypeptide chains are held together by hydrogen bonds which form between the amine ($-N-H_2$) group and the carbonyl group ($-CO-$) of the amide links which is present in all the amino acids making up each polypeptide chain. Using enzymes such as collagenase, the alpha helix structure of the collagen is separated into polypeptides by breaking the hydrogen bonds which hold together the polypeptide chains in the helical conformation. These polypeptides could then be further hydrolyzed into shorter chains by breaking some of the amide bonds using enzymes like proteolytic enzymes such as papain and alkaline proteinase (Felician et al. 2019).

Collagen is characterized by five main peaks on the FTIR spectrometry; the amide A, B, I, II and III at a wavelength of 3433, 2926, 1641, 1549 and 1240 cm^{-1}, respectively. The amide III bands are present when the collagen triple helix structure is still intact (Chi et al. 2014). This can therefore be used to distinguish between whole collagen and hydrolyzed collagen.

Solubility is an important factor in the applicability and bioactivity of collagen. This impacts their ability to dissolve under physiological conditions and interact with the cells to carry out their biological activity. An example of such is in transdermal delivery of therapeutics using microneedles, where the drug is loaded onto a microneedle patch made from hydrolyzed collagen. Upon insertion into the skin, these hydrolyzed collagen microneedles dissolve in the fluid within the skin releasing the compounds loaded within (Olatunji et al. 2014; Olatunji and Olsson 2015). Lower molecular weight collagen from fish skin of Spanish mackerel for example show faster solubility in neutral pH and at acidic pH but will however decrease in solubility at alkaline pH (Chi et al. 2014). The increased solubility at lower molecular weight is attributed to the shorter polypeptide chains having better ability to be surrounded by the water molecules and form hydrogen bonds. This molecular weight and pH dependence of collagen solubility prove useful in designing products which are targeted to be released at specific points along the alimentary canal since the pH varies at the different parts.

12.4 Availability of Raw Materials

FAO reported 171 million tonnes of fish produced in 2016, 54.1 million tonnes of this being farmed finfish (valued at about 138.50 billion USD). It was stated that 88% of this was utilized for direct human consumption (FAO 2018). Of these percentage consumed as food, some of these also include the waste generated from the fish food waste. The term fish here also includes other aquatic animals such as shrimps, mollusks and lobsters but excludes aquatic plants and algae. It has also been observed that as the world population grows the consumption of fish grows twice as much. Since the global population is ever on a constant rise so will the demand for fish, and hence the availability of the fish waste from which collagen can be produced.

Due to its presence in a broad range of organisms, collagen is relatively abundant. One of the issues with availability of raw materials for commercial collagen production is the fact that some of the sources of collagen raw materials are consumed as food. Those which are not eaten are usually sold alongside with the food for different reasons like preserving quality and reducing the cost of processing. For example, fin fish are often sold and served whole, such that obtaining the waste bones along the value chain might prove difficult. Fish waste such as bones is usually discarded alongside organic waste, they also degrade relatively fast, post-consumer collection of such wasted is not as feasible as collecting non-biodegradable wastes such as plastics. One way to counter this limitation is to collect from fish processors who descale, fillet or extract oils from the fish. Coproduction of fish collagen can also be carried out alongside the aforementioned processes.

Fish production either from fishing from the wild or fish farming using aquaculture forms an important part of the human diet. Aquatic sourced food is a major source of protein in many diets across the world. A long-term sustainable production of fish and fisheries products is a key part of sustainable development. Fish processing results in the generation of a considerable amount of waste, about 50–70% of the fish results in waste which includes the fins, the scales and in some cases the skin and guts. These wastes comprise up to 30% collagen which is of high commercial value. Aquatic-based raw material for collagen production is therefore promised to be sustainable.

The availability of aquatic raw materials for collagen production however faces some challenges. One of which is the challenge of overfishing and irresponsible fishing which has resulted in decline of some fish species and threatens the long-term availability of such resources. About 33.1% of fish available in the waters are fished beyond biologically sustainable limits. Commercial extraction of collagen from fish scales should therefore not seek to exert further strain on the rate of fishing but rather align the production rate with the waste generated from a sustainable level of fishing. The process of extraction should also not impact adversely on the aquatic environment within which these resources exist.

The main competition for collagen production is the fishmeal production. The fish wastes are also used for the production of fishmeal which is high in protein and minerals. Since in the wild, the big fishes tend to fish on other little fishes (as well as other aquatic organisms such as planktons), these fish by-products are ideal sources for fish meal. It is also particularly important that the waste is returning into the food cycle to further ensure continuous production of fish. However, fishmeal production has since been on a general decline since 1994 when global fishmeal production was at an all time high of 30 million tonnes. Of the 171 million tonnes of fish produced in 2016, 91 million tonnes are captured from natural stock while the rest are farmed in aquaculture (FAO 2018). This implies a demand for fishmeal for aquaculture is relatively high. On the other hand, collagen has a broader market which spans pharmaceuticals, food, biomedical and cosmetics. Other than fish meal, fish skin, particularly of larger fish, is used for producing leather for a range of products such as shoes, bags, garments and furniture.

As of 2016, China was the world's highest producer of fish, followed by Norway, Vietnam and Thailand. Much of these fish are being exported to Europe or the USA as they are the greatest consumers of fish (FAO 2018). Fresh whole fish is usually the most preferred form of fish such that the fish moves down the distribution channel with very little processing or removal of any part of it. Much of the cost goes into transportation and preservation. This means that much of the fish waste is not generated until the point of consumption when the fish is cut and cooked. Therefore, although China is the biggest producer of fish, it might not necessarily be the region where fish waste for collagen production is highest. The countries where the fish is being cooked will therefore be more likely generators of fish waste.

Collagen is also sourced from other aquatic animals such as the jellyfish (Cheng et al. 2017), squids (Dai et al. 2018) and sea cucumbers (Liu et al. 2019), and these are captured in lower quantities from natural stock and hardly farmed. They are recorded as other sea animals make up about 938,500 tonnes annually as of 2016. However, recent giant jellyfish blooms mean there is an increasing amount of these aquatic nuisances to be utilized.

12.5 Extraction of Collagen

There are various methods which exist for the extraction of collagen from different sources. A single source of collagen might have more than one method of extraction of the collagen. Here, we review some of the existing extraction methods for some specified sources. Some of these methods lead to the production of collagen peptides as the process of extraction results in the breakdown of the collagen triple helix structure. This could be desirable where the end goal is to obtain peptides since the two steps of extraction and formation of polypeptide are achieved in one process.

Collagen content varies from species to species and also varies in different parts of the animal. It also varies for different studies. Extraction yields of collagen from the skin of *C. catla* and *C. mrigala* are 13 and 11.2%, respectively, while that from the scales were 9.5 and 8.3, respectively, and 13 and 13.1%, respectively, from the fins (Mahboob 2015). In jellyfish *R. esculentum* recent studies report a yield of 4.31% (weight of dry collagen to weight of wet jellyfish) compared to other studies which reported lower yields (Felician et al. 2019). A yield of 13.6% (dry collagen per wet weight of skin) was obtained from acid extraction of collagen from Spanish mackerel fish skin (Chi et al. 2014) while a 30% yield (dry weight of hydrolyzed collagen per dry weight of fish scales) was obtained from hydrothermal extraction of hydrolyzed collagen from fish scales of croaker fish (Olatunji and Denloye 2017). In another study (Ahmed et al. 2019), a yield of 16.7% was reported for enzyme extraction of collagen from skin of bigeye tuna using pepsin enzyme. Using acid extraction, the acid-soluble collagen yield from the skin was 13.5% while yield of pepsin soluble collagen from the scale and bones were 4.6 and 2.6%, respectively. The acid-soluble collagen obtained from scales and bones of the tuna fish was negligible.

12.5.1 Extraction of Collagen from Combined Fish Waste

Collagen has been extracted from fish skins, scales and bones of a variety of fish such as bigeye snapper, red snapper, carp, black drum, Nile perch and yellowfin tuna (Ahmed et al. 2019; Olatunji and Denloye 2017; Mahboob 2015; Chi et al. 2014). Fish skins can be sourced from fish processing factories where skinless fish fillets or other fish products are made. Processing the skin separately from the bones and other parts since the composition of bones and skin significantly varies. Separate extraction of the different parts of the fish waste allows for better process optimization by tailoring the operating conditions to the respective composition. However for practicality and reduction of labor and process complexity, it is desirable to combine all the fish wastes and extract collagen from the mixtures in one batch.

To extract collagen from combined fish waste both the chemical and enzymatic method follows the same steps for pretreatment of the fish waste parts. Here, an example method (Mahboob 2015) which was used as a general method to extract collagen from scales, fin and fish skin is presented. Although the parts were separated, the same process conditions were applied to all parts. The different parts were separated prior to extraction in order to analyze them separately for collagen contents and structure in the different parts of the fish. The pretreatment stage is targeted at removing non-collagenous proteins so that these are not extracted alongside the collagen and contaminate the extract. To achieve this, the mass is treated with an alkali, most commonly sodium hydroxide at 0.1 M using a 10 ml solution per gram of fish waste mass. This is carried out at a temperature of 4 °C under constant stirring for 6 h with the alkali solution changed every 2 h after which the spent alkali solution must have been saturated with non-collagenous proteins that have been removed. This is then washed with water until all the alkali with the non-collagenous proteins has been washed off and the system is neutral. The type of water used depends on the level of purity required. Clean tap water could be used, or distilled water or ultrapure water.

For the acid-based extraction, the fish waste was treated with acid in order to dissolve out the acid-soluble collagen from the rest of the tissue. The first step was to remove the fat using 10% butyl alcohol at a 1:10 ratio of solid to alcohol for a duration of 8 h. The fats are alcohol soluble so it will be removed from the fish waste. The fish waste is then washed with cold water to remove any residue of alcohol and fat from the surface. The next stage is then to extract the collagen in acetic acid at a concentration of 0.5 M and 1:15 solid to liquid ratio. The extraction is allowed a duration of 24 h. The extraction time can be varied for different samples since the composition of collagen and other components varies for different types of fish. Defatting prior to extraction prevents the fat from being released alongside the collagen during the extraction. The fat will dissolve in alcohol but collagen will not, however the fat could be released into the acetic acid solution hence the order is also important. The collagen dissolved in acetic acid is then precipitated out in a 0.05 M tris (hydroxymethyl) aminomethane containing sodium chloride at a concentration of 2.6 M. To separate the precipitated collagen from the solution, centrifugation is used at 20,000 g for 1 h. A refrigerated centrifuge is used in order to maintain the required temperature. To further purify the collagen sample, it can then be redissolved in 0.5 M acetic acid and then separated by repeated dialysis. The collagen extract now needs to be dried for more effective storage and packaging. For this mild drying method is required as excessive heat will destroy the triple helix structure and the resulting extract will be hydrolyzed collagen. Freeze drying can be employed here. It is commonly used in such cases where a delicate sample needs to be dried. A temperature between 4 and 8 °C is maintained throughout the pretreatment and extraction to prevent degradation of the collagen.

The enzyme extraction can be carried out using the residue from the acid extraction (Mahboob 2015). The residue contains the collagen proteins which do not dissolve in acid but can be dissolved in pepsin. This method allows obtaining two types of collagen from the same source thus making optimal use of the fish waste. The pH is adjusted to that suitable for the enzyme by soaking in 0.5 M acetic acid in a solid to liquid ratio of 1:15. The pepsin is then added and continuously stirred for a duration of 48 h. A low temperature is also maintained at 4 °C. At the end of the extraction, the solid residue is then separated from the liquid which contains the collagen. The collagen is precipitated using sodium chloride, and the precipitated collagen is then separated from the salt solution and further purification by repeated dialysis then follows.

The type of extraction process has some effect on the collagen yield and properties. For example, acid extraction from scales of goatfish gave collagen yield of 0.46% while pepsin extraction gave a yield of 1.20% (dry weight basis). The acid-soluble collagen had slightly higher glass transition temperature of 41.58 °C while that of pepsin soluble collagen was 41.01 °C (Matmaroh et al. 2011). Sodium hydroxide extraction results in mostly denatured collagen or hydrolyzed collagen which is a mixture of polypeptides of different chain lengths. The extraction process therefore

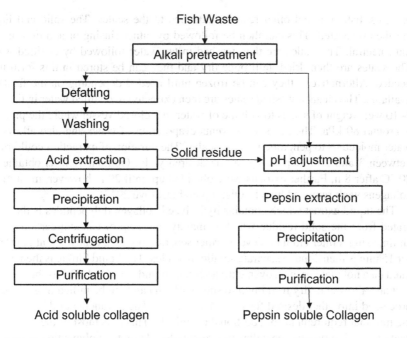

Fig. 12.2 Process for acid and pepsin soluble collagen extraction from fish waste

depends on the form in which the collagen is required. Figure 12.2 summarizes the extraction process for acid-soluble and pepsin soluble collagen extraction.

12.5.2 Hydrothermal-Based Extract of Hydrolyzed Collagen from Fish Scales

Fish scales are particularly attractive source for collagen as it is a food source, and it is the part of the food that is discarded as it is often not used for fishmeal or fish oil. The extraction of hydrolyzed collagen from fish scales could be carried out using the chemical or enzymatic method described for combined fish waste; however, another alternative is the hydrothermal method. This method requires minimal use of acids or alkali. It uses a combination of heat and pressure for a prolonged period of time (Olatunji et al. 2014; Olatunji and Olsson 2015; Olatunji and Denloye 2017). In the said method the fish scales are generally collected at the end of the day from fish traders who sell and scale a large amount of fish daily. The scales are transported to the laboratory where they are washed with tap water to get rid of debris. They are then washed with sodium hydroxide with a concentration of 0.1 M in a ration of 1:15 (weight of dry scales to volume of alkali solution) at room temperature for 1 h and stirring at intervals. This step is carried out to get rid of non-collagenous proteins, fish skin and other residues attached to the scales. The solid and liquid are then separated. This can then be followed by either rinsing or addition of acid until neutral. The scales are then washed in tap water followed by distilled water. The scales are then dried until bone dry and they can be stored in this form until needed. Alternatively, they can be frozen until needed depending on the facilities available. The cleaned washed scales are then extracted in distilled water in the ratio 1–10 wet weight of scales to volume of water in a closed vessel where the pressure is around 80 kPa. The pressure prevents evaporation of water and also allows the water molecules to penetrate into the scales. The duration of extraction could range between 3 and 20 h depending on the desired yield. Optimal yield is obtained at 80 °C after 8 h. Further extracts were obtained up until 20 h, however, most of the collagens were removed by 8 h (Olatunji and Denloye 2017).

The liquid extract which contains hydrolyzed collagen polypeptides is then separated from the solid residue which is mainly made up of calcium carbonate and chitin, using a filter cloth or sieve. Further separation with a centrifuge at 2500 rpm for 15 min is then done to separate smaller particles. The liquid filtrate is then evaporated on a hot plate or in an oven until a viscous liquid is formed. This is then solvent cast and allowed to dry forming into sheets which can then be milled in granules or processed into other desired forms such as pellets. The advantage of this method is the minimal requirement for additional chemicals. This puts most of the cost of the heat required to rupture the collagen structure to release the polypeptides.

12.5.3 Extraction of Collagen from Jellyfish

Collagen obtained from jellyfish is of particular interest due to recent findings that polypeptides from these collagens have some antioxidant, hemostatic and antihypertensive bioactivities (Zhuang et al. 2009, 2012; Cheng et al. 2017). Extraction of collagen can be carried out using the enzyme pepsin, and this has been used for extraction of collagen from aquatic sources such as jellyfish and starfish (Felician et al. 2019; Tan et al. 2013) used pepsin enzyme. 203 g of jellyfish filaments which had been washed with ultrapurified water was then homogenized in a 50% w/v acetic acid for 10 min. The extraction is then carried out using 1% pepsin under constant stirring at 4 °C. The extraction was carried out for a duration of 72 h. Centrifugation at 8000 rpm for 15 min was then used to separate the extracted collagen from the other components. The collagen was then precipitated out by adding sodium chloride at 2 and 0.05 mol/L tris to get a pH of 7. Further centrifugation was then carried out at 10,000 rpm for 20 min to separate the precipitate. This is then dissolved in 50% acetic acid. Dialysis against acetic acid is at 0.1 mol/L and ultrapure water for 2 h and 2 days, respectively. Advantage of this method for extraction of collagen is that it is done at low temperature of 4 °C such that the collagen can be obtained in its whole form rather than degraded into collagen peptides. Such method is desirable when the whole collagen is needed, the main cost therefore goes into use of the enzyme and maintaining the conditions required for the enzyme to function. To obtain collagen peptides, another enzyme is then used to break the bonds between the polypeptide chains. Since these enzymes act on specific sites, a different enzyme is required for this. Here, the collagenase enzyme is used at 5% w/w composition.

12.6 Environmental Implication

This section discusses some key issues associated with collagen extraction from aquatic resources. Some of these are positive environmental impacts while some are adverse. The overall environmental impact is therefore a sum of the positive and negative impacts for the specific source and type of extraction process.

12.6.1 Jellyfish Blooms

In recent years, rapid rise in population of jellyfish has become an environmental nuisance in various parts of the world including Asia, the USA, Europe and Australia (Dong 2019). This has had devastating effects and economic losses in the fisheries industry and other related industries. Fishermen have reported frequent fishing trips resulting in the fishing nets filled with large amounts of giant jellyfish with little or no fish. Most of the jellyfish caught results in wasted efforts and resources and in some cases poses risk to the lives of the fishermen. One of the measures to address this is to utilize the jellyfish for production of much-needed biopolymer products like collagen. These jellyfish feed on plankton at a relatively faster rate than small fish. Most population blooms as seen with the algae blooms, result in an eventual decline as the food runs out and the jellyfish begin to die out and decay. This is likely to result in another severe environmental problem in the aquatic environment. Therefore, capturing them and utilizing them for the production of biopolymers such as collagen is one way of abating the problem until a more permanent solution os developed for reversing the bloom.

12.6.2 Fisheries Waste Management

Processing of fish waste for the extraction of collagen is beneficial to the environment in the sense that it prevents the accumulation of fish waste which results in environmental deterioration and health hazard. One issue with processing of fish is the fishy smell associated with fish. This is largely due to the presence of trimethylamine oxide TMAO (Subramaniam 2018). One of the advantages of collagen as a naturally sourced biopolymer is that it can utilize fish waste and does not require special conditions for the cultivation of fish for collagen production. Collagenous waste makes a good fraction of the fish waste which would otherwise be discarded as waste if not used for fishmeal. Collagen production from fish waste therefore serves a significant role in helping to manage the waste generated from fisheries whether captured or farmed.

12.6.3 Water Consumption

The water consumed is used in different aspects of collagen production, in the washing of the fish biomass prior to pretreatment, washing the pretreated biomass where it has been pretreated with alkali and the water used in the main extraction. The purity of water varies from clean tap water to ultrapure water which requires additional energy to clean. In the washing stage using the hydrothermal method for extraction of collagen from fish scales (Olatunji et al. 2014) as an example, water is used in an estimated ratio of 13:1 (volume of water in ml to mass of fish waste in grams), in the washing following pretreatment this usually requires repeated washing before the water becomes neutral after sodium hydroxide pretreatment (in the case of extraction from fish scales). If we assume it is washed six times after pretreatment in the same water to fish waste ratio, this means for the cleaning and pretreatment stage about 91 ml of water is required per gram of fish scales. In the extraction process, more water is used as the solvent for the extraction. For each gram of fish scales, 10 ml of water is used. This makes an estimated 10 ml of water for every gram of collagen produced from fish scales. The water from washing could be recycled either on site or sent to water treatment plant. The water used in the extraction could be recovered during the drying process by condensation, although this will then require further energy and material consumption for these additional operations.

12.6.4 Energy Consumption

In the enzyme extraction process, less energy is consumed in heating as the extraction is carried out at low temperature. However, energy is required for cooling. Where the average room temperature in cold climates is around 16 °C and in hotter climates could reach much higher temperatures of around 30 °C, energy input is required to remove heat from the system such that at least 50 J is required per gram of water being heated. Where hydrothermal extraction at 80 °C is used, energy is required to heat the system from room temperature (assuming this to be about 25 °C) to 80 °C. This heating will require an input of about 230 J per gram of the water being heated for the entire duration of the extraction. Additional energy is then utilized in drying; this can be done either by freezing and sublimation in freeze drying or by evaporation. Other operations requiring more energy input include operating the centrifuge for separation at about 2500 rpm for about 15 min, the power consumed per centrifuge cycle depends on the volume of the centrifuge and the power efficiency of the particular unit.

Energy is also consumed in transporting the fish waste from the point of collection to the processing factory. Additional transportation is then required to get the products to the distributor or to other manufacturers for further processing into more refined forms or for use in other products. Renewable sources of energy could be considered for each of these stages to minimize the environmental impact.

12.6.5 CO_2 Emission

In the extraction of collagen from fish scale, the step before the extraction with either an acid, alkali or hydrothermal method is the removal of the calcified minerals from the scale. This is mainly composed of calcium carbonate ($CaCO_3$) and some calcium hydroxyapatite ($Ca_{10}(PO_4)_6(OH)_2$). The decalcification of scales, fins and bones is a chemical reaction between the calcium compound and acid. For example, the reaction between hydrochloric acid and calcium carbonate is expressed in Eq. (12.1).

$$CaCO_3 + 2HCl \rightarrow CaCl_2 + H_2O + CO_2 \tag{12.1}$$

Given a fish scale of for example the Northern Pike fish (*Esox lucius*) made of 20% calcium salt (Sionkowska and Kozlowska 2014), that means for every gram of fish scales xxx grams of CO_2 is released.

12.6.6 Salts, Acids and Alkali

Where the acid-based or alkali-based extraction is used for the extraction of acid-soluble collagen, for example, acetic acid at a concentration of 50% w/v is required for the extraction medium in the enzyme extraction of collagen from jellyfish. More of the same acid is used to redissolve during purification and a lower concentration of 0.1 mol/L of acetic acid is used for the dialysis such that the acid is still required in the enzyme extraction. In the chemical extraction, either acid or alkali can be used for the extraction of collagen, depending on the type of collagen desired.

In the extraction of collagen from fish, skin and fins, pretreatment is usually required to get rid of debris and non-collagenous proteins. This usually requires treatment with 10 ml of 0.1 M sodium oxide for every gram of fish waste. To remove fat, particularly in fish skin, alcohol is required. 10 ml of a 10% concentration of butyl alcohol per gram of fish scales are used for this purpose. Sodium chloride is generally used for precipitation of collagen when either low-temperature extraction of intact collagen with acid or pepsin is being used.

Therefore, the extraction process of collagen whether enzymatic, chemical or hydrothermal requires varying amount of additional salts, acid and alkali. Some salts are also produced in the process of decalcification of the scales, fins and bones when these parts are used. If these salts, acid and alkali are released to the environment they

Table 12.1 Consumptions for extraction of acid-soluble collagen from fish skin

Consumption	Amount
Butyl alcohol	10 ml of 10% v/v per gram of fish skin
Acetic acid	30 ml of 0.5 M per gram of fish skin
Tris (hydroxymethyl) aminomethane	10 ml of 0.05 M per gram of crude extract
NaCl	2.6 M 10 ml per gram of crude extract

contribute to altering the salinity and pH of the water. There is therefore a need for careful regulation of these processes involving large-scale use of these compounds for collagen extraction. Table 12.1 gives a summary of typical consumption for acid extraction of collagen from fish skin (Mahboob 2015; Matmaroh et al. 2011).

12.7   Applications

Beyond it is mainly structural role within the tissue, when orally consumed, implanted, digested, injected or transdermally applied collagen can present a broad range of benefits. Here we explore some of the notable once from recent literature. These applications highlight the relevance of collagen to the economy, environment and health.

Collagen is most often used in its partially degraded form as polypeptides. In the triple helix form collagen play mostly structural roles, however, when unraveled into polypeptides by breaking the secondary bonds between the peptide chains, these have more bioactivities. The peptide chains also have higher bioavailability when applied to the skin or ingested. Their respective bioactivity is dependent on the amino acid composition of the polypeptide chains. The length of the polymer chains also affects their bioactivity and bioavailability. Generally, lower molecular weights more easily penetrate membranes and this is important in for example delivering collagen polypeptide-based formulation into the skin. When orally ingested polypeptides are easier to break down and transported into the circulatory system from where they can elicit their biological activity.

12.7.1   Skin Mimicking

Testing of products which require understanding or measuring how they interact with the skin such as topical skin care, adhesives and physical devices requires the use of skin samples. These are either obtained from cadaver, grown in tissue culture or obtained from other animals whose skin closely resembles that of humans in terms of porosity, thickness, mechanical properties and moisture content. Recently, artificial skin based on hydrolyzed collagen in the form of gelatin has been developed by different researchers (Dabrowska et al. 2017). Aside from reducing the complexity of growing skin in tissue culture, it also addresses ethical concerns which surround the use of human or animal skins for testing of products. Some of these artificial skins are designed to mimic different aspects of the skin, porosity, water content and mechanical properties. The artificial skin reported by Dabrowska et al. (2017) was made using a composite of cross-linked collagen and cotton fibers. This resulted in a skin-like material which mimicked the friction properties of the skin. The friction properties of the skin are important when designing products such as textiles and adhesive patches which are designed to interact physically with the skin. It seems not surprising that collagen would play a role in such material since it already serves as one of the key components of human skin.

12.7.2 Wound Healing

Collagen aids in wound healing by protecting the wound from infection and physical abrasion, keeping it moist thereby allowing more effective and fast healing. It can be used in combination with other wound healing aids such as antimicrobial agent, and it can also serve as a delivery agent for drugs which support healing and pain relief. In addition to this, collagen peptides can induce skin rejuvenation and healing by signaling the fibroblast cells to produce new collagen cells. They can also promote overall cell turnover by increasing the rate of cell proliferation and migration, such that more old worn-out skin cells are removed and newer skin cells are formed.

Complete wound healing requires reformation of new blood vessels, new skin cells and reconstruction of the different layers of the new skin tissue. Collagen deposition is part of the key processes in wound healing and skin repair. The other processes are angiogenesis, granulation tissue formation and re-epithelialization. A wound healing aid should either act as a barrier to prevent further infection of the wound, serve as a surface for cell adhesion to promote cell proliferation and differentiation or serve as a scaffold for rebuilding if the new tissue, or all of the aforementioned. Such material must be biocompatible and biodegradable should not induce any adverse immune response and must have the right mechanical properties (Muthukumar et al. 2014; Elango et al. 2018; Jeevithan et al. 2015; Chattopadhyay and Raines 2014).

When tested on human embryonic vein cells and in vivo on adult male mice, collagen polypeptides obtained from the jellyfish *R. esculentum* and hydrolyzed into different molecular weight showed significant wound healing bioactivity. Effective and rapid acceleration of wound healing rate was also observed in rat skin cells treated with collagen extracted from the skin of tilapia fish (Chen et al. 2019).

In improving the effectiveness of wound healing, collagen-based wound healing material contribute significantly by reducing the hospital stay and cost of treatment.

It also reduces the period over which the patient has to withstand the pain and discomfort from the wound hence improvement in the quality of health care (Hochstein and Bhatia 2014).

For effective wound healing the form in which the collagen is applied to the wound is also important. The physical environment should be one which provides the right architecture that promotes the forming of new skin cells. Collagen wound healing materials have been made in the form of sponges produced using collagen obtained from scales of mrigal fish (Pal et al. 2016), dressings made from electrospun fibers of collagen extracted from tilapia fish (Moura et al. 2014; Zhou et al. 2016). When tested in rat models, these different forms of collagen-based wound healing materials showed enhancement of wound healing processes such as proliferation of keratinocytes, fibroblast migration and collagen deposition.

12.7.3 Bone Regeneration

Collagen forms the matrix within which the bone mineral, hydroxylapatite is dispersed. Damage to the bone often requires a membrane to hold the bone in place during healing after which the material is removed. The membrane serves to guide the direction of bone regrowth ensuring that the bone returns to its original shape and form. The removal of the membrane often requires an additional surgical procedure which implies more healthcare cost and further discomfort for the patient. Non-biodegradable bone guiding membranes are made of materials such as polytetrafluoroethylene (Fiorellini et al. 1998), and this has the right mechanical properties to support the bone regeneration and is unreactive enough not to induce an undesirable immune response by the body. It will also not be degraded by the matrix metalloproteinase enzymes (MMPs) which are responsible for bone protein degradation, these enzymes get rid of damaged old bone tissue and replace with the newly regenerated tissue (Armstrong and Jude 2002). In collagen-based scaffolds used in bone regeneration, the activity of these MMPs enzymes is inhibited by treating the scaffold with a tetracycline which inhibits the activity of MMPs (Moses et al. 2008).

Scaffolds are used where starter cells as well as a structural guide are required to aid tissue regeneration. Collagen-based scaffolds have been developed with the goal to replicate the macro- and microstructure of the bone tissue. Using collagen and hydroxyapatite composites for such scaffolds further better mimics the biological tissue physically and chemically. Different forms of these collagen/hydroxyapatite scaffolds have been presented for bone regeneration applications (Zhang et al. 2018; Xia et al. 2013; Yunoki et al. 2006, 2007). However, there is limited understanding of the complex mineralization process and reproducing the intricately ordered nanocrystalline structure of the collagen/mineral bone composite is not yet achievable. Further, development is still required in the area of tissue engineering to develop tissue scaffolds which better mimic the tissue structure.

The area of bone regeneration has further advanced toward development of 3D printed scaffolds (Govindharaj et al. 2019). The ability to 3D print tissue scaffolds offers the advantage of making more complex and more detailed and dimensionally precise design of these scaffolds more possible. This will result in bone regrowth which is as close to the original tissue as possible. Materials used for these include polycaprolactone and hydroxyapatite composite (Liu et al. 2019). Such scaffolds have been tested in vitro and in vivo on laboratory animals (commonly regeneration of bones in damaged rat skull regions). The use of hydroxyapatite ($Ca_{10}(PO_4)_6(OH)_2$) is with the aim of using a material which is the same as that found in the bone tissue. It acts as the dispersed phase in the composite while the polymer acts as the matrix.

The membranes or scaffolds could be required to remain in place for different periods of time; therefore, the collagen material needs to be designed for specific applications. This is done by for example varying the degree of cross-linking which in turn affects the rate of degradation (Moses et al. 2008). Composites of collagen with varying degree of degradation could also be used. Most of the commercially available tissue scaffolds are based on bovine and porcine-based collagen, and fish waste could potentially replace or augment the use of bovine or porcine collagen in tissue regeneration. Many aquatic vertebrates including fish collagen, like human collagen, are mostly of type 1 (Zhang et al. 2018). Collagen alone usually does not possess sufficient mechanical properties to act as a scaffold or membrane for tissue regeneration; it is therefore often used as a composite alongside other materials such as carbon, minerals or other polymers. The mechanical properties can also be further enhanced by cross-linking with crosslinkers such as glutaraldehyde or ribose (Bai et al. 2018).

12.7.4 Biomedical Implants

The biocompatibility of collagen makes it an excellent material for implants and tissue replacement. An example of such is in corneal implants into the eye. Corneal repair generally requires replacement of the damaged cornea with an artificial one which is implanted into the eyes to replace the old one. The material from which such is made needs to be biocompatible and be retained in the eye and allow growth and proliferation of new corneal cells. Fish scales have been tested as ideal candidate as a biocompatible corneal replacement. Scales from the collagen-based artificial cornea have the advantage of not requiring human corneal donor, and it eliminates the chance of an undesired immune response. Obtaining this from a widely available resource such as fish scales which is not limited to only a particular region also offers the prospect of having multiple producers across the world. The arrangement of the collagen structure is quite important in tissue regeneration as the micro- and nano-structure must allow sufficient space for oxygen transport to the cells growing within and also allow for cell adhesion. To address this, a similar material which already possesses an extracellular matrix structure similar to the microporous structure of the cornea is used. This approach was used by Lin et al. (2010). Scales from tilapia fish were decalcified using nitric acid treatment and living behind the organic part which comprises of collagen. With the calcified part of the scales removed, what is left behind is a cellular architecture which has the right cellular architecture in terms of pore size and shape to allow for effective cell growth.

12.7.5 Microneedles

Microneedles are transdermal patches which comprise of tiny projections with micrometer dimensions. They are designed to pierce through the upper layer of the skin without reaching the dermis where the pain receptors exist and in so doing open up the skin in a painless manner and creating microconduits through which therapeutics can be delivered. The lower layer of the skin (dermis) is easy reach to the blood vessels from which the drugs can be absorbed into the bloodstream and to target cells. These microneedles can be designed for delivery via different modes. One way is to either pierce the skin, after which it is removed and replaced by a drug loaded patch which then enters into the lower layer of the skin through the holes created by the microneedles. These types are referred to as solid microneedles which could be microfabricated using biocompatible materials such as metals such as stainless steel or polymers such as silicone and polycarbonate. These solid microneedles could also be coated with the drugs in the form of a dry coating which will then dissolve into the skin when inserted and comes in contact with the skin fluid. Microneedles can also be designed to dissolve into the skin upon piercing and release the drugs loaded within into the lower layers of the skin from where it can be absorbed. These types of microneedles are strong and rigid when in the dry state; however, they are soluble in physiological fluid. Such that when they are inserted into the skin and come into contact with the moisture in the skin, they dissolve off. The production process usually involves loading the drug within the polymeric formulation in the wet state followed by forming into microneedles such that the drug is entrapped within the microneedle structure. These are referred to as dissolving microneedles and are made from soluble polymers such as hydrolyzed collagen (Olatunji et al. 2014), polylactic acid (Adamczak et al. 2011) and polyvinyl alcohol (Sullivan et al. 2010). A third mode is where the microneedle with the drug loaded is inserted into the skin, the drug is released from the microneedles after which the microneedle is then removed. These are referred to as hydrogel microneedles. They are made using cross-linked polymers with the drug embedded within. Cross-linked polymeric structures have the ability to absorb water and swell. As the water is absorbed and the cross-link network expands, the drugs which are trapped within the polymer chains are released into the skin while the microneedle stays intact and can be removed after the drug is released. Another type of microneedles exists which are the porous microneedles. These are designed to inject the drug in liquid form into the skin such that it can be absorbed from the dermis. However, these have been of much less interest due to the complexity of their production process and high possibility of blockage of the micron-sized holes upon insertion.

Recently, microneedles produced using fish scale-derived hydrolyzed collagen hydrogels have been developed (Olatunji and Denloye 2019). These are a form of hydrogel microneedles which in the dry state are rigid enough to pierce the skin; however, once inserted into the skin it will absorb water and release the entrapped loaded compound into the lower layer of the skin from where it can be absorbed into the bloodstream more easily. In vitro studies showed that these hydrogel microneedles have sufficient strength to pierce the skin and can swell and release loaded drug compound when in contact with water. The hydrogels were made using hydrolyzed collagen extracted from fish scales of croaker fish using the hydrothermal extraction method.

Delivery of collagen into the skin using microneedles also serves cosmetic purpose of boosting the skin collagen content. This could be a means to address the decreased collagen production rate associated with aging and by so doing decrease the appearance of wrinkles. Due to its large molecular weight collagen or hydrolyzed collagen cannot pierce the upper layer of the skin; therefore, delivering the collagen through topical means might not achieve the same goal as the collagen will not penetrate the impermeable top layer of the skin (*Stratum corneum*). Another option is the use of hypodermic injections to deliver collagen underneath the skin. However, this method comes with the disadvantage of discomfort and risk of accidental injury from the injection. Dissolving microneedles made of lower molecular weight hydrolyzed collagen can therefore be used as painless means to deliver collagen into the skin. This also has the advantage of self-administration by the user without requiring medical expertise unlike with injection using hypodermic needles. The microneedle could either be made entirely of collagen (Olatunji et al. 2014) or it could be blended with another polymer such as polyvinylpyrrolidone (Sun et al. 2014).

12.7.6 Cosmetics and Skin Care

The skin is the outermost layer of the body and by virtue of this it has a significant role to play in appearance. People are constantly seeking to smoothen, moisturize, color, bleach and dye or improve the appearance of the skin in one way or the other. This makes collagen also suitable as an active compound in cosmetics and skin care products. Collagen has become a key part of the billion dollar cosmetics industry. They have been applied in a range of products from topical skin care to cosmeceutical oral gels to injectables (Alves et al. 2017; Devgan et al. 2019).

Collagen plays a significant role in cosmetics either in attaining the desired physical or rheological properties in terms of thickening and gelling property, and it also plays a role as an active ingredient, mostly in the form of collagen hydrolysate where it is intended to penetrate into the epidermis and improve the skin mechanical properties which result in smoother and more youthful looking skin (Savary et al. 2016). Oligomeric forms of collagen which are very short polypeptide fragments are used in skin care products such as skin lotions where they are thought to easier penetrate the epidermis from where they can reach the dermis and induce skin rejuvenating effect. There are further studies into confirming the exact manner or extent to which this happens (Bradley et al. 2015). Collagen, like other polymers, is usually used in low concentration in cosmetics products, usually below 1% weight fraction.

12.7.7 Food

In food, collagen is used mainly in the hydrolyzed form. One of the more common applications is its use in jellies. In the presence of water at a temperature of around 4 °C, gelatin forms a soft aqueous jelly. However, this jelly is temperature sensitive and melts at higher temperatures. Although not yet a commercial reality, hydrolyzed collagen is also used in the production of food packaging films (Hanani et al. 2014).

Like all polymers when dissolved in a solvent, it increases the viscosity of the solvent such that there is a correlation between its concentration and the thickness of the solution in a molecular weight-dependent manner. For this reason, hydrolyzed collagen can be used as a thickening agent in food. In this dissolved form, when the temperature is reduced to around 4 °C, the polypeptide chains take on a coiled helix conformation and form a three-dimensional network trapping water within the network of polymer chains. This manifests physically as a gel. This gel is thermoreversible, and this comes an advantage when the melt in mouth feeling is required for the food product (Rayner et al. 2016).

The gel properties vary for different types of hydrolyzed collagen. Gel properties can be characterized by the gel strength, bloom, gel transparency and gel elasticity. Other gel-forming polymers such as carrageenans tend to form stronger gels that gelatin; however, in some applications a softer gel is required while in others a strong gel is required. Hydrolyzed collagen from aquatic sources such as fish skin has been shown to be also suitable in the production of food jellies similar to those made using bovine and porcine sourced hydrolyzed collagen (Gamarro et al. 2013).

Hydrolyzed collagen is used in the production of soft gels which are used in the production of some nutritional supplements such as omega-3 and production of herbal capsules. These gels and capsules make it easier to ingest nutrients or herbs which are either not taken in the diet or for easier intake where the taste or flavor is not desirable.

12.7.8 Antioxidant Activity

Collagenous proteins obtained from edible jellyfish known as *R. esculentum* have been shown to possess antioxidant activity (Zhuang et al. 2009), and collagens from fish also show some antioxidant property (Chi et al. 2014). This antioxidant property however is dependent on the structure of the collagen.

Antioxidant property could be demonstrated by a molecule or compounds ability to scavenge free radicals, donate protons or electrons, deactivate the reactive oxygen species or chelate transition metals which promote oxidation. Collagen shows better antioxidant activity in the hydrolyzed form. Further, lower molecular weight hydrolyzed collagen shows better antioxidant properties than higher molecular weight once. For example, when tested in vitro using hydrogen peroxide and 2,2-diphenylpicrylhydrazyl (DPPH) radicals, hydrolyzed collagen from the skin of Spanish mackerel fish with molecular weight of 47.82 and 5.04 kDa showed 35.82 and 65.72% DPPH scavenging activity, respectively. The scavenging activity against hydrogen peroxide was 41.72 and 76.99%, respectively (Chi et al. 2014).

The antioxidant ability of lower molecular weight hydrolyzed collagen is attributed to the reducing property of the peptide groups present in the hydrolyzed forms. In the helical form, the amine and carbonyl groups of the different polypeptide chains are engaged in intermolecular secondary bonds; however, once hydrolyzed these groups become available to other bonds, hence reducing tendencies of hydrolyzed collagen over the intact form of collagen. Further, hydrolysis of the polypeptides in lower molecular weight further frees up more peptide functional groups which can have reducing effects by donating electrons to free radicals.

12.7.9 Anti-inflammatory Activity

The skin responds to external stimuli which are harmful to the body through inflammation. However in certain cases, the process of inflammation could be exaggerated or lead to inflammatory diseases such as psoriasis, rheumatoid arthritis and atherosclerosis. Anti-inflammatory agents are used as a therapeutic measure to treat such inflammatory-related diseases. Fish scale collagen polypeptides have been used to demonstrate anti-inflammatory activity of collagen extracted from fish. The collagen peptides interfered with inflammatory response by preventing the secretion of proteins associated with inflammatory response (Ahn et al. 2017). Collagen extracted from squid also shows anti-inflammatory effects in the treatment of osteoarthritis. The collagen type II extracted from squid offers superior therapeutic effect against osteoarthritis by eliminating the immunogenicity associated with treatment using collagen sourced from terrestrial animal (Dai et al. 2018).

12.8 Commercial Production

Collagen is a protein of much commercial value. Bones and flesh of bovine and porcine have been the main commercial source of commercial collagen production. With the outbreak of bovine spongiform encephalopathy and the foot and mouth disease, for health and economic reasons have led to the rising demand for alternative collagen sources. The aquatic environment offers such alternatives. Both marine and freshwater are a source of diverse range of collagen-producing organisms. Some of these are bones, skins and scales of fishes, jellyfish, starfish and sponges (Felician et al. 2019; Olatunji et al. 2014; Khong et al. 2016).

The global collagen market is expected to reach 4.6 billion USD by 2023. In 2018, the market was valued at 3.5 billion USD with a 5.2% estimated growth rate annually. This includes collagen and hydrolyzed collagen, hydrolyzed collagen making a larger share of the market. The marine sourced collagen shows constant growth as the industry seeks alternatives to bovine and porcine sourced collagen. Main drivers of growth for the collagen market are the increased use in the pharmaceutical, cosmetics and food industry (Research and Markets 2019).

Companies which are commercially producing collagen and/or collagen peptides from aquatic sources include Nitta Gelatin which has been in the business of gelatin since 1979. Gelita AG, Weishardt Group, Darling Ingredients, Connoils, Lapi Gelatine S.p.a and Gelnex are some other companies which are key producers of gelatine. The produce gelatin and collagen peptides for various target markets include food, cosmetics and pharmaceutical. They produce from both aquatic and non-aquatic sources.

Collagen plays a significant role in sustaining and improving human quality of life. Such includes wound healing, tissue repair, disease prevention and cosmetics. As human civilization expands and we develop new technologies to keep people alive for longer, the tendency for injuries, diseases and need to improve appearance increases. Therefore, an ever-growing market exists for collagen in its wide range of applications.

12.9 Conclusion

Collagen is the most abundant protein in nature. It is present in both aquatic and terrestrial animals. Fish is the most common source of aquatic collagen, other sources are squid and jellyfish. Aquatic sourced collagen offers the advantage of avoidance of the risk of mad cow disease and an alternative collagen source where ethics and religious beliefs prohibit the use of collagen from pork or animal skin. Collagen from aquatic source also has bioactive, mechanical and rheological properties which differ from those of the conventional porcine collagen. This allows more diverse forms of collagen for a wider range of applications. Most of the aquatic sourced collagens are from the waste by-products of these aquatic resources; therefore, the extraction of a valuable biopolymer such as collagen significantly contributes to optimal utilization of and value addition to aquatic resources.

References

Adamczak M, Ścisłowska-Czarnecka A, Genet MJ, Dupont-Gillain CC, Pamula E (2011) Surface characterization, collagen adsorption and cell behaviour on poly(L-lactide-co-glycolide). Acta Bioeng Biomech 13(3):63–75

Aerssens J, Dequeker J, Mbuyi-Muamba JM (1994) Bone tissue composition: biochemical anatomy of bone. Clin Rheumatol 1:54–62

Ahmed R, Haq M, Chin BS (2019) Characterization of marine derived collagen extracted from the by-products of bigeye tuna (*Thunnus obesus*). Int J Biol Macromol 135:668–676

Ahn MY, Hwang JS, Ham SA, Hur J, Jo Y, Lee SY, Choi M, Han SG, Seo HG (2017) Subcritical water-hydrolyzed fish collagen ameliorates survival of endotoxemic mice by inhibiting HMGB1 release in a HO-1-dependent manner. Biomed Pharmacother 93:923–930

Alves AL, Marques APL, Martins E, Silva TH, Reis RL (2017) Cosmetic potential of marine fish skin collagen. Cosmetics 4(4):39

Armstrong DG, Jude EB (2002) The role of matrix metalloproteinases in wound healing. J Am Podiatr Med Assoc 92:12–18

Bai X, Gao M, Syed S, Zhuang J, Xu X, Zhang X (2018) Bioactive hydrogels for bone regeneration. Bioact Mater 3:401–417

Benedetto CD, Barbaglio A, Martinello T et al (2014) Production, characterization and biocompatibility of marine collagen matrices from an alternative and sustainable source: the sea urchin *Paracentrotus lividus*. Mar Drugs 12:4912–4933

Bradley E, Griffiths C, Sherrat M, Bell M, Watson R (2015) Over the counter anti aging topical agents and their ability to repair and protect photo aged skin. Maturitas. https://doi.org/10.1016/j.maturitas.2014.12.19

Chang S, Shefelbine SJ, Buehler MJ (2012) Structural and mechanical differences between collagen homo- and heterotrimers: relevance for the molecular origin of brittle bone disease. Biophys J 102(3):640–648

Chattopadhyay S, Raines RT (2014) Review collagen-based biomaterials for wound healing. Biopolymers 101:821–833

Chen J, Gao K, Liu S, Wang S, Elango J, Bao B, Dong J, Liu N, Wu W (2019) Fish collagen surgical compress repairing characteristics on wound healing process in vivo. Mar Drugs 17(33):1–12

Cheng X, Shao Z, Li C et al (2017) Isolation, characterization and evaluation of collagen from jellyfish *Rhopilema esculentum* Kishinouye for use in hemostatic applications. PLoS One 12:e0169731

Chi CF, Cao ZH, Wang B, Hu FY, Li ZR, Zhang B (2014) Antioxidant and functional properties of collagen hydrolysates from Spanish mackerel skin as influenced by average molecular weight. Molecules 19:11211–11230

Dabrowska A, Rotaru GM, Spano F, Affolter C, Fortunato G, Lehmann S, Derler S, Spencer ND, Rossi RM (2017) A water-responsive, gelatine-based human skin model. Tribol Int 113:316–322

Dai M, Liu X, Wang N, Sun J (2018) Squid type II collagen as a novel biomaterial: isolation, characterization, immunogenicity and relieving effect on degenerative osteoarthritis via inhibiting STAT1 signaling in proinflammatory macrophages. Mater Sci Eng C 89:283–294

Delphi L, Sepehri H, Motevaseli E et al (2016) Collagen extracted from Persian Gulf squid exhibits anti-cytotoxic properties on apple pectic treated cells: assessment in an in vitro bioassay model. Iran J Public Health 45:1054–1063

Devgan L, Singh P, Durairaj K (2019) Minimally invasive facial cosmetic procedures. Otolaryngol Clin North Am 52(3):443–459

Dong Z (2019) Blooms of the moon jellyfish *Aurelia*: causes, consequences and control. In: World seas: an environmental evaluation, vol III, 2nd edn., pp 163–171

Elango J, Lee JW, Wang S, Henrotin Y, de Val J, Regenstein JM, Lim SY, Bao B, Wu W (2018) Evaluation of differentiated bone cells proliferation by blue shark skin collagen via biochemical for bone tissue engineering. Mar Drugs 16:350

FAO (2018) The state of world fisheries and aquaculture 2018—meeting the sustainable development goals. Rome. Licence: CC BY-NC-SA 3.0 IGO. ISBN 978-92-5-130562-1

Felician FF, Yu RH, Li MZ, Li CJ, Chen HQ, Jiang Y, Tang T, Qi WY, Xu HM (2019) The wound
   healing potential of collagen peptides derived from the jellyfish Rhopilema esculentum. Chin J
   Traumatol 22:12–20

Fiorellini J, Engebretson SP, Donath K, Weber HP (1998) Guided bone regeneration utilizing
   expanded polytetrafluoroethylene membranes in combination with submerged and nonsubmerged
   dental implants in beagle dogs. J Periodont 69:528–535

Gamarro EG, Orawattanamateekul W, Sentina J, Gopal TKS (2013) By-products of tuna processing.
   GlobeFish 112(48):1–18. FAO, Rome

Govindharaj M, Roopavathi KU, Rath SK (2019) Valorization of discarded Marine Eel fish skin
   for collagen extraction as a 3D printable blue biomaterial for tissue engineering. J Clean Prod (in
   press)

Hanani ZAN, Roos YH, Kerry JP (2014) Use and application of gelatin as potential biodegradable
   packaging materials for food products. Int J Biol Macromol 71:94–102

Hochstein AO, Bhatia A (2014) Collagen: its role in wound healing. Podiatry Manage 103(106):109–
   110

Jeevithan E, Zhang JY, Wang NP, He L, Bao B, Wu WH (2015) Physico-chemical, antioxidant and
   intestinal absorption properties of whale shark type-II collagen based on its solubility with acid
   and pepsin. Process Biochem 50:463–472

Khong NM, Yusoff FM, Jamilah B et al (2016) Nutritional composition and total collagen content
   of three commercially important edible jellyfish. Food Chem 196:953–960

Lin CC, Ritch R, Lin MS, Ni M, Chang Y, Lu YL, Lai HJ, Lin F (2010) A new fish scale-scaffold
   for corneal regeneration. Eur Cells Mater 19:50–57

Liu D, Nie W, Li D, Wang W, Zheng L, Zhang J, Zhang J, Peng C, Mo X, He C (2019) 3D printed
   PCL/SrHA scaffold for enhanced bone regeneration. Chem Eng J 362:269–279

Mahboob S (2015) Isolation and characterization of collagen from fish waste material—skin, scales
   and fins of *Catla catla* and *Cirrhinus mrigala*. J Food Sci Technol 52(7):4296–4305

Matmaroh K, Benjakul S, Prodpran T, Encarnacion AB, Kishimura H (2011) Characteristics of acid
   soluble collagen and pepsin soluble collagen from scale of spotted golden goatfish (*Parupeneus
   heptacanthus*). Food Chem 129(3):1179–1186

Moses O, Vitrial D, Aboodi G, Sculean A, Tal H, Kozlovsky A, Artzi Z, Weinreb M, Nemcovsky CE
   (2008) Biodegradation of three different collagen membranes in the rat calvarium: a comparative
   study. J Periodontol 79(5):905–911

Moura LIF, Dias AMA, Suesca E, Casadiegos S, Leal EC, Fontanilla MR, Carvalho L, de Sousa
   HC, Carvalho E (2014) Neurotensin-loaded collagen dressings reduce inflammation and improve
   wound healing in diabetic mice. Biochem Biophys Acta 1842:32–43

Muthukumar T, Anbarasu K, Prakash D, Sastry TP (2014) Effect of growth factors and proinflammatory cytokines by the collagen biocomposite dressing material containing *Macrotyloma
   uniflorum* plant extract—in vivo wound healing. Colloids Surf B Biointerfaces 121:178–188

Olatunji O, Denloye A (2017) Temperature-dependent extraction kinetics of hydrolyzed collagen
   from scales of croaker fish using thermal extraction. Food Sci Nutr 5:1015–1020

Olatunji O, Denloye A (2019) Production of hydrogel microneedles from fish scale biopolymer. J
   Polym Environ 27(6):1252–1258

Olatunji O, Olsson RT (2015) Microneedles from fishscale-nanocellulose blends using low
   temperature mechanical press method. Pharmaceutics 7:363–378

Olatunji O, Igwe CC, Ahmed AS, Alhassan DOA, Asieba GO, Das DB (2014) Microneedles from
   fish scale biopolymer. J Appl Polym Sci 131:40377–40388

Pal P, Srivas PK, Dadhich P, Das B, Maity PP, Moulik D, Dhara S (2016) Accelerating full thickness
   wound healing using collagen sponge of mrigal fish (*Cirrhinus cirrhosus*) scale origin. Int J Biol
   Macromol 93:1507–1518

Rayner M, Ostbring K, Purhagen J (2016) Application of natural polymers in food. In: Olatunji O (ed) Natural polymers: industry techniques and applications. Springer, Switzerland. https://doi.org/10.1007/978-3-319-26414-1_5

Research and Markets (2019) Collagen market by product type—global forecast to 2023, Feb 2019. Report ID: 4756592

Robinson RA (1979) Bone tissue: composition and function. Johns Hopkins Med J 145(1):10–24

Savary G, Grisel M, Picard C (2016) Cosmetics and personal care products. In: Olatunji O (ed) Natural polymers: industry techniques and applications. Springer, Switzerland, pp 219–261

Shoulders MD, Raines RT (2009) Collagen structure and stability. Annu Rev Biochem 78:929–958

Sionkowska A, Kozlowska J (2014) Fish scales as a biocomposite of collagen and calcium salts. Eng Mater 587:185–190

Subramaniam S (2018) Trimethylamine oxide (TMAO): a new toxic kid on the block. J Biomol Res Therapeut 7(1)

Sullivan SP, Koutsonanos DG, del Pilar M et al (2010) Dissolving polymer microneedle patches for influenza vaccination. Nat Med 16(8):915–920

Sun W, Inayathullah M, Manoukian MAC, Malkovsky AV, Manickam S, Marinkovich MP, Lane AT, Tayebi AM, Rajadas J (2014) Transdermal delivery of functional collagen via polyvinylpyrrolidone microneedle. Ann Biomed Eng 43(12):2978–2990

Tan CC, Karim AA, Latiff A et al (2013) Extraction and characterization of pepsin-solubilized collagen from the body wall of crown-of-thorns starfish (*Acanthaster planci*). Int Food Res J 20:3013–3020

Xia Z, Yu X, Jiang X, Brody HD, Rowe DW, Wei M (2013) Fabrication and characterization of biomimetic collagen-apatite scaffolds with tunable structures for bone tissue engineering. Acta Biomater 9(7):7308–7319

Yunoki S, Ikoma T, Monkawa A, Ohta K, Kikuchi M, Sotome S, Shinomiya K, Tanaka J (2006) Control of pore structure and mechanical property in hydroxyapatite/collagen composite using unidirectional ice growth. Mater Lett 60(8):999–1002

Yunoki S, Ikoma T, Monkawa A, Marukawa E, Sotome S, Shinomiya K, Tanaka J (2007) Three-dimensional porous hydroxyapatite/collagen composite with rubber-like elasticity. Mater Sci Eng C 18(4):393–409

Zhang D, Wu X, Chen J, Lin K (2018) The development of collagen based composite scaffolds for bone regeneration. Bioact Mater 3:129–138

Zhou T, Wang NP, Xue Y, Ding TT, Liu X, Mo XM, Sun J (2016) Electrospun tilapia collagen nanofibers accelerating wound healing via inducing keratinocytes proliferation and differentiation. Colloids Surf B 143:415–422

Zhuang YL, Zhao X, Li BF (2009) Optimization of antioxidant activity by response surface methodology in hydrolysates of jellyfish (*Rhopilema esculentum*) umbrella collagen. J Zhejiang Univ Sci B 10:572–579

Zhuang Y, Sun L, Zhang Y et al (2012) Antihypertensive effect of long-term oral administration of jellyfish (*Rhopilema esculentum*) collagen peptides on renovascular hypertension. Mar Drugs 10:417–426

Chapter 13
Starch

Abstract Starch plays a significant role in the economy as it serves as a source for two products; food and bioethanol, both of which are essential commodities. It is also used in several other industries and discussed in this chapter. In the aquatic environment, starch can be sourced from aquatic plants and algae. Starch-producing aquatic organisms have a significant role to play in the production of third-generation biofuel which does not require already limited arable land for the cultivation. The unique chemistry of some aquatic starch forms makes them attractive for specific industrial applications.

Keywords Starch · Polysaccharides · Carbohydrates · Polymers · Aquatic

13.1 Introduction

Starch is a carbohydrate which is more commonly associated with terrestrial food crops such as cassava and corn which forms staple foods and biofuel feedstock in many parts of the world. Some aquatic organisms also contain starch; for example, aquatic sources of starch are algae, duckweeds and water hyacinths. Although starch is a biopolymer that is not exclusive to the aquatic environment, aquatic-sourced starch plays a significant role as an alternative source of starch in many industries.

One of the potential economic importance of aquatic-sourced starch is the potential for the use in biofuel production. While the first-generation biofuels created the promise of a renewable source of energy (Eichelmann et al. 2016), the second-generation biofuel promises a renewable source of energy that does not compete with food (Negm et al. 2018), and the promise of the third generation of biofuel from aquatic resources such as algae (Yew et al. 2019) is a supply of fermentable biomass for ethanol production which does not threaten food security or the environment and does not compete with terrestrial animals and humans for land space. Third-generation biofuel from algae and aquatic plants can achieve the goal of carbon-neutral biofuel production. This is mainly owing to the fact that the process of cultivation of algae and aquatic plants removes carbon dioxide from the environment. This is unlike in the case of fossil fuel extraction and refinery which leads to further carbon emissions.

© Springer International Publishing 2020

287

O. Olatunji, *Aquatic Biopolymers*, Springer Series on Polymer and Composite Materials, https://doi.org/10.1007/978-3-030-34709-3_13

Food security is increasingly becoming a global concern. Starch forms an important part of the human diet and making up over half of the global calorie intake. It is a major source of carbohydrate and energy food. Although approximately 65 million tonnes of starch are produced annually in the world and constantly increasing at an annual rate of 2–3% (Sullivan-Trainor 2013; McWilliams 2017), much more starch would be required if it is to serve as a sustainable source of biofuel as well as a component of food and other products used by humans.

Starch is one of the major polymers being explored to solve one of the biggest global crises of single-use plastic pollution through the development of starch-based bioplastic packaging. It is therefore pertinent to seek alternative, preferably more abundant source of starch which do not interfere with the human consumption of starch and starch-based products but rather augments it where possible. Conventional starch sources are maize, rice, wheat, potato, tapioca and cassava. From these, starches for biofuel, bioplastic, food and other starch-based products must be sourced.

In addition to the existing and potential application of aquatic-sourced starch to create products, the process of cultivating and processing of aquatic-sourced starch should also be considered in order to understand the overall impact of aquatic-sourced starch as a biopolymer resource. This chapter therefore discusses the sources, availability, production process, chemistry, potential applications and the impact on the environment and global economy of starch sourced from aquatic ecosystems.

13.2   Occurrence in Nature

Starch, a major energy storage carbohydrate, occurs naturally in plants. It is present in the leaves, seeds, fruits, stems, roots and tubers. It is also present in red, brown and green macroalgae and microalgae (Prabhu et al. 2019). These photosynthetic organisms absorb light energy from the sun and store it in chemical form as starch. Aquatics plants like the water hyacinth, duckweed and *Azolla* are examples of aquatic sources of starch. These starch-producing organisms can be found in both marine and freshwater (Zhang et al. 2018) in varying amounts. The starch content of any given organism depends on factors such as species and abiotic factors of the aquatic ecosystem, and it is growing within.

Microalgae contain a considerably large amount of starch. On average, they contain around 37% starch by weight. Strains such as Chlamydomonas, Chlorella, Spirulina, Dunaliella, Scenedesmus have the most abundant starch content which could be as much as 50% carbohydrates most of which is starch (Rehman and Anal 2018). At optimized conditions, the cultivation of the green marine microalgae strain Tetraselmis subcordiformis with starch content of up to 62.1% dry weight has been achieved (Yao et al. 2012). These findings suggest that not only is starch present in microalgae but also that the rate of starch accumulation in the microalgae biomass can be controlled and thus optimized through the limitation of nutrients.

The rate of accumulation of starch and biomass within the aquatic organism is affected by the environmental conditions as well as the nutrients available for growth.

For example, marine red microalgae, *Porphyridium marinum*, grow optimally at a light intensity of 100 μmol photons $m^{-2} s^{-1}$, sodium nitrate ($NaNO_3$) concentration of 1 g/L and a sodium chloride concentration of 20 g/l (Himaa et al. 2019). At this optimal condition, these algae can accumulate up to 140.21 μg per ml which is 13% higher than when grown in suboptimal conditions. The lower the light intensity and salinity, the higher the starch accumulated within the microalgae.

Starch is also present in a considerable amount in aquatic plants such as water hyacinth and duckweed. The composition of starch in aquatic plants varies from species to species and is also affected by seasonal variation and growth conditions. In *Azolla*, for example, filiculoides produces up to 36.4 g/L more sugars than another strain of *Azolla pinnata*. Filiculoides biomass contains 6.05% starch while pinnata contains 4.7%. Pinnata also has higher lignin content (Miranda et al. 2016). *Azolla*, in general, has the highest lignin content of all the common aquatic plants explored in the literature; however, this lignin content (13.2% for pinnata and 10.3% for filiculoides) is still lower than those of terrestrial lignocellulose crops such as corn and sugar beets.

In algae, the location of starch on the organism could vary. The starch could be present in the chloroplast, cytoplasm or pyrenoids. They could also be present in the form of plates or granules. For example, in the red algae Gracilariopsis, starch is present in the form of granules within the cytoplasm; in the green algae Cladophora, starch is present within the chloroplasts in form of granules; and in the green algae ulva, starch is present in the form of plates around the pyrenoids and as granules in the chloroplast thylakoid membranes.

Starch content could vary depending on the environmental conditions. Growing in the open environment weather conditions significantly affect the starch content at any period. Starch content also varies with the organism's reproductive cycle. Studies on the green algae *Ulva ohnois* indicated a minimal starch accumulation at the initial growth phase; in the particular study, this occured between March and May and maximum starch content was recorded in June. Note that the biomass accumulation rate does not necessarily correlate with the starch accumulation rate. A faster growth rate is not necessarily implicative of a higher starch content, and the organism could be accumulating other compounds such as cellulose, proteins and lipids.

There are also instances where starch is degraded into other polymers within the aquatic organism, and this has been observed in red algae. Degradation of starch into carrageenan was observed in *S. cordalis*, a unicellular red algae. It appears that under various conditions such as nutrient supply and irradiance, and the starch within the organism is converted into carrageenan (Fournet et al. 2000). This is attributed to the need for the organism to switch to a biochemical pathway which favors the surrounding condition. Starch also exists in the aquatic environment in the form of floridean starch (Himaa et al. 2019). This refers to starch sourced from a particular strain of red algae known as the florideophyceae. It is mainly characterized by the absence of the branched polymer of starch, amylopectin. Floridean starch is discussed in more detail in a separate section within this chapter.

13.3   Availability of Raw Material

Availability of starch is limited to some starch-producing aquatic plant and algae biomass. Within this selection lies thousands of species and these include duckweeds, *Azolla*, water hyacinth macroalgae and microalgae. Duckweed, for example, comprises of 37 species (Appenroth et al. 2013). Here we look at the availability of starch from the aquatic environment based on the availability of these sources and their starch component.

Duckweeds grow almost all year round with a growth season lasting between 9 and 12 months (Ziegler et al. 2015). Duckweed of up to 39.1–105.9 tonnes per hectares annually is achievable with a starch content of up to 31.0–45.8%. 94.7% of this starch can be converted into bioethanol, and duckweed is therefore a relatively abundant source of starch. Methods such as selectively breeding high-performance duckweed strains and improving starch enrichment in the plant are among the area of research focus toward advancing duckweed for commercial production of starch for biofuel application (Xu et al. 2014).

Duckweeds are exceptionally high in starch compared to other aquatic and non-edible plants in general. Duckweed species include *Wolffia arrhiza* (rootless duckweed), *Spirodela polyrhiza* (greater duckweed), *Lemna gibba* (fat duckweed), *Lemna minuta* (least duckweed), *Lemna minor* (common duckweed), *Lemna trisulca* (ivy-leaved duckweed). Duckweed annual accumulation estimated to 39.1–105.9 tonnes of starch per hectare annually (Xu et al. 2012). They also have a relatively high protein content. Their low lignin content is also a desirable feature as it makes the extraction process less demanding. Accumulation of starch in duckweed is affected by factors such as light intensity, temperature, nitrogen content and phosphorus content of the water within which they are growing. Adequate monitoring and control of these parameters can improve the rate of starch accumulation.

Availability of biomass for starch production can be further improved by practicing a mixed culture system of aquatic farming. The productivity of biomass of some aquatic plants can be improved by polyculture method. This involves growing more than one strain of the plant within the same growth space. Duckweed polyculture comprising of *Lemna aequinoctialis* 5505, *Landoltia punctata* 5506 and *S. polyrhiza* 5507 attained a starch composition of up to 28.78 g m^{-2}. When compared to monocultures of each species alone, the combination of all three gave a higher starch biomass yield. However, the use of a polyculture does not always guarantee an improvement in starch accumulation. This highly depends on the combination of the species. While one combination of species in a polyculture could result in improved starch content compared to monoculture, in a different combination of species the starch accumulation could be less. For example, a monoculture of *L. aequinoctialis* is grown at a temperature of 20 °C, light intensity of 105 μmol m^{-2} s^{-1}, nitrogen concentration of 35 mg L^{-1} and potassium concentration of 15 mg L^{-1} which are accumulated with 14.22 g of starch per m^2 of duckweed growth area. At the same growth condition, the combination of all three species *L. aequinoctialis*, *L. punctata* and *S. polyrhiza* attained a starch accumulation of 13.96 g m^{-2}.

Aquatic plants which produce starch can accumulate biomass at a faster rate than terrestrial plants. Duckweeds, for example, can double in biomass between 1 and 3 days (Li et al. 2016). The effect of factors such as light intensity and temperature on the rate of accumulation from biomass has been studied by different researchers; however, the effects of these factors depend on other complex factors such as origin of the plant and adaptation to the previous environment, such that a particular trend cannot be recommended. For example, that an increase in light intensity results in higher starch accumulation for one species sourced from a particular region might not mean the same species from a different origin will respond in the same manner.

Starch production by some aquatic plants can be further optimized by limiting the growth of the plant biomass to only the starch yielding parts of the plant. This results in the faster harvest and minimizes the growth of non-starch materials. For example, in an earlier study (Fujita et al. 1999), optimal turion growth is achieved at low nutrient concentration and low spacing between plants that have been demonstrated in another duckweed strain, *W. arrhiza* (rootless duckweed). The turion is a dormant tissue and can remain dormant for long periods (up to a month) in the absence of nutrients. The turion is rich in starch (65.65%) and relatively low in lignocellulosic (12.82%), and this is desirable for isolation of the starch component and can be grown with less nutrient compared to the whole plant. Duckweed turion can achieve a starch productivity of up to 2.90 g/m^2 daily from a turion biomass accumulation rate of 3.78 g/m^2 daily. The turion of *S. polyrhiza*, also commonly known as rootless duckweed or spotless watermeal, is another highly productive starch-producing aquatic plant. 0.34 g of ethanol is obtainable per gram of its turion. The saccharified starch achieved a sugar to ethanol conversion of 91.67% (Xu et al. 2018).

The water hyacinth is another aquatic plant with considerable starch content. Starch is present within different parts of the plant with the highest starch content present in the rhizomes while the roots have the least. There is also a moderate amount of starch present in the stolons, leaves and peduncles (Penfound and Earle 1948). Starch content in green macroalgae ranges between 1.59 and 21.44%. The starch content is affected by growth conditions and varies from season to season. Up to 3.43 tonnes of starch can be obtained per hectare annually from offshore cultivation of *Ulva ohnoi* green algae (Prabhu et al. 2019).

Several reports have shown that starch content in aquatic plants and algae can be significantly increased through nutrient starvation (Prabhu et al. 2019; Rehman and Anal 2018; Korzen et al. 2016; Andrade et al. 2004). Such potential for yield improvement through biotechnology and agricultural technology makes aquatic-sourced starch very promising for applications such as biofuel production, a process which requires large quantities of starch. Starch content in algae can also be boosted by controlling the wavelength of the light under which they grow. For example, *Ulva pertusa* has been shown to have improved starch content in blue light (Muthuvelan et al. 2002). Starch content in algae could be as much as 32% on a dry weight basis as it is found in *Ulva rigida* (Korzen et al. 2016). Table 13.1 gives the starch content from some aquatic sources, and Table 13.2 lists starch production rate in terrestrial crops.

Table 13.1 Starch content of some aquatic plants and algae

Source	Starch content (% weight)	References
Duckweed turion	65.86	Xu et al. (2018)
Green macroalgae	1.59–21.44	Prabhu et al. (2019)
Filamentous green algae *Oedogonium nodulosum*	46.29	Zhang et al. (2016)
L. aequinoctialis (duckweed)	21.70	Li et al. (2016)
S. polyrhiza (duckweed)	23.23	Li et al. (2016)
Coculture of *L. aequinoctialis* and *S. polyrhiza*	23.78	Li et al. (2016)
Landoltia punctata duckweed	60.03	Liu et al. (2018)

Table 13.2 Starch accumulation rate in some terrestrial plants and green algae

Plant	Starch (tonnes ha^{-1} yr^{-1})	References
Wheat	1.84	Beloshapka et al. (2016)
Rice	1.79	Beloshapka et al. (2016), Reddy and Bhotmange (2014)
Maize	1.56	Beloshapka et al. (2016)
Potato	5.46	Rahman et al. (2016)
Cassava	10.39	Edison and Srinivas (2016)
Ulva ohnoi	2.01–3.38	Prabhu et al. (2019)

13.4 Chemistry of Aquatic Starch

Starch comprises of two polymers, amylose and amylopectin, and the monomeric unit of these polymers is D-glucose. The glucose monomers are either linked together in linear structure forming amylose chains or in branched structure as seen in the amylopectin fraction of the starch. In the amylose chains, the glucose monomers are linked together in a alpha 1–4 linkage; while in the amylopectin, there is about 5% branching with glucose 1–6 linkages forming branches along with the amylopectin chain. Both amylopectin and amylose are combined within a starch granular structure with a varying level of crystallinity. The ratio of amylose to amylopectin and the level of crystallinity of the starch granules vary for different starch sources (Robyt 2008). Figure 13.1 shows the chemical structure of starch represented by (a) 1–6 linked alpha glucose units and (b) a linear alpha 1–4 linked alpha glucose repeating units.

One of the important properties of starch for applications in, for example, food and pharmaceutics is the gelatinization temperature. The gelatinization temperature of starch from green algae *U. ohnoi* is reported as 69 °C. Starch is insoluble in water and will gelatinize when exposed to water at a relatively high temperature. This gelatinization temperature varies for different starch sources. Gelatinization of starch is an important process, whereby starch granules swell in the presence

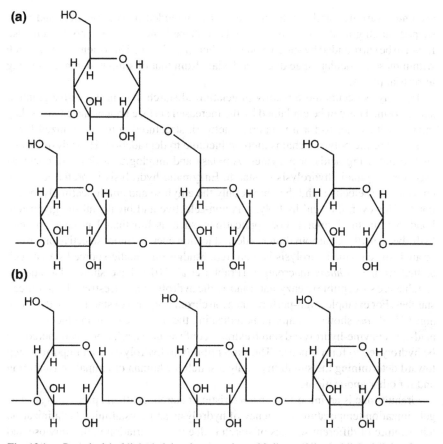

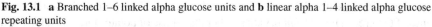

Fig. 13.1 a Branched 1–6 linked alpha glucose units and **b** linear alpha 1–4 linked alpha glucose repeating units

of excess water at the gelatinization temperature. This property is important in the application of starch in, for example, film formation and uses as a thickening agent. Gelatinization of the red algae *Gracilariopsis lemaneiformis* is 55.1 °C, and that of red algae *Gracilaria chilensis* is 52.7 °C, while that of green algae *U. ohnois* is 118.4 °C. Gelatinization temperature of starch from terrestrial plants is generally higher, for example the gelatinization temperature of potato starch is 120.8 °C (Prabhu et al. 2019) lower values for the gelatinization temperature of potato starch of 66.2 °C have been reported in other studies (Yu et al. 2002). This suggests that the gelatinization temperature from algal source is possibly generally lower than that from terrestrial plants.

In one of the rare studies to observe the starch structure in algae cells, TEM and confocal imaging revealed starch granules between 5 and 7 μm present within the cell cytoplasm and thylakoids of the chloroplasts in *U. ohnoi* green algae. TEM imaging presented various starch conformations which included ovoid, spherical and pear-shaped starch granules as well as irregularly shaped ones. The pyrenoids were also found to be surrounded by starch plates (Prabhu et al. 2019). This orientation of starch within these particular algae does not deviate from that of similar starch-containing terrestrial plants.

Hydrolysis occurs more readily in gelatinized starch than in the native granular starch form. This can be explained by the increased surface area of the chain as they interact with water to form temporary network structures in the gelatinized form, thus leaving the polymer chains more vulnerable to degradation. Hydrolysis can be carried out using acids or enzymes. Amylase and amyloglucosidase are enzymes used for enzymatic hydrolysis of starch. Enzymatic hydrolysis is preferred as the enzymes are selective, and they break only the amylose and amylopectin glycosidic bonds. However, the acid hydrolysis is non-selective and breaks all the glycosidic bonds within the biomass. For applications, such as bioethanol production acid, hydrolysis of all the carbohydrates is desirable; however, for more selective processes or analysis, enzyme hydrolysis is preferred. Studies on aquatic-sourced starch such as that from *U. ohnois* macroalgae (Prabhu et al. 2019) have shown that aquatic starches are susceptible to enzymatic and acidic hydrolysis as terrestrial plant-sourced starches. For example, high-performance ion chromatograms of starch from the green algae *U. ohnois* showing major peak indicating the presence of main glucose in the acid- and enzyme-hydrolyzed starch extracts confirm that starch from such source can be hydrolyzed to form glucose. The susceptibility to hydrolysis is an important step toward determining the suitability of algal starch for human or animal consumption and for other applications.

Granule size is an important characteristic of starch. It affects properties such as gelatinization temperature, efficiency of hydrolysis and crystallinity. Gelatinization temperature of different sources of starch is due to the variation in the amylose and amylopectin content and the size of the starch granules. For instance, the starch granules of the green algae *U. ohnois* are smaller than those of potato starch, resulting in a higher gelatinization temperature of potato starch. Starch from microalgae *Chlorella sorokiniana* has granule size of 1 μm. In addition to this, its molecular weight, A type crystallinity pattern (crystallinity 30%) as well as amylose content is similar to starch from cereal plants (Gifuni et al. 2017). When SEM images of starch granules extracted from aquatic plant *U. ohnoi* were compared, the surfaces of the algal starch appeared smoother compared to that of the potato-sourced starch. Typical size of red macroalgae starch granule is ~2 to 5 μm (Yu et al. 2002) that of *U. ohnoi* was observed to range an average of 6.6 μm with a relatively large standard deviation as the granule sizes ranged from 0.7 to 27.4 μm with multiple peaks in the size distribution characterization. Although the size of *U. ohnoi* starch granules is larger than that of typical red macroalgae, this variation in the size distribution of starch granules within species of aquatic plants is not too far from those of other typical terrestrial-sourced starch. For instance, potato starch granules have an average size of 45.5 μm while those of rice starch are between 2 and 7 μm (Le Corre and Brass 2010). This implies that the applicability of starch extracted from aquatic source is not limited by the granule size. Although the cause of this variation in starch granule size is not fully understood, it is hypothesized to be due to the timing of the granule initiation process during starch synthesis within the cell. Other studies have established a relationship between the size of granules, period of formation and shape of the granules and the amylose content of the starch (Ziegler et al. 2015).

The crystallinity of starch also varies for different sources. FTIR spectroscopy showed that starch from green algae has a more amorphous structure, indicated by more intense peaks at the wavelength of $1014.5 \, cm^{-1}$. Starches from green algae *U. ohnoi* and potato starch show the same peaks for starch in the fingerprint region of the FTIR and thus confirming that the starch from this aquatic source has identical chemical composition to that of terrestrial-sourced starch. Decomposition temperature is another important parameter to consider in polymeric materials, and the decomposition temperature informs processing and handling thermal parameters. At $283.7 \, °C$, starch of green algae *U. ohnois* had began to degrade. This decomposition temperature is similar to that of potato starch.

The physicochemical properties of the starch from aquatic source show many similarities to those of common terrestrial sources such as rice, potato and cassava. This similarity in the physical and chemical properties of starch from aquatic source to those of terrestrial source suggests similar industrial applicabilities. Future studies on toxicology and biocompatibility, it is required to confirm the possible range of application to include direct consumption in food, cosmetics and medicine.

13.4.1 Biodegradation of Starch

Starch is a digestible polymer. It is broken down into its glucose units which can then be converted into ATP to meet the energy requirement of the organisms. Looking at the chemical structure of starch as shown in Fig. 13.1, it comprises of carbon, hydrogen and oxygen, all of which can be safely absorbed into the environment. Starch-producing organisms have biomechanisms for degradation of starch which they accumulate and convert to energy when required. Humans also have biomechanics to degrade the starch they consume from plants and algae as food. Edibility depends on the safety of other molecules attached to the starch. Since starches from terrestrial and aquatic sources are chemically identical, no distinction is made between their respective degradation processes.

Starch-based products, such as bioplastics not consumed as food, need to be disposed of following their usage lifespan. One method of disposal is in landfill or buried in soil or compost heaps where they degrade and serve as a carbon source for the soil. Biodegradation of starch buried in soil in the forest occurs within several days. Within 15 days, fungal hyphae colonize starch films and pronounced degradation of starch granules is detectable by 45 days based on the observations using scanning electron microscopy (SEM) (Lopez-Llorca and Valiente 1993).

The chemical properties of starch and the degradation conditions affect the degradation process. Properties such as crystallinity, molecular weight and granule size affect the degradation rate and mechanism. Higher molecular weight starch degraded faster; however, degraded rate does not seem to be affected by granule size. Retrogression decreases the starch degradation rate. This is due to retrograded starch having a more ordered structure which inhibits degradation (Li et al. 2015). These findings suggest that the rate at which starch is broken down varies for different types of starch since as discussed in the earlier section, the molecular weight, granule size and crystallinity vary for starch from different sources. This is important in the design of products such as bioplastic films where the rate of degradation needs to be slow enough to retain product stability during storage and usage life but fast enough to allow safe degradation in the environment without accumulation.

Starch is broken down by amylase enzymes which can cleave alpha 1–4 and alpha 1–6 bonds in amylose and amylopectin. This can occur in nature either by microorganisms or within the digestive systems of animals and humans which metabolize the required enzymes. Organisms which carry out degradation of starch in natural environment include fungi such as penicillium Appressoria and chlamydospores (Lopez-Llorca and Valiente 1993), *Trichoderma viride* I (Schellart et al. 1976) and bacillus bacteria (Wang et al. 2019). These organisms produce amylase and glucoamylase enzymes which catalyze starch degradation and therefore serve as sources of enzymes for commercial processes involving starch degradation such as hydrolysis of starch to fermentable sugars for bioethanol production. Degradation of starch also occurs through acid hydrolysis and heat treatment. This can either be desirable or undesirable. The acid hydrolysis of starch is, for example, desirable in the production of bioethanol where the starch is hydrolyzed into fermentable sugars. In the processing of starch into plastic films, undesirable thermal degradation of starch occurs during the process of extrusion (Liu et al. 2010). This is considered undesirable and is prevented by careful control of the process to ensure proper temperature distribution.

13.5 Floridean Starch

While starch is generally defined as consisting of amylose and amylopectin in different ratios, a form of starch exists where the amylose content is zero. This is known as floridean starch and found in red algae as a highly branched phosphorylated amylopectin chain; floridean starch is a short-chain starch with a degree of polymerization of 18 and a branching of 4.8 (Yu et al. 2002). The highly branched structure and low molecular weight make it less stable than the long-chain starch with linear structure.

Ordinarily starch is found in land plants and green algae within the plastids. A different form of starch, floridean starch is found in the cytosol of red algae and glaucophytes, cryptophytes, dinoflagellates and apicomplexa parasites. These organisms use a different pathway to synthesize and store starch in a form called floridean starch (Dauvillee et al. 2009). The storage of floridean starch within the cytosol in red algae is similar to the manner in which fungi and animals have their storage carbohydrate (glycogen) stored in the cytosol. Accumulation of starch within the plastid in photosynthesizing plants is more conventional since this is the part associated with manufacture and storage. Although it was recently thought that floridean starch also contained amylose (McCracken and Cain 1980), it was later found that floridean starch contains only amylopectin and is limited to a specific group of red algae so named as Florideophyceae (Dauvillee et al. 2009). The other group of red algae, Bangiophyceae, does not contain floridean starch (McCracken and Cain 1980). The composition of floridean starch can be up to 80% in red algae.

Floridean starch extracted from red algae in one study show granules structures of mostly spherical orientation, although some other granule shapes are also observed (Yu et al. 2002). The granule length was measured to be in the range 1.7–3.4 μm and was species dependent. For example, the red algae species *G. chilensis* has an average granule size of 3.4 μm while that from the red algae Gracilariopsis sp. had an average granule size of 1.7 μm.

Although floridean starch is relatively newer concept, the Florideophyceae red algae are widely commercially explored for food and extraction of agar and carrageenan. There is therefore room to integrate the extraction of floridean starch alongside the extraction of these other polymers from the red algae. The highly branched structure of floridean starch means they are less stable and more easily hydrolyzed than linear form. This has significance in the area of biofuel production.

13.6 Extraction Process

Extraction of starch from the natural aquatic source generally involves separation from the other components such as lignin, cellulose, proteins, lipids and other components. In plants and algae with no lignin, the complexity of the extraction process is reduced to a significant extent. Starch can also be obtained as a by-product of other extraction processes. It is more efficient to be able to obtain multiple biopolymers from the same source, particularly one that can be done in a single process concurrently.

In the following subsections, we review the extraction processes for starch from various aquatic sources. Some similarity exists in some of these processes, while the variations should have some significant economic and environmental implications. The variations in the processes are in most cases which are due to the difference in the structural composition of the plant or algae feedstock.

13.6.1 Extraction of Starch from Microalgae

Microalgae are an important source of biopolymers, and one of the reasons for this is the ability to control and influence the metabolism of microalgae compared to macroalgae. Here, the extraction of starch from the microalgae *C. sorokiniana* is used as an example of the extraction of starch from microalgae. The microalgae are cultivated in a photobioreactor. It is also important to consider the process for extraction of starch from microalgae harvested from the open aquatic ecosystem as this is a way to utilize the microalgae generated from green tides. The limitation of harvesting microalgae from the open aquatic ecosystem includes risk of contamination and uncontrolled growth condition therefore the starch content might not be optimized.

In the closed harvest system, the microalgae are cultivated in photobioreactors in Bold's Basal algae growth medium. The growth conditions are maintained at 25 °C, illumination of 300 μE m^{-2} s^{-1}, aerated at 0.02 vvm with air containing 2% CO_2. The system was allowed to run until it reached nitrogen starvation to optimize starch accumulation by the microalgae. The microalgae biomass is then harvested and taken through cell disruption. This process of cell disruption disintegrates the cell wall freeing up the components which include starch. Cell disruption can be achieved by methods such as enzyme or alkali pretreatment of mechanically by bead beating with 0.5 mm glass beads while suspended in ethanol. Further, cell disruption is achieved by heating at a temperature of 75 °C. The starch is then separated from the other component of the microalgae by centrifugation at 10,000 g for 20 min. The starch is recovered as the solid residue and is resuspended in distilled water and centrifugation is repeated and separation carried out through Percoll gradient to achieve purer starch. Figure 13.2 summarizes the extraction process for obtaining starch from microalgae.

Fig. 13.2 Process flowchart for the extraction of starch from microalgae

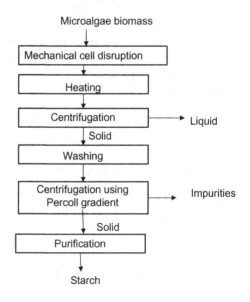

13.6.2 Extraction of Starch from Duckweed

Removal of lignin is carried out by treating the biomass with 25 mM of sodium acetate at a pH of 5.5 and autoclaving at 121 °C for 20 min. The residue from this is the starch and cellulose component of the biomass. The starch can then be separated from the cellulose by methods such as hydrolysis whereby the starch is removed as soluble sugar. This can be enzymatic hydrolysis with amylase or amyloglucosidase. In this process, 500 and 50 μL of alpha-amylase and alpha-amyloglucosidase, respectively, are added per gram of dried biomass and heated under continuous agitation of 250 rpm for a duration of 5 h after in which hydrolysis is completed. The hydrolyzed starch in the form of glucose can then be separated from the cellulosic residue. This method of the extraction of saccharified starch is used, for example, as stage in the conversion of aquatic plant biomass for bioethanol production. This particular method has been adapted for use in the extraction of starch for use in the production of biofuel from duckweed (Miranda et al. 2016). The starch can also be extracted in the whole form.

13.6.3 Extraction of Starch from Macroalgae

Extraction yield from algae varies depending on the type of algae and efficiency of the process. Here, we review a typical method used for extraction of starch from the green macroalgae *U. ohnoi*. Extraction yield of 50.37% has been reported for the green macroalgae *U. ohnoi*. The starch extract had 75.45% starch content per extract mass.

In the method reported for extraction of starch from the green seaweed, *U. ohnoi*, the cultivated seaweed was washed and added to distilled water in the ratio 1:20 dry weight of biomass to water volume. The mixture of water and seaweed was then homogenized until it forms a fine suspension. This was then followed by filtration using nylon filters with pore sizes of 100 μm and then through 50 μm and finally through 10 μm filters. This allowed for more effective filtration. The filtrate obtained was then centrifuged at 4000 rpm for 6 min. The solid fraction from the centrifugation contains mainly starch. The residual pigments and lipids are removed by washing three times with absolute ethanol.

The extraction process can be scaled up for mass production. The process does not deviate much from the typical process used for industrial production of starch from terrestrial sources. Extraction yield of 50.37% has been reported for starch from the green algae *U. ohnoi*. The purity obtained using this particular method was 75.45% (Prabhu et al. 2019). This is similar to the purity of other non-aquatic starches such as cassava, rice and potato which are 80, 75–86.8 and 78.6%, respectively (Bhat and Riar 2016; Sjöö and Nilson 2018).

13.6.4 Extraction of Floridean Starch from Red Algae

To extract floridean starch from red algae, the algal biomass is pulverized to smaller particle sizes. This can be achieved using equipment such as a hammer mill- or motor-driven mortar and pestle. The pulverization is done under liquid nitrogen to prevent oxidation. 50 mM of citrate buffer is then added to the biomass at a pH of 6.5. At this stage, the floridean starch is dispersed as smaller particles and the protein dissolved in the liquid state since acid-soluble proteins will dissolve out of the algae tissue at low pH. These are separated from the larger solid residues of the algae tissue by filtration. The filtrate is then centrifuged at $13,000 \times$ g for 10 min. This separates the dispersed starch granules which settle at the bottom from the dissolved proteins which is decanted off. Further, purification is achieved by subsequent addition of distilled water and centrifugation. Using this method of extraction, a yield of 1% starch has been reported for starch extracted from the Gracilariales red algae species (Yu et al. 2002). The process for extraction of floridean starch from red algae is therefore similar to that used for the extraction of starch.

13.7 Environmental Impact

As a biopolymer, starch is a benign material. Within the organisms from which it is sourced, starch serves as a source of energy and is accumulated in different parts of the organism. The edible plants and algae can be consumed whole as dietary starch sources with less input of energy or materials. In order to make use of starch in its isolated form for more advanced applications such as additives in food and pharmaceuticals, further consumption other than the biomass is involved. In this section, the environmental impact of some processes involved in the extraction of starch from aquatic biomass is discussed. Table 13.3 gives a summary of the consumption in a typical process for extraction of starch from microalgae (Gifuni et al. 2017). Some

Consumption	Amount
Microalgae biomass	2.5 g per gram starch
Water	30 ml per gram[a]
Ethanol	30 ml[a]
CO_2	2% at 0.02 vvm
Percoll reagent	15 ml/g[a]
Energy	Mechanical disruption Heating 75 °C for ~1 h Centrifugation—$10,000 \times$ g for 40 min

Table 13.3 Consumptions for extraction of starch from microalgae

[a]Estimation based on typical volume to mass ratios

of the environmental issues associated with the process of starch production from aquatic resources are then discussed.

13.7.1 Waste Utilization

In the extraction of other biopolymers from algae, the starch is usually seen as a waste. This starch could be further processed and purified for other applications such as bioethanol production, where the starch content is relatively lower, and this is combined with starch from other extraction process to augment the amount of starch in the feedstock. The biochemistry of starch from aquatic plants and algae is similar to that of terrestrial crops used in bioethanol production; therefore, combining multiple sources of starch for bioethanol production should not pose difficulty in processing.

13.7.2 Cultivation of Aquatic Plants and Algae for Starch Production

Cultivation of some aquatic life forms for the production of biopolymers interferes with the balance in the food chain as some of these biomass serve as food for other life within the aquatic system. For example, some fish feed on algae such as *U. ohnoi* (Ingle et al. 2018). It is important to have a consideration for the replacement cycle of these organisms for a truly sustainable marine biorefinery. Furthermore, to optimize the bioaccumulation of starch, it is often required to vary the environmental condition to stimulate optimal starch accumulation by the aquatic plant or algae. This could limit the production of starch from aquatic sources to indoor cultivation in controlled environment which could be at increased capital and running costs. For example a typical indoor photobioreactor for cultivation of microalgae *C. sorokiniana* for production of starch under nitrogen starvation using a photobioreactor requires indoor temperature maintained at 25 °C, light intensity at 300 μE m^{-2} s^{-1} air flow of 0.02 vvm with 2% CO_2 composition in an inorganic Bold basal algae growth medium (Gifuni et al. 2017). If grown in mariculture, altering the natural environmental conditions in the open marine or freshwater ecosystem could result in blooms which have a detrimental effect on the entire aquatic ecosystem, and this should be prohibited.

13.7.3 Water Consumption

Aquatic plants and algae also have an advantage of, unlike terrestrial crop plants, not requiring freshwater for cultivation. This is particularly important as freshwater is increasingly scarce, and where available is required for consumption by humans and farm animals. The water consumption given in Table 13.3 is the water consumed during the extraction process. While algae do not require freshwater to grow in and can fix nitrogen and other elements from the water for their growth, the extraction process requires distilled water to extract purified starch.

13.7.4 CO_2 Emission

One key advantage of starch source from aquatic plants and algae is that the emission during extraction is balanced by the CO_2 that the plant or algae removes from the environment; it can be assumed to be a zero emission process. The process of cultivating microalgae is CO_2 negative since CO_2 is consumed in the process. For example, in the cultivation of *C. sorokiniana*, the system is aerated with air containing 2% carbon dioxide at 0.02 vvm (Gifuni et al. 2017). Starch production from terrestrial plants also involves the plan removing CO_2 from the environment; however, aquatic plants grow at a much faster rate than some of the terrestrial plants which commonly serve as sources of starch (Table 13.2) some also have higher starch content (Table 13.1); therefore, the cultivation of aquatic plants and algae for starch production results in more CO_2 removal per gram of starch produced.

For systems, where the cultivation is carried out in a bioreactor or in aquaculture system which are within the factory where the starch is extracted, CO_2 emission from running vehicles on fossil fuel to transport the biomass to the factory for extraction is minimized. This is particularly for starch which the bioaccumulation of starch is optimized by using carefully controlled growth conditions and nitrogen starvation.

13.7.5 Water Remediation

Aquatic plants and algae are able to extract impurities, mainly compounds of nitrogen and phosphorus from wastewater. Water recycling systems which implement the cultivation of macrophytes for production of biochemicals (Muradov et al. 2014) such as starch for bioethanol production are an efficient and low impact means of providing freshwater for other applications such as terrestrial crop farming which requires clean water. This provides a solution to water pollution and contributes to ensuring food security through providing water for plant cultivation and also a source of third-generation renewable energy.

13.8 Applications of Aquatic-Sourced Starch

Applications of starch include textiles, food, biomedical, pharmaceutical and energy industries. Here, we take a look at some recent developments in the application of starch, where aquatic-sourced starch could play a particular role.

13.8.1 Third-Generation Biofuel Production

In terms of biomass for bioethanol production, duckweed takes the lead of all the aquatic plants studied so far. With a starch composition of up to 64% and the ability to grow to twice its size with 5–6 days, this aquatic mosquito fern is an attractive source of bioethanol production. For this reason, it has generated a lot of research interest, and this includes how to best optimize its growth to selectively boost the rate of production of desired metabolites and making use of its water remediation properties in the treatment of industrial waste. Duckweed is also used in the recovery of nutrients from water.

Algae can be used to produce either bioethanol or biodiesel as they contain lipids, starch and other polysaccharides. Recent studies have experimented with carbon switching in algae to direct the metabolism of the fixed carbon toward either production of lipids or starch (Zhang et al. 2018). This is primarily achieved through controlling the salinity of the water. This carbon switching occurs in both freshwater and marine algae. *C. sorokiniana* is one of the species which demonstrates this property.

Starch from algae has gained increasing attention for its potential in serving as the key to low-cost, low-impact and commercially profitable biofuel that is globally available.

13.8.2 Bioplastic Production

One of the global problems facing the world today is pollution caused by the accumulation of non-biodegradable plastic. Between 2015 and 2017, an estimated of 6.3 billion tonnes of plastics have been accumulated in the ocean globally and that number continues to increase at an annual rate of 10–20 million tonnes (Urbanek et al. 2018). In the great pacific between California and Hawaii, 79 thousand tonnes of plastics are accumulated covering an area of 1.6 million km^2 based on models developed from the data collected from vessels and aircraft surveys (Lebreton et al. 2018). The approaches to addressing the plastic pollution problem include reducing the production and use of non-biodegradable plastics, reusing the plastics and hence keeping them away from being accumulated in the environment as waste and recycling the plastics into other long-lasting products and keeping them in use for longer periods of time. In reducing the production of non-biodegradable plastics, other replacements which serve the same purpose are increasing in demand. Starch is one of the most widely used alternatives to non-biodegradable plastics. Starch can be processed into plastic films which have the same mechanical properties as non-biodegradable plastic films made up of synthetic, fossil-derived plastics such as polyethylene. However, these starch-derived plastics degrade into compounds that are absorbed into the environment and utilized by living organisms without causing harm.

One of the limitations of bioplastic applications is their ability to retain their properties at the service conditions. Properties such as water absorbance and thermomechanical properties in bioplastics vary more with environmental conditions compared to the non-biodegradable plastics. Plastics made from aquatic-derived starch could potentially have higher temperature tolerance.

13.8.3 Waxy Starch

Starches with high amylopectin content and little amylose content are desirable in the formation of waxy starch. Waxy starch is used as a food thickener, bread making, emulsion stabilizers, coating and film forming. It is preferred for its improved digestibility due to its branched structure. Branched polymers tend to be less stable than linear polymers, thus making waxy starch more prone to break down by digestive enzymes. The branched nature of the waxy starches also makes then suitable for nanoparticles formation. These nanoparticles can then be used in a variety of applications such as drug delivery and scaffold formation (Sarka and Dvoracek 2017). Waxy starch has preferred gelatinization properties compared with starch with higher amylose content. It is also more resistant to retrogation and hence acts as a more stable viscosity enhancer (Sarka and Dvoracek 2017).

Floridean starch which is found in the cytoplasm of red algae is similar in structure to these waxy starches (Dauvillee et al. 2009) since they contain only amylopectin with no amylose content (McCracken and Cain 1980). Although waxy starch is most commonly sourced from maize, maize is also largely used in other applications; the unique aquatic floridean starch could potentially replace or augment the use of maize for production of waxy starch.

13.8.4 Food

Starch is the major staple food across the world. About 35% of the daily calorie intake comes from starch (Gifuni et al. 2017). Starch is also widely used as a thickening agent in food products. Isolated starch is also on its own a meal such as cornmeal or tapioca. The gelatinization property in hot water makes it a good viscosity enhancer. Combination of different types of starch results in a hybrid material with new properties which leads to improved quality or a new variety of products. Therefore, aquatic starch combined with terrestrial-sourced starch has potential for novel formulations for food applications. With an ever-increasing global population and decreasing arable land for food cultivation, augmenting the already existing, mostly land-based starch sources with aquatic-derived starch would be a significant contribution to global food security. Some studies have explored the production of edible starch from algae. When not acting as a direct food for human consumption, aquatic-sourced starch could also be used as a source of animal feed (Prabhu et al. 2019; Smith et al. 2010) as texture enhancers or stabilizers for animal food. As discussed in other chapters in this book, some aquatic-sourced biopolymers which do not meet the organoleptic or purity requirements to be used for direct human consumption are approved for the use for use in pet food. The feed for pets which are not consumed as food is different for those of animal feed which will eventually be consumed by humans.

13.8.5   Textile and Paper Industry

In the textile industry, starch is applied as a thickener in aiding the textile printing process, and it is also used in fabric treatment to improve finish (Teli et al. 2009). These applications are attributed to the coating property of starch and its ability to adhere to dyes. Application of starch from aquatic resource is particularly attractive toward developing a textile industry which integrates an aquatic-based water remediation system, whereby the aquatic source of starch-based raw material could also be used in the wastewater treatment. This allows the source of the raw material to also be used to clean the effluent from the industrial process.

Furthermore, the use of non-edible aquatic-sourced starch would make more edible starch from other sources such as corn, cassava and potato more available for food-based applications. This is of particular significance as the world faces the risk of global food shortage, and it becomes more important to limit the use of food crops for non-food applications.

Similarly, the papermaking is a non-food application of starch. Papermaking makes use of starch as a coating binder, to aid ink retention and in finishing (Maurer 2009). In such application, starch is most commonly used in the modified form. Starch for papermaking can be sourced from extraction from high yielding aquatic plants such as duckweed and as a by-product from the extraction of other aquatic biopolymers such as carrageenan.

13.8.6   Pharmaceutics

In the pharmaceutical industry, starch is used as a thickening and bulking agent in syrups, tablets and capsules. It protects the active ingredient from the first-pass metabolism in the digestive system when administered orally such that the active ingredient gets to the walls of the small intestine in its intact form. Usually in combination with other polymers and compounds, for example, floating gel tablets made from starch and cellulose blend in the weight ratio 3:7 used in gastric drug delivery. Starch is also used in wound dressings and tissue engineering (Liu et al. 2017). Although starch from aquatic plants and algae is yet to be explored in pharmaceutical applications, they have the same structure as the starch from terrestrial plants used in these applications, the aquatic-sourced starch can potentially be used for such. This is highly dependent on the aquatic-sourced starch meeting the same safety requirement.

13.9 Commercial Production and Applications

Present rate of biofuel production from algae cannot replace fossil fuels in terms of cost of production. Microalgae and seaweeds are a potential source of commercial-scale production of starch. The most economic way of commercial production of algal starch is the use of wastewater which contains the nutrients required for algae growth and are already the cause of uncontrolled algal pollution anyway. Current challenges in this area are the fluctuation of wastewater which depends on the source. As outlined in Fig. 13.3 starch production from algae can be integrated into biodiesel and bioethanol production by using the lipids and starch from the algae. This could be further expanded to utilize the other biopolymers, obtainable from algae. These biopolymers are discussed in other chapters of this book.

The bioplastic market also forms a great opportunity in the production of starch from aquatic biomass, particularly those that are not primarily consumed as food. The annual production rate of plastics is estimated to be >320 million tonnes (Lebreton et al. 2018). With the increasing concern and evidence of the adverse impact of the use of the non-biodegradable plastics on the environment, this means there is an already existing market for biodegradable plastics which can offer a better alternative to the conventional plastics.

Presently, the commercial exploration of the production of starch from aquatic biomass is largely limited by the fact that there still exist cheaper alternatives in

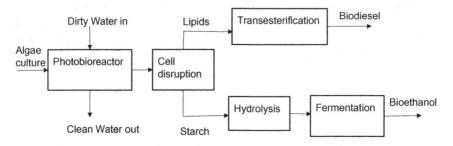

Fig. 13.3 Starch and lipids extraction to obtain biofuels from algae

terms of economic feasibility (Chia et al. 2018). However, as fossil fuels continue to deplete further against unrelenting increasing global consumption closely associated with population growth, alternative sources of versatile resources such as starch become more important; therefore, all alternative sources become important.

Commercial production of starch from algae would benefit much from the technique of nutrient starvation which results in optimal yield. This would mean less expense on nutrients unlike in terrestrial crop farming where fertilizers are constantly required for plant growth. Nutrient is required to reach the desired growth phase, and this is then followed by nutrient starvation which yields an increase in starch production. This is accompanied by reduced biomass production and as much of the nutrient available in this stressful period is then directed toward energy production for basic survival. This increased production of starch by plant during nutrient starvation accompanied by reduced biomass production could further make the starch extraction process more effective as more starch is present per unit mass.

Starch accumulation rate in green algae growing in the wild varies with season. An example of such variation has been detected in green algae *U. ohnoi* (Prabhu et al. 2019) where winter conditions seem to favor optimal starch production in Tel Aviv, Israel. This seasonal variation in starch content implies fluctuation in the quality of raw materials sourced from wild at different periods; therefore, commercial starch production from aquatic source requires cultivation in controlled photobioreactors or aquaculture or storage facilities for those sourced from the wild to ensure constant availability throughout the year.

13.10 Conclusion

A variety of sources of starch exist in the aquatic environment. The growth rate and starch accumulation rate of these organisms can be further optimized by adequate control of their growth system in controlled cultivation. They can also be sourced from natural stocks as part of efforts to utilize natural resources which otherwise cause environmental nuisance. To date, starch remains one of the algae resources which is yet to be industrially explored. Although aquatic plants and algae have been explored for other biopolymers such as alginate and proteins, when considering a zero-waste marine biorefinery, it is important to have a refinery process which integrates the extraction of starch alongside the other extractives from the aquatic biomass since it makes up a significant weight fraction. The low lignin content of aquatic sources of starch makes them a promising source of fermentable starch for bioethanol production. Therefore, despite not having as diverse applications as other polymers reviewed within this book, its potential application in a low-cost, high-volume and highly valued product such as biofuel gives it high commercial significance.

References

Andrade LR, Farina M, Amado GM (2004) Effects of copper on *Enteromorpha flexuosa* (Chlorophyta) in vitro. Ecotoxicol Environ Saf 58:117–125

Appenroth KJ, Borisjuk N, Lam E (2013) Telling duckweed apart: genotyping technologies for the Lemnaceae. Chin J Appl Environ Biol 19:1–10

Beloshapka A, Buff P, Fahey G, Swanson K (2016) Compositional analysis of whole grains, processed grains, grain co-products, and other carbohydrate sources with applicability to pet animal nutrition. Foods 5:23–32

Bhat FM, Riar CS (2016) Effect of amylose, particle size & morphology on the functionality of starches of traditional rice cultivars. Int J Biol Macromol 92:637–664

Chia SR, Ong HC, Chew KW, Show PL, Phang SM, Ling TC, Nagarajan D, Lee DJ, Chang JS (2018) Sustainable approaches for algae utilisation in bioenergy production. Renew Energy 129(Part B):838–852

Dauvillee D, Deschamps P, Ral J, Plancke C, Puteaux J, Devassine J, Durand-Terrasson A, Devin A, Ball SG (2009) Genetic dissection of floridean starch synthesis in the cytosol of the model dinoflagellate *Cryptecodinium cohnii*. PNAS 106(50):21126–21130

Edison S, Srinivas T (2016) Status of cassava in India an overall view. Crops 46:7–172

Eichelmann E, Wagner-Riddle C, Warland J, Deen B, Voroney P (2016) Comparison of carbon budget, evapotranspiration and albedo effect between the biofuel crops switchgrass and corn. Agr Ecosyst Environ 231:271–282

Fournet I, Zinoun M, Deslandes E, Diouris M, Yves Floch J (2000) Floridean starch and carrageenan contents as responses of the red alga *Solieria chordalis* to culture conditions. Eur J Phycol 34:125–130

Fujita M, Mori K, Kodera T (1999) Nutrient removal and starch production through cultivation of *Wolffia arrhiza*. J Biosci Bioeng 87(2):194–198

Gifuni I, Oliveri G, Krauss IR, D'Errico G, Pollio A, Marzocchella A (2017) Microalgae as new sources of starch: isolation and characterization of microalgal starch granules. Chem Eng Trans 57:1423–1428

Himaa HB, Dammak M, Karkouch N, Hentati F, Laroche C, Michaud P, Fendri I, Abdelkafi S (2019) Optimal cultivation towards enhanced biomass and floridean starch production by *Porphyridium marinum*. Int J Biol Macromol. https://doi.org/10.1016/j.ijbiomac.2019.01.207 (in press)

Ingle KNKN, Polikovsky M, Chemodanov A, Goldberg A (2018) Marine integrated pest management (MIPM) approach for sustainable agriculture. Algal Res 29:223–232

Korzen L, Abelson A, Israel A (2016) Growth, protein and carbohydrate contents in *Ulva rigida* and *Gracilaria bursa-pastoris* integrated with an offshore fish farm. J Appl Phycol 28:1835–1845

Lebreton L, Slat B, Ferrari F, Sainte-Rose B, Aitken J, Marthouse R, Hajbane S, Cunsolo S, Schwarz A, Levivier A, Noble K, Debeljak P, Maral H, Schoeneich-Argent R, Brambini R, Reisser J (2018) Evidence that the great pacific garbage patch is rapidly accumulating plastic. Nat Sci Rep 8:4666

Le Corre BAD, Bras J (2010) Starch nanoparticles: a review. Biomacromolecules 11:1139–1153

Li M, Witt T, Xie F, Warren FJ, Halley PJ, Gilbert RG (2015) Biodegradation of starch films: the roles of molecular and crystalline structure. Carbohyd Polym

Li Y, Zhang F, Daroch M, Tang J (2016) Positive effects of duckweed polycultures on starch and protein accumulation. Biosci Rep 36(00380):1–8

Liu W, Halley PJ, Gilbert RG (2010) Mechanism of degradation of starch, a highly branched polymer, during extrusion. Macromolecules 43(6):2855–2864

Liu G, Gu Z, Hong Y, Cheng L, Li C (2017) Electrospun starch nanofibers: recent advances, challenges and strategies for potential pharmaceutical application. J Controlled Release 252:95–107

Liu Y, Wang X, Fang Y, Huang M, Chen X, Zhang Y, Zhao H (2018) The effects of photoperiod and nutrition on duckweed (*Landoltia punctata*) growth and starch accumulation. Ind Crops Prod 115:243–249

Lopez-Llorca LV, Valiente MFC (1993) Study of biodegradation of starch plastic films in soil using scanning electron microscopy. Micron 457–463

Maurer HW (2009) Starch in the paper industry. In: Starch, 3rd edn. Food science and technology, pp 657–713

McCracken DA, Cain JR (1980) Amylose in floridean starch. New Phytol 88:67–71

McWilliams A (2017) Biodegradable polymers. BCC Research, Wellesley, MA, USA

Miranda AF, Biswas B, Ramkumar N, Singh R, Kumar J, James A, Lal B, Subudhi S, Bhaskar T, Mouradov A (2016) Aquatic plant *Azolla* as the universal feedstock for biofuel production. Biotechnol Biofuels 9(221):1–17

Muradov N, Taha M, Miranda AF, Kadali K, Gujar A, Rochfort S, Stevenson T, Ball AS, Mouradov A (2014) Dual application of duckweed and *Azolla* plants for wastewater treatment and renewable fuels and petrochemicals production. Biotechnol Biofuels 7(30):1–17

Muthuvelan B, Noro T, Nakamura K (2002) Effect of light quality on the cell integrity in marine alga *Ulva pertusa* (Chlorophyceae). Indian J Mar Sci 31:21–25

Negm NA, Zahran MK, Elshafy MRA, Aiad AI (2018) Transformation of Jatropha oil to biofuel using transition metal salts as heterogeneous catalysts. J Mol Liq 256:16–21

Penfound WT, Earle TT (1948) The biology of the water hyacinth. Ecol Monogr 18(4):447–472

Prabhu M, Chemodanov A, Gottlieb R, Kazir M, Goldberg A (2019) Starch from the sea: the green macroalga *Ulva ohnoi* as a potential source for sustainable starch production in the marine biorefinery. Algal Res 37:215–227

Rahman M, Roy TS, Chowdhury IF (2016) Biochemical composition of different potato varieties for processing industry in Bangladesh. Agric Sci Pract 2:81–89

Reddy DK, Bhotmange MG (2014) Viscosity of starch: a comparative study of Indian rice (*Oryza sativa* L.) varieties. Int Rev Appl Eng Res 4:397–402

Rehman ZU, Anal AK (2018) Enhanced lipid and starch productivity of microalga (Chlorococcum sp. TISTR 8583) with nitrogen limitation following effective pretreatments for biofuel production. Biotechnol Rep 20. https://doi.org/10.1016/j.btre.2018.e00298

Robyt J (2008) Starch: structure, properties, chemistry and enzymology. In: Fraser-Reid BO, Tatsuta K, Thiem J (eds) Glycoscience. Springer, Berlin, Heidelberg

Sarka E, Dvoracek V (2017) Waxy starch as a perspective raw material (a review). Food Hydrocolloids 69:402–409

Schellart JA, Visser FMW, Zandstra T, Middelhoven WJ (1976) Starch degradation by the mould *Trichoderma viride* I. The mechanism of starch degradation. Antonie van Leeuwenhoek 42:229

Sjöö LM, Nilson L (2018) Starch in food: structure, function and applications, 2nd edn. Elsevier, p 58

Smith JL, Summers G, Wong R (2010) Nutrient and heavy metal content of edible seaweeds in New Zealand. N Z J Crop Hortic Sci 38:19–28

Sullivan-Trainor M (2013) Starches/glucose, global markets. BCC Research, Wellesley, MA, USA

Teli MD, Rohera P, Sheikh J, Singhal R (2009) Application of germinated maize starch in textile printing. Carbohyd Polym 75(4):599–603

Urbanek AK, Rymowicz W, Mironczuk AM (2018) Degradation of plastic-degrading bacteria in cold marine habitats. Appl Microbiol Biotechnol 102:7669–7678

Wang Y, Pan S, Jiang Z, Liu S, Feng Y, Gu Z, Li C, Li Z (2019) A novel maltooligosaccharide forming amylase from *Bacillus stearothermophilus*. Food Biosci 30:100415

Xu J, Zhao H, Stomp AM, Cheng JJ (2012) The production of duckweed as a source of biofuels. Biofuels 3:589–601

Xu J, Stomp A, Cheng J (2014) The production of duckweed as a source of biofuels. Biofuels 3(5):589–601

Xu Y, Fang Y, Li Q, Yang G, Guo L, Chen G, Tan L, He K, Jin Y, Zhao H (2018) Turion, an innovative duckweed-based starch production system for economical biofuel manufacture. Ind Crops Prod 124:108–114

Yao C, Ai J, Cao X, Xue S, Zhang W (2012) Enhancing starch production of a marine green microalga *Tetraselmis subcordiformis* through nutrient limitation. Bioresour Technol 118:438–444

Yew GY, Lee SY, Show PL, Tao Y, Law LC, Nguyen TTC, Chang J (2019) Recent advances in algae biodiesel production: from upstream cultivation to downstream processing. Bioresour Technol Rep 7:10027

Yu S, Blennow A, Bojko M, Madsen F, Olsen CE, Engelsen SB (2002) Physico-chemical characterization of floridean starch of red algae. Starch/Staerke 54:66–74

Zhang W, Zhao Y, Cui B, Wang H, Liu T (2016) Evaluation of filamentous green algae as feedstocks for biofuel production. Bioresour Technol 220:407–413

Zhang L, Pei H, Chen S, Jiang L, Hou Q, Yang Z, Yu Z (2018) Salinity-induced cellular cross-talk in carbon partitioning reveals starch-to-lipid biosynthesis switching in low-starch freshwater algae. Bioresour Technol 250:449–456

Ziegler P, Adelmann K, Zimmer S, Appenroth KJ (2015) Relative in vitro growth rates of duckweeds (Lemnaceae)—the most rapidly growing higher plants. Plant Biol 17:33–41

Chapter 14
Cellulose

Abstract The most abundant polymer on earth is also present in the aquatic environment, produced by aquatic plants, algae and cellulose-producing bacteria. Cellulose can be modified into various forms such as cellulose nanofibers and methyl cellulose, and it finds diverse applications in various industries which include textiles, paper and energy. The feasible commercial production of bioethanol from cellulosic biomass is limited due to the stable structure of cellulose which makes it less susceptible to hydrolysis. However, cellulose from aquatic biomass may serve other industries such as textiles where the unique chemistry of aquatic-derived cellulose has some desirable properties. Extraction of cellulose from biomass reduces the methane released into the environment from the degradation of dead aquatic plant and algae biomass.

Keywords Cellulose · Polysaccharide · Polymers · Hydrolysis · Cellulase

14.1 Introduction

Cellulose occurs in the aquatic ecosystem in aquatic plants, algae and cellulosic bacteria. Aquatic plants which serve as sources of cellulose include water hyacinth, duckweed, *Azolla* and many other aquatic plants whose cell wall comprises of cellulose, a prominent feature in plants. Algae sources include red, brown and green macroalgae, where cellulose also plays a structural role in the cell walls. While bacteria such as acetobacter are also sources of high-quality cellulose, this chapter is focused on cellulose from aquatic sources. Therefore, much of the discussion shall be on plant and algae sourced cellulose.

Cellulose is a biopolymer of huge industrial significance. Cellulose is the most abundant polymer on earth, and this extends to cellulose sourced from the aquatic ecosystem. Exploring the availability of cellulose in the aquatic environment would further expand the ubiquity of cellulose. More importantly, the aquatic environment serving as an additional or alternative source of cellulose for certain industrial application could pose some significant advantage to the environment and commerce by reducing the strain on land-based cellulose sources and potentially offering lower-cost processing. Cellulose in aquatic organism is generally free from lignin and hemicellulose, which are either completely absent or present in much lower amounts

© Springer International Publishing 2020   311

O. Olatunji, *Aquatic Biopolymers*, Springer Series on Polymer and Composite Materials,
https://doi.org/10.1007/978-3-030-34709-3_14

in aquatic organisms compared to terrestrial sources such as wood. This alters the extraction process in ways we shall explore within this chapter.

As the world population continues to increase annually, more land space is required for housing, animal grazing and growing of plants. It becomes important to understand the role the aquatic environment plays in replacing or augmenting the cellulose from terrestrial plants. Cultivation and harvesting of aquatic-based cellulosic biomass take a different process compared to that of terrestrial plants. The fact that the land requirement is omitted means the production of aquatic-based cellulose does not interfere with the occupation of land by humans for habitation, animal grazing and crop production.

This chapter therefore explores the economic impact in terms of availability of feedstock and applications in different industries. The environment impact is also assessed through reviewing the extraction processes and occurrence in nature. Additionally, the chapter also explores the chemistry of cellulose and peculiar characteristics of cellulose from aquatic sources. This is aimed to guide and inform exploration of cellulose production from aquatic sources and encourage the conservation of such resources as well as further research in the area.

14.2 Occurrence in Nature

Cellulose is the most abundant polymer on earth. It is present in all plants and algae as a structural polymer and in bacteria as protection against UV radiation and harsh chemicals. It is indigestible by humans and however can be digested by ruminant animals and some bacteria. Within the aquatic environment, aquatic plants and algae serve as the sources of cellulose. The rate of generation if cellulosic biomass by aquatic plants and algae is much faster than that it terrestrial plants. Non-competition with space used for cultivation of edible crops and for human habitation and the bioremediation properties of some aquatic plants and algae make aquatic plants and algae serve as particularly appealing option for cellulose-based products.

Aquatic plants have the advantage over algae as a source of cellulose being free-floating plants. This makes their harvesting and cultivation much simpler and less expensive. Aquatic plants which have drawn much interest over the years are *Azolla*, duckweed and water hyacinth. Each one is of interest for different reasons, due to either the resource potential or the adverse impact on human activity and the environment. The occurrence of these plants in nature is also affected by human activities such as introducing pollutants into the water. The environmental and economic impact of these plants can therefore be controlled by understanding their resource potential and growth parameters.

The *Azolla* plant makes an interesting source of cellulose because of the plant's remarkable ability to grow in wastewater. *Azolla* metabolizes contents of wastewater such as nitrogen and phosphorus and produces metabolites for its growth and survival. Metabolites include cellulose, hemicellulose, starch and lipids. *Azolla* grows well in rural lagoons, rivers, ponds, irrigation channels, ditches and wetlands under a wide range of temperature conditions. It required little or no maintenance or care, is endemic to most parts of the world and therefore occurs in abundance in nature. *Azolla* can contain up to 35% cellulose by weight (Miranda et al. 2016), and it has a symbiotic relationship with nitrogen-fixing bacteria known as *Anabaena azollae*. This symbiotic relationship with *A. azollae* makes it possible for *Azolla* to grow in water where nitrogen is not present (Kollah et al. 2016). Two most common *Azolla* species are *A. filiculoides* and *A. pinnata*, and 7 species of *Azolla* have been identified. *Azolla* has a rather unique composition of cellulose, hemicellulose starch and lipids which is being explored as a universal biofuel feedstock (Miranda et al. 2016). This implies that pyrolysis and hydrothermal liquefaction of *Azolla* biomass could result in good yield of hydrocarbons, and the transesterification of the lipids would yield biodiesel while the enzymatic hydrolysis of starch and cellulose would yield sugars for bioethanol production (Miranda et al. 2016).

Duckweed is relatively well-explored aquatic plants, and they have been used for over two decades for the remediation of industrial and municipal wastewater in countries such as the USA, Bangladesh and Israel and in biofuel production. They grow in slow flowing or stagnant water as floating plant. They grow best in warm waters where they can grow all year round in some areas. Duckweeds can double in size within 2–7 days and in as little as 20 h at optimal growth conditions. Duckweeds have the unique property of being able to accumulate high levels of microelements and heavy metals from wastewater (Basile et al. 2012). Although starch makes up a higher percentage by mass in duckweed, they do contain a considerable amount of cellulose and are therefore worth considering as a source of cellulose. Total carbohydrate content of duckweed of 51.2% has been reported from composition analysis (Zhao et al. 2014) an estimated 35% of this being cellulose. Table 14.1 lists the chemical composition of duckweed.

Another abundant floating aquatic plant is the water hyacinth (*Eichhornia crassipes*). Water hyacinth contains 25% cellulose, 33% hemicellulose and 10% lignin (Thiripura and Ramesh 2012). It grows rapidly in lakes, basins and rivers in tropical and subtropical regions throughout the world and has been identified as far back as the 1940s (Penfound and Earle 1948). Water hyacinth has been categorized as the worst-growing weed in the world, due to its ability to cover vast areas of water in a relatively short period of time causing damaging effects to aquatic life and water

Table 14.1 Duckweed composition (Zhao et al. 2014)

Duckweed component	% w/w
Pectin	20.3
Starch	19.9
Total carbohydrates	51.2
Hemicellulose	3.5
Phenolics	0.03%
Essential fatty acids	0.6% alpha-linolenic and linoleic/linoelaidic acid 0.015% p-coumaric acid

Table 14.2 Cellulose content is some aquatic plants and algae

Aquatic resource	Cellulose (%)	References
Water hyacinth	25	Thiripura and Ramesh (2012)
Duckweed	55.2	Yadav et al. (2017)
Azolla	21.8–12.8	Miranda et al. (2016)
Green algae	20–30	Mihranyan (2010)
Red algae (*Gelidium elegans*)	17.2	Chen et al. (2016)
Brown algae (*Sargassum tenerrimum*)	11	Siddhanta et al. (2011)

transportation. Water hyacinth has a higher cellulose content than *Azolla*. It therefore is an important aquatic source of cellulose.

Macroalgae cell walls consist of layers of cellulose in relatively large amounts (Munoz et al. 2014). Green algae Cladophora is a seasonal light-sensitive aquatic filamentous macroalgae that are submergent aquatic organisms. They bloom in a variety of temperatures on rough rock surfaces. Brown algae are widely used as a source of alginate, and the residue from the alginate extraction can be further treated to isolate cellulose. Table 14.2 gives yield of cellulose from different aquatic organisms.

14.3   Chemistry of Aquatic Cellulose

The chemistry of cellulose is well established. It is a linear polymer made up of glucose monomers in 1–4 glycosidic bonds. Cellulose is identified by a color change when in contact with iodine and sulfuric acid. This is characteristic of cellulose as other polymers will not give the same response, for example, alginate from marine algae tends to be more crystalline, forming thick microfibrils when in contact with the same (Koyama et al. 1997).

Although having the same unit structure, glucose ring, the secondary structures of the cellulose from algae are different from that of higher plants. The cellulose in algae is more tightly packed than in terrestrial plants. A density of 1.64 g/cm^3 for cellulose is obtained from green algae Cladophora while that of terrestrial plant is around 1.56 g/cm^3. Furthermore, cellulose from algae distinguishes itself from terrestrial and aquatic plants by its lower level of hornification. This is thought to be associated with its well-ordered structure. This property allows formation of well-dispersed microfibrillated cellulose without agglomeration. Cellulose from green algae can go through repeated cycles of hydration and dehydration with its structure intact.

Crystallinity of cellulose varies in different algae depending on what group the algae fall into. This grouping is based on the nature of cellulose that makes up the algae cell wall. Group 1 comprises of algae which have mainly native cellulose within their cell walls. Algae belonging to this group are the Cladophorales and some Siphonocladales. Most algae fall into Group 2 in terms of their cell wall structure. Group 2 comprises of algae whose cell walls are made up of mercerized cellulose derived from native cellulose. This form of cellulose has low crystallinity. Example of such is the Spongomorpha. The third group in this classification is Group 3 algae. These are characterized by heterogeneous cell wall structure in which cellulose is not the major component of the cell walls. Spirogyra and Vaucheria fall into this group. The variation in crystallinity in cell walls is attributed to the cellulose synthase complex responsible for determining the structure of the cellulose microfibrils (Brown and Saxena 2000). Selectivity of species for cellulose extraction is therefore necessary in order to obtain cellulose of desired crystallinity.

Hornification occurs to a much lower degree in crystalline cellulose (Fernandez Diniz et al. 2004) due to the high level of orderliness in the microfibrils which hinder water absorption. Such property aids formation of uniform dispersion in micro- and nanoformulations of cellulose. This further adds to the appeal of cellulose from green algae. Cellulose from Cladophora has a surface area of 95 m^2/g, a relatively high surface area compared to that of industrial adsorbents which could be around 100 m^2/g. This surface area is about 100 times higher than that of pharmaceutical grade microcrystalline cellulose which is around 1 m^2/g. This high surface area coupled with their very low level of hornification makes green algae sourced cellulose a potentially better candidate for production of cellulose-based aerogels that are more resistant to humidity than those produced from native cellulose.

Cellulose has particularly excellent mechanical properties due to its linear orientation, microfibrillar structure and high degree of polymerizations. Linear polymers are able to form well-ordered structures resulting in densely packed secondary structure which prevents penetration of heat, moisture or other molecules. The superior mechanical properties of starch fit well with the structural role it plays in the plant cell wall. Cellulose provides rigidity and strength to the plant. The strong hydrogen bonds present within the cellulose structure also contribute to the strong mechanical structure of cellulose. This property also makes the hydrolysis of cellulose much more difficult than that of starch.

Such high mechanical strength of cellulose does not seem so pertinent in algae compared to land plants which need to grow higher above the ground and require strong cell walls for strength. Hypothesis developed to explain the need for such crystalline cellulose in the Cladophora algae is to retain turgor pressure where salinity fluctuates, as is common in aquatic environment and/or withstand the drag flow of water (Johnson and Shivkumar 2004). The cellulose from aquatic source can therefore be said to have similar chemical structure as terrestrial cellulose but for differences in secondary structure in some cases such as in Cladophora.

14.3.1 Decomposition of Cellulose

Cellulose has a very simple linear polymer structure made up of glucose repeating units. The absence of branching gives cellulose a very stable crystalline structure making it resistant to degradation. This makes it an excellent structural component of the cell wall. While in plants cellulose plays mainly structural role and not utilized for energy or cell metabolism, bacteria are able to decompose cellulose and use as a carbon source. Humans do not produce any known enzyme for digestion of cellulose. However, cellulose is important as a fiber in the human diet. Degradation of cellulose by cellulose-degrading organisms involves a cocktail of enzymes working in synergy. These enzymes are referred to as cellulases (Beguin and Aubert 1994). Industrial production of enzymes for degradation of cellulose involves isolation of these enzymes from cellulose degrading organisms. Ruminant animals are able to digest cellulose at relatively faster rate due to symbiotic relationship with bacteria present in the rumen of these animals. The rate is further assisted by the process of regurgitation, a process whereby the food is returned to the mouth and further particle size reduction is achieved by chewing before being returned to the rumen for digestion (Russel et al. 2009). Cellulose can also be degraded by thermochemical process through the use of high temperature and pressure in the absence of oxygen. This is used in the production of biofuel from cellulose in the process of pyrolysis.

The chemical structure of cellulose from both aquatic and terrestrial sources is identical. Therefore, the degradation process is expected to follow the same mechanism. The difference in the degradation of cellulose within aquatic biomass lies in the difference in the other components of the cell walls. In the isolated form, degradation depends on the level of modification and the form in which it exists in the product.

14.4 Availability of Raw Materials

The growth rate of most aquatic plants is much faster than most non-edible and edible crops being used or considered for biofuel production. They can therefore generate biomass at a faster rate than the terrestrial crops. One of the limitations facing commercial utilization of cellulosic biomass for biofuel production is the availability of raw materials. With the faster biomass accumulation rate of aquatic plants and algae, availability of cellulosic biomass can be significantly increased. Cultivated *Azolla* plant in aquaculture can grow at a rate of 2.9–5.8 g/m^2 in dry weight per day, and in the wild, it can grow at a rate of 25.6–27.4 g dry weight per m^2 per day (93.4–100 t/ha-year dry weight). *Azolla* is one of the fastest-growing aquatic plants with the ability to double its weight within 5–6 days. On average, cellulose content in algae is between 20 and 30% dry weight (Mihranyan 2010), and however, there have been reports of cellulose content in filamentous algae as high as 45%.

Duckweeds can grow at a rate of between 39.2 and 44 tonnes per hectare annually. Duckweed contains up to 45.7% starch and for over two decades has been explored for application in nutrient recovery from wastewater which can then be used as animal feed. It has also been largely explored for production of biofuel through fermentation of its non-edible starch (Miranda et al. 2016). Another cellulose contains aquatic plants. Water hyacinth can cover a whole hectare of water space within 6 months growing to an estimated 125 tonnes wet weight (Istirokhatun et al. 2015).

Algae blooms have been reported in different parts of the world such as in the southern coast of California (Smith et al. 2018). This has been accompanied by a loss of devastating amount of fish and negative impact of the aesthetics of the location. Algae biomass for cellulose extraction is therefore beyond abundant. It is accumulated to devastating degree. Conversion of such waste into useful polymer products such as cellulose is beyond its economic returns but also the gains to the environment and all life concerned.

An estimated 100–150 billion tonnes of cellulose is produced annually by cellulose synthesizing plants, bacteria and algae (Hon 1994). Most plants contain cellulose, lignin and starch, and however, they contain these in varying compositions. The composition of each plant will determine if the extraction of cellulose is economical, or it might be more economical to extract other carbohydrates from them. *Azolla*, for example, can accumulate 34 tonnes of cellulose in dry weight of biomass per hectare of land annually. Although not polymers, lipids are also potential by-products from cellulose extraction from aquatic plants. *Azolla*, for example, can be a source of up to 8 tonnes of lipids per hectares annually. This is higher than other sources of lipids of terrestrial plants such as oil palm, soy and rapeseed. *Azolla* has been described as a universal biofuel crop as its chemical composition mimics that of a combination of terrestrial and macroalgae. However, as a source of cellulose, it has moderate cellulose composition.

Cellulose can therefore be said to be relatively abundant in the aquatic environment since it is present in fast-growing plants and aquatic algae whose biomass accumulation rate is faster than those of terrestrial plants. Actual availability of aquatic biomass for cellulose production depends on several other factors such as demand-driven expansion of cultivation of these sources and the cost of processing compared to the typical sources of cellulose. Future availability of abundant aquatic plants and algae could reduce as other applications are being discovered. For example, farmers in tropical regions had adopted water hyacinth for use as compost fertilizer (Polprasert et al. 1994). This value addition to such aquatic plant which was previously seen as a nuisance could therefore lead to an increase in the price of the aquatic plant as a feedstock for cellulose production.

14.5 Extraction of Aquatic Cellulose

The chemical resilience of cellulose compared to other components of the cell wall of plants and algae serves as a basic for its extraction. The chemical-based process for extraction involves dissolving off the other components of the cell wall in strong acids and bases under high temperature followed by the recovery of cellulose from the residual material. It is important to maintain the structural integrity of the cellulose during the process of extraction as this eventually affects the quality and applicability of the cellulose obtained. Properties such as crystallinity, degree of polymerization, microfibril structure, wettability and mechanical properties are used to measure the quality of cellulose extracted. Here, we consider extraction methods that have been reported for the main cellulose sources in the aquatic ecosystem.

14.5.1 Extraction of Cellulose from Aquatic Plants

High moisture content of water hyacinth means the process of drying may require more time and energy input compared to other plants. Water hyacinth has a moisture content of around 93–96%. Following the drying, the water hyacinth fibers are heated at 115 °C in toluene solvent for a duration of 3 h. The next step is the removal of color by bleaching in sodium hypochlorite NaClO (3%) under heating for 2 h at 80 °C. The hemicellulose within the cell wall is removed using the process of hydrolysis with sodium hydroxide (1%) at a temperature of 60 °C for 2 h. Further treatment is carried out with sodium hypochlorite at 1% at 75 °C while stirring for a duration of 3 h, to remove the residual lignin (Istirokhatun et al. 2015).

To extract cellulose from *Azolla* plant, in the method by Miranda et al. (2016), the *Azolla* plants harvested were washed and dried in hot air oven at 70 °C overnight. This was followed by grinding to reduce particle size. Lignin was removed by autoclaving in sodium acetate at 121 °C for 20 min. The remaining mass contains a mixture of polysaccharide starch and cellulose. The starch can be removed by chemical treatment with sodium hydroxide and heat or enzymatic treatment with amylase and amyloglucosidase which reduces the starch to glucose. Where the goal is to obtain cellulose, the glucose can then be washed off as it is water soluble leaving behind cellulose fibers. This process is summarized in Fig. 14.1.

14.5.2 Extraction of Cellulose from Algae

One key advantage of cellulose extraction from algae over aquatic plants such as *Azolla* is the very little to no lignin content in algae. This eliminates the high energy and chemically demanding delignification process. Although cellulose is not a main structural polymer in brown algae, some have been reported to have no cellulose content, for example, brown algae is made up of 32% alginate, 15% laminarin which is the glucose polymer in brown algae, 18% mannitol and the rest ash or salts This composition was based on average values from three harvests of *S. latissima*, a strain of brown algae commonly used in the industry (Konda et al. 2015). Red algae on the other hand contain cellulose, glucans and galactan (Wi et al. 2009; Yoon et al. 2010). Nanocellulose extraction from red algae (*Gelidium elegans*) involved alkaline treatment, bleaching and acid hydrolysis. The nanocellulose extracted had diameter

Fig. 14.1 Extraction of
cellulose from aquatic plant

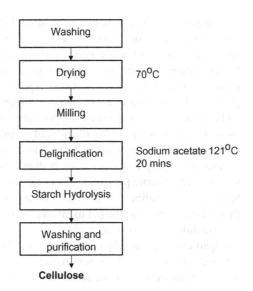

Cellulose

of around 21.8 nm and a length of around 547 nm (Chen et al. 2016). Marine green algae contain 9% cellulose, 21% hemicellulose and 1.9% lignin (Yaich et al. 2011).

Microalgae also contain cellulose (Chen et al. 2009). Extraction of cellulose from the microalgae *N. oceanica* yielded cellulose nanofibrils with mechanical properties superior to nanofibril cellulose from wood. The nanofibrils obtained had a diameter of 9 nm and tensile strength between 3 and 4 GPa. This process of extraction from microalgae required no delignification process as microalgae do not contain lignin. The process of extraction is involved in deproteinization and removal of lipids followed by purification and TEMPO-mediated oxidation under gentle mixing (Lee et al. 2018).

Most commercially used biopolymers in algae are the phycocolloids alginate, agar and carrageenan. The residue from the extraction of these phycocolloids contains significant amount of cellulose. For example, the cellulose obtained from the residue of red algae from which agar had been extracted yielded high-quality cellulose nanocrystals (Achaby et al. 2018). This use of the cellulosic residue reduces the waste generated from the processing of red algae for production of agar and hence optimizes the utilization of the aquatic resource. Although discarding might seem the least energy-intensive option following agar extraction, the conversion of the cellulose from the waste residue into high-value products used in, for example, biomedical industry could compensate for the additional processing cost.

Likewise, cellulose can be extracted from the residue from alginate extraction of brown algae. Cellulose with molecular weight of 2.69×10^5 was obtained with an average fiber length of 1.1 μm and fiber width of 4 nm. The cellulose obtained using this method showed good biocompatibility and potential for application in food industry demonstrated by its superior thickening property due to the ability to bind to milk (Gao et al. 2018).

Compared to the extraction from aquatic plants, the extraction of nanocellulose from terrestrial wood is more intensive. The pulp used for this process contained lignin of 0.7% and hemicellulose 13.8%. The pulp had been pretreated to reduce lignin content. The pulping process requires the use of either the kraft process (Lahnalammi et al. 2018) or the acid sulfite process (Hanhikoski et al. 2019). The kraft process involves treatment of wood with high concentrations of sodium hydroxide and sodium sulfide at elevated temperatures while the acid sulphite process makes use of high concentration of sulfuric acid at high temperatures. The process to convert pulp to nanocellulose involved suspension of wood pulp in water at a ratio of 1:100 (grams wood pulp to water). This is then followed by oxidation with 10 mmol of sodium hypochlorite per every gram of cellulose mediated by a mixture of 2,2,6,6-tetramethylpiperidine-1-oxyl (TEMPO) and sodium bromide. Sodium hydroxide is then added to increase the pH to 10 and the reaction allowed to take place for 5 h to form cellulose nanoparticles, which was then washed with water and the cellulose nanoparticles separated by centrifugation at 14,000 rpm (Olatunji and Olsson 2015). The extraction of cellulose from wood requires a more rigorous process to remove the lignin and hemicellulose which are present at a higher content.

14.6 Environmental Implications

Here, some environmental issues arising from the extraction of cellulose from aquatic plants and algae are discussed. In this section, more focus is directed at the process of extraction as discussed in the previous section. Further environmental issues arise in additional processes such as transportation of the raw materials to the factory and the packaging process. These issues have been discussed in other sections of the book and will not be discussed here to avoid repetition.

14.6.1 Energy for Drying

Most of the aquatic plants contain a considerable amount of water. Water hyacinth contains the most amount of water ranging between 93 and 96% (Penfound and Earle 1948). Drying could be done prior to transportation to factories where the biopolymers are extracted or the biomass could be transported before drying. Spoilage could be prevented by initial drying, and the reduced moisture content cold also reduces transportation cost. To dry a sample of duckweed, for example, required drying at 60 °C for 2 days (Chen et al. 2012) while *Azolla* required drying at 70 °C for several hours overnight (Miranda et al. 2016). Drying of biomass is usually done using electric-powered hot air ovens. This could be from hydroelectricity, nuclear or renewable energy-powered source. Drying could also be achieved using open air drying in hotter climates, where high level of purity is not required at this stage.

14.6.2 Chemical Consumption

Depending on the source of cellulose, the types and quantity of chemicals consumed could vary. From plants with lignified cell walls, mineral acids and alkali are often required for delignification. The amount of chemical used also depends on the grade of cellulose required. Cellulose with all lignin, lipids and extractives such as proteins removed and cellulose with all the hemicellulose removed have different requirements.

Sodium hydroxide is commonly used for removal of the non-cellulosic components, and toluene has also been used for removal of lipids while depigmentation and further delignification are achieved using sodium hypochlorite. Sodium acetate at high temperature can also be used for delignification. Sodium hydroxide and sodium acetate can be neutralized during wastewater treatment, thereby limiting the impact on environment pH. Toluene however is an organic solvent derived from fossil fuel during the production of gasoline and other products. Exposure to toluene either through the respiratory tract, gastrointestinal tract through contamination of soil, air or drinking water as a result of various forms of release into the environment. Toluene has been associated with neurological, reproductive and developmental adverse effects (Vulimiri et al. 2017). Being a volatile organic compound, use of toluene in the production of cellulose from aquatic biomass will require additional processes for treatment such as adsorption, condensation and catalytic oxidation (Martinez de Yuso et al. 2013). Exposure limits and safety of the workers coming in contact with the harmful chemical will also require additional precautions hence additional cost.

Cellulose production in general requires use of a wide variety of chemicals in large amounts for delignification. The process of extraction results in generation of solids and some other chemicals as effluent, and some of these chemicals are unknown due to the complex structure of lignocellulosic biomass. One advantage of cellulose from algae and aquatic plant is the lower lignin content compared to, for example, wood. For example, duckweed has a lignin content of 12.2% (Yadav et al. 2017). This results in reduced requirements for the heavy chemicals and energy used in delignification.

14.6.3 Methane Emission

While the process of cultivation of aquatic plants and algae can be considered as CO_2 emission negative, the process of decomposition of the biomass in nature leads to the emission of methane, another greenhouse gas of concern (Heilig 1994). Controlled degradation of isolated cellulose and utilization of the decomposition products as fuel are the ways to minimize the release of methane to the environment from the decomposition of aquatic plants and animals. Cellulose has been shown to have a higher biomethane potential of the three lignocellulosic compounds (Li et al. 2018).

More methane is produced in combined biodigestion than when isolated anaerobic digestion of cellulose, hemicellulose and lignin.

The environmental significance of this is that allowing aquatic plants and algae to degrade in the environment results in more methane emission into the environment. One large-scale example of this is the production in paddy fields where the decomposition of the straws in the paddy fields results in methane release. A proposed solution to this is the collection of these straws and utilizing them (Fusi et al. 2014). One way of utilizing this is extraction of the cellulose content for applications such as those discussed in the following section. Therefore, the controlled extraction and utilization of cellulose from these sources could contribute to reducing the release of methane into the environment.

Furthermore, one of the solutions proposed and being adopted for addressing the water hyacinth bloom in tropical regions is the harvesting of the water hyacinth from the water and use as compost in farming (Polprasert et al. 1994). This is a solution that makes use of the aquatic plant biomass in large quantities and significantly contributes to reducing the environmental nuisance. If the composting is done properly with adequate aeration to encourage aerobic decomposition, the process does not result in methane emission. On the other hand, if anaerobic decomposition occurs, this results in further release of methane. New methods have therefore been proposed to minimize anaerobic decomposition in compost pile using semipermeable membrane that aid aeration of the pile (Ma et al. 2018). This can be applied to the composting of water hyacinth in order to ensure that the positive benefits of this approach to the management of the water hyacinth is not overshadowed by the adverse impact of methane emission.

14.6.4 Land Space Occupied

The main advantage of aquatic biomass-sourced cellulose is that there is no competition with land occupied by humans and that used for cultivation of food crops. It also does not lead to depletion of nutrients from the soil. However, the process of extraction of cellulose requires land space. The use of aquatic environment for cultivation of cellulosic biomass for commercial applications such as biofuel production could save thousands of acres of land space which would be required to cultivate the required amount of terrestrial-based cellulose crops.

14.6.5 Aquatic Plants and Algae Bloom

Considering the devastating effect uncontrolled growth that some aquatic plants such as water hyacinth and algae have caused in recent times, harvesting of these aquatic organisms goes beyond the financial gains or economic boost. Effective system to convert these presently unmanaged resources would be of immense benefit to the environment and recovery from the financial loss to the fisheries and associated businesses. Algae blooms have been discussed in the previous chapters in this book. One of the solutions is to harvest the abundant algae and aquatic plant and extract biopolymers such as cellulose from them.

Industries which are affected by uncontrolled growth of aquatic plants and algae include tourism, seafood, water transportation, aquatic farming, fishing and water sports. Water hyacinth also affect activities on land by preventing access to water as a result of blocking if irrigation channels and rivers. This blockage or reduced access to water affects biodiversity as many animals on dry and wetlands time their migrations, reproduction and other activities crucial to their survival in tune with the aquatic cycle. For example, water turtles lay eggs on land and the timing is such that the egg hatches just in time as the water level rises to carry the hatched turtles into the water (Steen et al. 2012). If the water flow is interrupted and the turtles hatch before the water level rises, the turtles have longer distances to reach water and may die from dehydration. This results in a drastic drop in population as the older turtles die out without offspring to maintain population balance.

Therefore industries gaining profits from extraction of biopolymers such as cellulose from these aquatic plants and algae is a bonus, since solving the problem they create otherwise is a global priority.

14.6.6 Wastewater Treatment

Aquatic plants and algae have water remediating effect. Duckweeds in particular have exceptional water remediating effect. They can remove heavy metals and microelements from water and retain them at high concentration within the plant up 100,000 times the concentration in the surrounding water. Water hyacinth can also remove heavy metals such as zinc, cadmium, chromium and copper from wastewater (Sarkar et al. 2017; Hasan et al. 2007). In the studies reported mainly, the shoot of the water hyacinth was utilized for wastewater treatment. This leaves the rest of the plant for extraction of biopolymers. The water hyacinth biomass used for water treatment can be recovered, and the biopolymer also extracted from them. The heavy metals adsorbed unto the surfaces can be removed by dissolving in acid during extraction while the cellulose is recovered.

14.7 Applications

In this section, we take a look at some significant application of cellulose, particularly cellulose from aquatic plant. The applications reviewed here are primarily based on those of current non-aquatic plant-based sources of cellulose. This provides some perspectives on the potential market for aquatic sourced cellulose in such applications. The goal is therefore to explore the potential application in the industries (where none is currently existing) and the demand for such production in the global market.

14.7.1 Textile

The textile industry is one of the earliest and one of the largest in the world. The textile industry formed the basis for the first industrial revolution. As of 2008, 24,025,100 tonnes of cotton, 3,097,000 tonnes of cellulosic fibers, 361,590,800 tonnes of wool, 790,100 tonnes of flax and 558,400 tonnes of synthetic fibers were consumed globally (FAO 2011). There is an increasing demand for more sustainable methods of fabric production. Aquatic plants are yet to be commercially explored for production of alternative cellulose-based fabrics on commercial scale.

Cotton has been the main cellulose-based fiber in the textiles and garment industry. In the past seven decades, for reasons such as yield, pest on cotton plantations and the vast land occupied for cotton plant cultivation, cheaper alternative fabrics from cellulose such as viscose, Tencel and Modal (Shen et al. 2010) have been introduced to the global market. Unlike cotton where the cellulose naturally grows into cellulose fibers which are then carded and spun into long yarns that are then woven into fabrics, man-made or semi-synthetic cellulosic fabrics are made from dissolved cellulose extract from lignocellulosic plants, mainly wood which are then taking through the wet fiber spinning process (Olatunji and Olsson 2016) to form cellulose fibers which are then spun and woven into yarns and fabrics, respectively. Although having other environmental concerns such as chemicals used in dissolving the fibers, production of these man-made cellulose fibers requires less water in production and requires less land space compared to cotton fibers. This could further be made more sustainable by using cellulose from aquatic plants rather than land plant.

The process of producing cellulose-based fabrics from lignocellulosic plants requires a variety of chemicals, and one of the most chemically and energy-demanding stages is the lignin removal. Aquatic plants such as duckweed and water hyacinth contain a considerable amount of cellulose, but little or no lignin duckweed, for example, contains 12.2% (Yadav et al. 2017) while water hyacinth contains around 10% (Thiripura and Ramesh 2012).

Cladophora, a filamentous green algae, has mercerized cellulose structure (Mihranyan 2010). This is a more amorphous, swollen form of cellulose. This makes the cellulose adhere better to dye and have a glossier appearance. This property of cellulose in algae presents some potential application in advanced textile technology development.

Production of viscose fiber could consume as much as 110,000 kJ per tonne of fiber produced from wood (Shen et al. 2010). An advancement in the man-made fiber in the textile industry should aim at reducing the energy requirement for production of such using aquatic sourced cellulosic fiber.

14.7.2 Bioethanol Production

Aquatic plants fall into the category of third generation energy crops. Not only are they non-fossil, but also they are sourced from biomass and even better non-edible biomass that is grown without competing with food or space for human and livestock habitation or food cultivation. In addition to this, their cultivation is beneficial to the environment as they have the capacity to remove nitrogen, phosphorous, carbon and microelements and heavy metal from wastewater. Third generation biofuels are characterized by their ability to yield high biomass with lower resource requirements than conventional feedstocks. Another advantage of aquatic sources such as algae is the lower lignin content which reduces the complexity of separating lignin from cellulose.

Different methods have been explored for the production of ethanol which takes advantage of the cellulose content of aquatic plants, duckweed and water hyacinth (Bayrakci and Kocar 2014). The key attraction here is the use of plants with bioremediating properties which contain polysaccharides which can be converted to glucose for fermentation. The challenge of this is the right cocktail of enzymes and conditions which can most effectively convert both the cellulose alongside other polysaccharides of the cell wall into glucose without costly separation processes. Cellulose from aquatic source can be saccharified into glucose using the enzyme cellulase (Shen and Xia 2003), and however, the cost of enzyme limits the commercial feasibility of such process. Thermochemical process such as pyrolysis is an alternative to enzyme-based process for bioethanol production from cellulosic biomass. However, this process requires high energy input which makes it commercially infeasible.

The high yield and relatively fast growth rate of algae compared to terrestrial plants sourced as biomass made it a desirable third generation biofuel feedstock option. However, a major drawback for biofuel production by algae is the high cost of cultivation and harvest relative to the present price of fuel. For biomass to make the cut as an ideal source of biofuel, it needs to be produced cheaply and contains sufficient amount of organic compounds which can be converted to biofuel such as ethanol or diesel.

As the world's fossil resources begin to dwindle and the impact of using vast land for cultivation of sources of second generation biofuels begin to manifest, third generation biofuels are fast emerging. Aquatic-sourced cellulose is one of such.

Cellulose can be hydrolyzed to cellulase of cellobiose using enzymes such as cellulase. In an example study, up to 65.9 g/L of glucose has been achieved after enzymatic hydrolysis of cellulose from *Azolla*. This required use of a combination of four different enzymes: cellulase and cellobiase for the hydrolysis of the cellulose and amylase and amyloglucosidase for the hydrolysis of the starch. Hydrolysis of cellulose is more difficult than that of starch as cellulose has a more resilient unbranched polymer structure.

Fermentation of glucose from *Azolla*-derived cellulose using *Saccharomyces cerevisiae* achieved an ethanol yield of 0.56 g/g. However, these yields vary for different species of the plant. The ethanol yield from aquatic plants is comparable to those from crops presently used for ethanol production.

For the production of bioethanol from biomass containing a mixture of cellulose and starch, the hydrolysis can be achieved simultaneously using a cocktail of different enzymes since the enzymes do not interfere with each other.

The hydrolysis and fermentation of cellulose biomass from *Azolla* plants can be used to achieve up to 11,700 L of ethanol per hectare of aquatic space annually (Miranda et al. 2016). Duckweed has also been successfully used at experimental stage as feedstock for biogas production via anaerobic digestion (Yadav et al. 2017). Although this is yet to be industrially adopted, a 1:1 mix of cow dung and duckweed biomass yields 12,070 mL of biogas when anaerobically digested at 37 °C over 55 days duration. The yield from combination of cow dung and duckweed was higher than when cow dung was digested alone (11,620 mL) with comparable methane contents.

14.7.3 Cellulose Filler in Composites

Cellulose extracted from green algae, Cladophora has been used as fillers in polyurethane foams. At 5–10% content by weight, cellulose improves the thermal properties, elastic modulus, color retention and biodegradability and reduces the polyurethane content by part substituting with cellulose. The functional groups of cellulose show good affinity to that of polyurethane resulting in good compatibility between the cellulose fiber and polyurethane matrix.

Cellulose has shown good compatibility with other hydrophobic fossil-based polymers and has been used as a filler for different purposes. This has significant environmental impact as it improves the biodegradability of the materials.

14.7.4 Cellulose Nanofilters

One of the challenges in developing cellulose-based nanofilters is the limitation if native cellulose in attaining stable microfibrils in the nanometer range. Membrane filters developed using green algae-derived cellulose have addressed this challenge (Mitsuo and Eisuke 1996). This is due to their highly ordered structures which make achieving well-dispersed nanoscale fibers more possible. Therefore, aquatic-sourced cellulose has some structural advantages which make them superior option in some applications. Nanocellulose has broad range of applications in various industries. These include drug delivery applications where they have been used to produce polymeric drug delivery devices like microneedles (Olatunji and Olsson 2015).

14.7.5 Drug Carrier

The high surface area and inert structure of cellulose from green algae make it a suitable drug carrier. A good drug carrier should have sufficient surface to embed the drug compound but without altering the integrity and activity of the drug. In addition to the crystalline structure, the low degree of hornification makes it a good material for producing stronger tablets than other plant-based cellulose (Stromme and Mihranyan 2002; Mihranyan et al. 2004). Nicotine loaded in liquid carrier formulation using cellulose from green algae showed higher loading capacity and stability compared to microcrystalline cellulose grade.

14.7.6 Papermaking

One of the oldest applications of cellulose is in papermaking. Paper is said to have been invented by Ts'ai Lun in 105 AD china from mixture of plant and textile fibers (Hunter 1974). Today, the papermaking process still follows the same fundamental principle of using the fibers from plants to create thin sheets for various purposes. Over the years, however, these clear sheets are known to be made mostly of cellulose and the techniques have been refined to obtain more pure forms of paper. As of 2016, FAO reports a global pulp for paper production of 107,042,000 metric tonnes which is 92% of a production capacity of 150,273,000 metric tonnes, projected to rise to 156,850,000 metric tonnes production capacity by 2021 (FAO 2017).

The papermaking industry is also a water-intensive industry (Shen and Qian 2012), and sourcing the cellulose from aquatic plants which have water remediating properties offers possibilities of a paper production system, whereby the aquatic plants from with the cellulose produced are used in the cleaning of the gray water generated from the paper production process. This is a mare concept, and however, as the paper industry develops toward more sustainable production processes, such should be explored.

14.7.7 Production of Cellulose Derivatives

Cellulose derivatives such as carboxymethyl cellulose and cellulose acetate have numerous applications which cut across various industries. CMC is used as a rheological property enhancer. Cellulose diacetate, for example, is used in the production of membranes for water filtration. Cellulose extracted from water hyacinth has been successfully used in the production of membranes which effectively filtered humic acid from water showing that such membranes can effectively filter surface water (Istirokhatun et al. 2015).

14.7.8   Conductive Cellulose Paper

Due to the large surface area of green algae sourced cellulose, it makes a good conductive flexible paper with numerous potential applications. Cellulose in general has good compatibility with polypyrrole (PPy). Cellulose-PPy composites possess the mechanical properties of paper and the conductive properties of metal. This has potential applications such as energy storage and sensors and a range of other low cost, mass reproducible conductive devices.

Green algae cellulose-PPy composites that have been produced have recorded a conductivity of 0.3 S/cm (Mihranyan 2010).

Cellulose extracted from water hyacinth has been used as a reducing agent and a capping agent for the green synthesis of silver nanoparticles. Monodisperse, spherical silver nanoparticles with size 2.68–5.69 nm are obtained using cellulose extracted from water hyacinth as reducing agent and a capping agent (Mochochoko et al. 2013). This further shows the broad range of applicability of cellulose in green extraction processes for high-value products such as silver nanoparticle.

14.8   Commercial Production

Cellulose is relatively well explored commercially. Since prehistoric times, cellulose-based materials have been used for paper and textiles. After paperboard, man-made cellulosic fiber is the next largest biopolymer commodity by volume (Shen and Patel 2010). In the earlier years of the textile industry, 70% of the textiles in the world were cotton based. The textile industry depending on one plant for raw material poses limitations such as nutrient consumption from soil and large areas of land required for cultivation. In addition to this, cotton being a plant that is endemic to only a few regions in the world meant dependency on importation for many and concentration of resources in only a few areas. On the contrary, aquatic plants such as duckweed and water hyacinth are endemic to almost all regions of the world in abundance.

The 1930s brought forth the innovation of man-made fibers sourced from cellulosic plants. This allowed cellulose to be extracted from, mainly wood and cotton lint, and then processed into fabrics (Albrecht 2004). The man-made cellulosic fibers have considerable market value, and as of 2002, global production of man-made cellulose reached 2800 kt annually (Aizenshtein 2004; Lenzing AG 2006). While the global production of cotton and petroleum-based synthetic fibers have both shown steep rises in the past decades, man-made cellulose fiber production rate has not shown as much increase. This can be attributed to the limited land-based resources. Use of aquatic plants could potentially boost commercial production of man-made cellulosic fibers.

The cellulosic fiber industry is of more significance now as land resource for growing cotton becomes scarce and the fossil resource for fossil-based synthetic fibers is depleting. Also important is the development of more environmentally friendly production process for production of cellulosic man-made textiles. These newer methods eliminate the need for toxic compounds such as CS_2 and minimize the total amount of chemicals such as sodium hydroxide required for the production process (Shen et al. 2010). Aquatic-sourced cellulose from aquatic plants which are growing abundantly and are seen as environmental nuisance means relatively low-cost feedstock for commercial cellulose production. Recent studies have shown that high-quality cellulose nanoparticles (Jaurez-Luna et al. 2019) with low aggregation can be extracted from the water hyacinth stem.

14.9 Conclusion

Cellulose is relatively abundant in the aquatic ecosystem as it can be obtained from a diverse range of plants and algae. Some cellulose-producing species of aquatic plants and animals are so abundant that they have become a nuisance to water transportation and coastal activities. Extraction of cellulose from aquatic source presents one of the solutions to address such challenges. There needs to be a balanced system in place, whereby the aquatic plants re-absorb the nutrients from the wastewater released by industries such as paper and textile, among others, and then these aquatic plants in return act as sources of polymers such as cellulose which in turn serves as a source of fuel and feedstock for these industries. The superior quality of cellulose obtained from aquatic plants and algae indicates future potential in their high-value application in biomedicine and pharmacology.

References

Aizenshtein EM (2004) World production of textile raw material. Fibre Chem 36(1):3–7

Albrecht W (2004) Regenerated cellulose in chapter "Cellulose". In: Ullmann's encyclopedia of industrial chemistry, 7th edn. Wiley-VCH Verlag GmbH & Co. KGaA. https://doi.org/10.1002/14356007.a05_375.pub2

Basile A, Sorbo S, Conte B, Cobianchi RC, Trinchella F, Capasso C, Carginale V (2012) Toxicity, accumulation, and removal of heavy metals by three aquatic macrophytes. Int J Phytorem 14(4):374–387

Bayrakci AG, Kocar G (2014) Second-generation bioethanol production from water hyacinth and duckweed in Izmir: a case study. Renew Sustain Energy Rev 30:306–316

Beguin P, Aubert JP (1994) The biological degradation of cellulose. FEMS Microbiol Rev 13(1):25–58

Brown RM, Saxena IM Jr (2000) Cellulose biosynthesis: a model for understanding the assembly of biopolymers. Plant Physiol Biochem 38(1–2):57–67

Chen P, Min M, Chen Y, Wang L, Li Y, Chen Q, Wang C, Wan Y, Wang X, Cheng Y, Deng S, Hennessy K, Lin X, Liu Y, Wang Y, Martinez B, Ruan R (2009) Review of the biological and engineering aspects of algae to fuels approach. Int J Agric Biol Eng 2:1–30

Chen Q, Jin Y, Zhang G, Fang Y, Xiao Y, Zhao H (2012) Improving production of bioethanol from duckweed (*Landoltia punctata*) by pectinase pretreatment. Energies 5:3019–3032

Chen WY, Lee HV, Juan JC, Phang S (2016) Production of new cellulose nanomaterial from red algae marine biomass *Gelidium elegans*. Carbohyd Polym 51:1210–1219

FAO (2011) A summary of the world apparel fibre consumption survey 2005–2008. International Cotton Advisory Committee

FAO (2017) Pulp and paper capacities. Rome, Italy. ISSN 0255-7665, ISBN 978-92-5-009846-3

Fernandez Diniz JM, Gil MH, Castro JAAM (2004) Hornification—its origin and interpretation in wood pulps. Wood Sci Technol 37:489–510

Fusi A, Bacenetti J, Gonzalez-Garcia S, Vercesi A, Bocchi S, Fiala M (2014) Environmental profile of paddy rice cultivation with different straw management. Sci Total Environ 494–495:119–128

Gao H, Duan B, Lu A, Deng H, Du Y, Shi X, Zhang L (2018) Fabrication of cellulose nanofibers from waste brown algae and their potential application as milk thickeners. Food Hydrocolloids 79:473–481

Hanhikoski S, Niemela K, Vuorinen T (2019) Biorefining of Scots pine using neutral sodium sulphite pulping: investigation of fibre and spent liquor compositions. Ind Crops Prod 129:135–141

Hasan SH, Talat M, Rai S (2007) Sorption of cadmium and zinc from aqueous solution by water hyacinth (*Eichhornia crassipes*). Bioresour Technol 98:918–928

Heilig G (1994) The greenhouse gas methane (CH_4): sources and sinks, the impact of population growth, possible interventions. Popul Environ 16(2):109–137

Hon (1994) Cellulose: a random walk along the historical path. Cellulose 1(1):1–25

Istirokhatun T, Rokhati N, Rachmawaty R, Meriyani M, Priyanto S, Susanto H (2015) Cellulose isolation from tropical water hyacinth for membrane preparation. Procedia Environ Sci 23:274–281

Jaurez-Luna GN, Favela-Torres E, Quevedo IR, Batina N (2019) Enzymatically assisted isolation of high-quality cellulose nanoparticles from water hyacinth stems. Carbohyd Polym 220:110–117

Johnson M, Shivkumar SJ (2004) Filamentous green algae additions to isocyanate based foams. J Appl Polym Sci 93(5):2469–2477

Kollah B, Patra AK, Mohanty SR (2016) Aquatic microphylla *Azolla*: a perspective paradigm for sustainable agriculture, environment and global climate change. Environ Sci Pollut Res 23(5):4358–4369

Konda M, Singh S, Simmons BA, Klein-Marcuschamer D (2015) An investigation on the economic feasibility of Macroalgae as a potential feedstock for biorefineries. Bioenergy Res 8:1046–1056

Koyama M, Sugiyama J, Itoh T (1997) Systematic survey on crystalline features of algal celluloses. Cellulose 4:147–160

Lahnalammi A, Sixta H, Jamsa-Jounela S (2018) Control strategy scheme for the prehydrolysis Kraft process. Comput Aided Chem Eng 44:643–648

Lee H, Kim K, Mun SC, Chang YK, Choi SQ (2018) A method to produce cellulose nanofibrils from microalgae and the measurement of their mechanical strength. Carbohyd Polym 180:276–285

Lenzing AG (2006) Sustainability in the Lenzing Group. Lenzing AG, Lenzing. http://www.lenzing.com/sites/nh/english/e_index.html

Li W, Khalid H, Zhu Z, Zhang R, Liu G, Chen C, Thorin E (2018) Methane production through anaerobic digestion: participation and digestion characteristics of cellulose, hemicellulose and lignin. Appl Energy 226:1219–1228

Ma S, Sun X, Fand C, He X, Han L, Huang G (2018) Exploring the mechanisms of decreased methane during pig manure and wheat straw aerobic composting covered with a semi-permeable membrane. Waste Manag 78:393–400

Martinez de Yuso A, Izquierdo MT, Rubio B, Carrot PJM (2013) Adsorption of toluene and toluene-water vapor mixture on almond shell based activated carbon. Adsorption 19:1137–1148

Mihranyan A (2010) Cellulose from Cladophorales green algae: from environmental problems to high-tech composite materials. J Appl Polym Sci 119:2449–2460

Mihranyan A, Llagostera AP, Karmhag R, Strømme M, Ek R (2004) Moisture sorption by cellulose powders of varying crystallinity. Int J Pharm 269:433–442

Miranda AF, Biswas B, Ramkumar N, Singh R, Kumar J, James A, Lal B, Subudhi S, Bhaskar T, Mouradov A (2016) Aquatic plant *Azolla* as the universal feedstock for biofuel production. Biotechnol Biofuels 9(221):1–17

Mitsuo Y, Eisuke Y (1996) (to Mitsubishi Paper Mills). Jpn Pat 08-229318

Mochochoko T, Oluwafemi OS, Jumbam DN, Songca SP (2013) Green synthesis of silver nanoparticles using cellulose extracted from an aquatic weed; water hyacinth. Carbohyd Polym 98(1):290–294

Munoz C, Hidalgo C, Zapala M, Jeison D, Riquelme C, Rivas M (2014) Use of cellulolytic marine bacteria for enzymatic pretreatment in microalgal biogas production. Appl Environ Microbiol 80(14):4199–4206

Olatunji O, Olsson RT (2015) Microneedles from fishscale-nanocellulose blends using low temperature mechanical press method. Pharmaceutics 7:363–378

Olatunji O, Olsson RT (2016) Processing and characterization of natural polymers. In: Natural polymers, industry techniques and applications. Springer, Switzerland. ISBN 978-3-319-26412-7

Penfound WT, Earle TT (1948) The biology of the water hyacinth. Ecol Monogr 18:447–472

Polprasert C, Kongsricharoern N, Kanjana Prapin W (1994) Production of feed and fertilizer from water hyacinth plants in the tropics. Waste Manag Res 12(1):3–11

Russel JB, Muck RE, Weimer PJ (2009) Quantitative analysis of cellulose degradation and growth of cellulolytic bacteria in the rumen. FEMS Microbiol Ecol 67(2):183–197

Sarkar M, Rahman AKML, Bhoumik NC (2017) Remediation of chromium and copper on water hyacinth (E.crassipes) shoot powder. Water Resour Ind 17:1–6

Shen L, Patel MK (2010) Life cycle assessment of man-made cellulose fibres. Lenzinger Ber 88:1–59

Shen J, Qian X (2012) Addressing the water footprint concept: a demonstrable strategy for papermaking industry. BioResources 7(3):2707–2710

Shen XL, Xia LM (2003) Studies on immobilized cellobiase. Chin J Biotechnol 19(2):236–239

Shen L, Worrell E, Patel KM (2010) Environmental impact assessment of man-made cellulose fibres. Resour Conserv Recycl 55(2):260–274

Siddhanta AK, Chhatbar MU, Mehta GK, Sanandiya ND, Kumar S, Oza MD, Prasad K, Meena R (2011) The cellulose contents of Indian seaweeds. J Appl Phycol 23:919–923

Smith J, Connell P, Evans RH, Gellene AG, Howard MDA, Jones BH, Kaveggia Palmer L, Schnetzer A, Seegers BN, Seubert EL, Tatters AO, Caron DA (2018) A decade and a half of *pseudo*-nitzschia spp. and domoic acid along the coast of southern California. Harmful Algae. https://doi.org/10.1016/j.hal.2018.07.007

Steen DA, Gibbs JP, Buhlmann KA, Carr JL (2012) Terrestrial habitat requirements of nesting freshwater turtles. Biol Conserv 150(1):121–128

Stromme M, Mihranyan A, Ek R (2002) What to do with all these algae? Mater Lett 57:569–572

Thiripura M, Ramesh A (2012) Isolation and characterization of cellulose nanofibers from the aquatic weed water hyacinth—*Eichhornia crassipes*. J Carbohyd Polym 87:1701–1705

Vulimiri SV, Pratt MM, Kulkarni S, Beedanagari S, Mahadevan B (2017) Reproductive and developmental toxicity of solvents and gases, chap 21. In: Reproductive and developmental toxicology, 2nd edn. Academic Press, pp 379–396

Wi SG, Kim HJ, Mahadevan SA, Yang DJ, Bae HJ (2009) The potential value of the seaweed Ceylon moss (*Gelidium amansii*) as an alternative bioenergy resource. Bioresour Technol 100:6658–6660

Yadav D, Barbora L, Bora D, Mitra S, Rangan L, Mahanta P (2017) An assessment of duckweed as a potential lignocellulosic feedstock for biogas production. Int Biodeterior Biodegradation 119:253–259

Yaich H, Garnaa H, Besbesa S, Paquot M, Beckerc C, Attia H (2011) Chemical composition and functional properties of *Ulva lactuca* seaweed collected in Tunisia. Food Chem 128(4):895–901

Yoon JJ, Kim YJ, Kim SH, Ryu HJ, Choi JY, Kim GS, Shin MK (2010) Production of polysaccharides and corresponding sugars from red seaweed. Adv Mater Res 93–94:463–466

Zhao X, Moates GK, Wellner N, Collins SR, Coleman MJ, Waldron KW (2014) Chemical characterisation and analysis of the cell wall polysaccharides of duckweed (*Lemna minor*). Carbohyd Polym 111:410–418

Chapter 15
Polyesters

Abstract Polyesters are characterized by a structural unit made up of ester-linked monomers. Cutin is the main polyester found in the aquatic ecosystem. It is present in aquatic plants where it plays a role in the regulation of water permeation. It has potential applications in packaging films, cosmetics and biopolyester production. The understanding of the structure, function and chemistry is also useful in developing products which mimic cutin. This chapter discusses cutin obtained from the environment. It covers the natural sources of cutin within the aquatic environment, its chemistry, extraction process, applications, environmental implications of cutin production and the present state of its commercial production.

Keywords Cutin · Polyester · Biopolymer · Aquatic plants · Cuticle

15.1 Introduction

The main polyesters found in aquatic environment are cutin and suberin. These polyesters are the most native feature in plants. Although previously thought to be absent in all aquatic plants, cutin is now known to be present in a number of aquatic plants and makes up the bulk of the dry mass of the cuticle. Traces of suberin have been found in the water facing part of the aquatic plant. The presence of traces of suberin in the water facing part of aquatic plant suggests it is playing a similar role as the terrestrial plant suberin present in the root. Suberin is present in a much lower quantity than cutin, and therefore, this chapter will focus more on cutin as the main polyester found in aquatic plants.

Cutin constitutes the bulk of the cuticle by mass and serves as a matrix holding the wax and other components of the cuticle. Based on the available findings from study of the biochemistry of the cuticle, the role of cutin in aquatic plant is thought to be UV protection and as a barrier to protect against pathogen attack and external environmental stress. The protection against pathogen is possible due to its physicochemical properties which discourage cell/microbe attachments and protection against microbe is possibly due to the mechanical properties and stability.

Although methods to isolate long-chain polyester cutin are underway, methods for extraction of cutin monomers and oligomers have been presented in literature and

© Springer International Publishing 2020

O. Olatunji, *Aquatic Biopolymers*, Springer Series on Polymer and Composite Materials,
https://doi.org/10.1007/978-3-030-34709-3_15

patents. Relatively non-complex methods for extraction of cutin, which require the use of acids and alkali treatment at high temperature, consume moderate quantities of water. The cultivation of the feedstock for cutin production has beneficial impacts on environment such as water remediation properties such that the process of production of cutin would include cultivation of aquatic plants which also serve to clean the aquatic environment.

Despite not yet being fully commercially exploited, cutin has some signifi-cant applications in, especially the food and the pharmaceutical industry where its mechanical properties as well as the role of the barrier/protective role of the cuticle can be mimicked for human use. Furthermore, the side products from cutin extrac-tion from aquatic plant cuticle are also useful. Compounds such as palmitic acid and linolenic acids find applications in areas such as soap making, biodiesel and antiinflammatory agent production.

In a global market that is increasingly demanding more plant-based alternatives to the dwindling fossil-derived products, particularly those that are either injected or indirect contact with the body or food, cutin is another one of the vast biopolymer resources the aquatic ecosystem offers.

15.2   Occurrence in Nature

Cutin occurs in nature as part of plant cuticle which is a protective coating on the sur-face plants. The cuticle protects plant from environmental stress and diseases and aids water retention. It is adhered to the plant epidermal wall through a proteinaceous layer with which it is interconnected. Cutin is a branched polymer, cross-linked through ester linkages between the chains forming the polymer network. Cutin content in most plants is between 40 and 80% dry weight of the cuticle; around 600 μg of cutin can be obtained per m^2 of a plant surface. The cuticle is measured to be between 0.5 and 40 μm thick. It is a composite of polysaccharides, biopolyester (cutin) waxes and phenolic compounds. The cutin served as the polymeric continuous phase, holding the other components of the cuticle. Figure 15.1 is a simplified representation of the cutin location on the plant surface. This is simplified since the pectinaceous layer is not always distinctly separated as the figure suggests.

Until the recent studies on cutin in aquatic plants, cutin has been widely regarded as a feature only in terrestrial plants with a sole purpose of aiding water retention. Studies

Fig. 15.1 A representation of cuticle layer next to the epidermal wall

in recent years confirm that cutin is not only a compound restricted to terrestrial plants but also is present in aquatic plants. This further revealed that the role of cutin in aquatic plants is not for water retention, since aquatic plants have always abundance of water and in fact have features to adapt them for growth in the environment. Cutin in aquatic plants serves as a protective layer against UV radiation, pathogens and environmental stress (Li et al. 2017). Because aquatic plants do not have the property of terrestrial plant of being able to direct parts of their leaves to grow away from the sun's rays, they use cuticular matrix as a protection against UV.

Aquatic plants generally have the water facing sides and the sun-facing side. Traces of suberin like compounds have been detected in the water facing side (Borisjuk et al. 2018). Suberin is the polyester synthesized in the roots of vascular plants which aids in water retention. It is therefore conceivable that they are present in the water facing sides of some aquatic plants. Further studies are yet to fully define the roles or suberin in aquatic plants.

Cutin does not occur in all aquatic plants as not all aquatic plants have need for it. Much is yet to be known about cutin in aquatic plants, and it has generated increasing interest. Duckweed is one aquatic plant containing cutin which has attracted growing interest for its potential commercial and ecological application. The cuticle contains between 40 and 80% cutin by weight.

15.3 Chemistry of Cutin

Cutin is a polymer made up of C16 or C18 fatty acids with hydroxyl or epoxy groups attached. The cutin polymer chains are cross-linked by ester linkages forming a rather complex structure. It has polyester structure formed from the ester linkage of C16 or C18 fatty acids binding with diols. The main monomer of cutin is the 10,16-dihydroxyhexadecanoic acid monomer (Graca and Lamosa 2010). Figure 15.2 shows ester linkages between this monomer and other hydroxyl compounds. Cutin consists of different monomers covalently linked together as well as cross-linking between different polymer chains. It is usually present in the form of a cutin matrix with waxes and other compounds embedded within the matrix or as a layer within the cuticle structure. Other components such as phenolic compounds could also be present within the cutin matrix (Heredia 2003).

In a recent analysis of the biochemistry of aquatic plant cuticle, GC-MS analysis of cutin extracted from duckweed fronds confirms the presence of a substantial amount of fatty acids, and up to 95% of the cutin monomers were fatty acids. These include saturated palmitic acid (C16) making up the bulk of the fatty acids. Cutin matrix monomer also consisted of linolenic acid, an unsaturated C18 fatty acid which is also an essential fatty acid for humans as it is not produced in the human body. Other fatty acid components of cutin include other long-chain fully saturated fatty acids as well as hydroxyl fatty acids, aromatic acids and phenolic ring-cinnamic acids (Borisjuk et al. 2018). The wax fraction from the cuticle contains other valuable compounds such as phytosterols and squalene which is a component of human skin

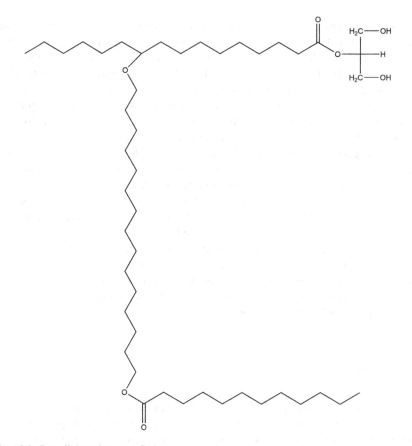

Fig. 15.2 Ester linkage between C16 monomers of cutin and a glycerol

that serves as the first point of protection against ultraviolet radiation and physical barrier from the environment. The cutin from aquatic plants comprises of the same compounds present as that of terrestrial plants, and however, the composition of each compound varied for the different species. This is expected as the cutin in aquatic plants serves different functions as those in terrestrial plants where the main function is water retention.

Although suberin is the plant polyester found mainly in the part of the plant below the ground, the root (Li et al. 2017), it is not expected to be present in aquatic plants. However, traces of suberin were detected in the water facing duckweed fronds (Borisjuk et al. 2018). Suberin from terrestrial plants is used in the production of cork used commonly in the wine bottling (Graca 2015). Their presence in aquatic plant could therefore be of commercial interest. However, more studies are required to fully understand their role in aquatic plants.

An important property of polymers is to have the mechanical properties required for their applications which may include load bearing, pulling and tearing, or serving as a barrier against water and air. Mechanical properties of cutin from plants at present state of the technology are still inferior to those of the fossil-based polymers widely used in industry. The elongation at break, young modulus and tensile strength measured for cutin is 27%, 45 and 12.3 MPa (Heredia-Guerrero et al. 2017a, b; López-Casado et al. 2007). Much lower than that of the petroleum-based polymers such as polyethylene, polystyrene and polypropylene where elongation at break is in the hundreds and the young modulus in multiple GPa. On the other hand, cutin has a relatively high thermal degradation temperature of 200 °C, similar to that of rubbers and polyethylene. Polymers such as polystyrene, polyvinyl chloride and polystyrene have lower degradation temperatures. Cutin shows no melting point in thermogravimetry analysis, thus confirming that it is a thermoset polymer. This is expected of its cross-linked structure. It has a transition temperature of −22 °C.

Cuticle composition and cutin content vary from plant to plant. Factors which determine cutin content include developmental stage of the plant, species, part of the plant and the environmental conditions within which the plant is grown (Yeats et al. 2012). Although there is a significant variation in the monomers making up the cutin from different species and different parts of the plant, the species and anatomy structure relation is well defined enough to allow obtaining a biomass with uniform cutin composition. This is important in order to achieve standard controlled processes and uniform products.

15.3.1 Biodegradation of Cutin

As a biodegradable polymer, it is essential that cutin decomposes into harmless monomers which further decomposes into smaller molecules and finally carbon, nitrogen and hydrogen compounds (Angst et al. 2016). Unlike other commonly used fossil-derived polyesters such as polyethylene terephthalate that take over three decades to degrade, cutin can be fully degraded within 3–8 months when buried in soil (De Vries et al. 1967). The degradation of cutin by microorganisms such as bacteria and fungi and other microorganisms involves the cleavage of ester linkages catalyzed by cutinases released by the organisms (Kolattukudy 2001a, b; Heredia-Guerrero et al. 2017a, b). Understanding of the mechanism of degradation of cutin by these organisms using cutinase is key to the development of processes for degradation of the fossil fuel-derived polyesters into monomers or oligomers with faster degradation rate or further applications.

15.4 Availability of Raw Materials

Although cutin which is present in the cuticle, is a more common feature in terrestrial plants, cutin is also present in a diverse range of aquatic plants such as duckweed, sea grasses and water hyacinth (Borisjuk et al. 2018). The cutin from these plants has been shown to contain the same monomeric unit as those in land plants. Therefore, the applications of cutin from land plants can also be extended to those of aquatic plants. In aquatic plants, the cuticle switches roles from primarily a water retention feature to serving structural protective roles. Furthermore, despite terrestrial plants being more abundant source of plant polyester, exploring cutin in aquatic plants provides improved understanding of the role of cutin beyond water retention. It is also an important additional source of plant biomass for cutin production, where aquatic plants grow in excess (e.g., water hyacinth growth in lagoons), and it serves as a means to manage aquatic plant growth through harvesting and utilizing them.

While some aquatic plants such as the water lotus are preserved, aquatic plant considered sacred in some parts of the world and valued for the aesthetic appeal of its flower. Other more abundant aquatic plants such as duckweed on the other hand are of more economic potential as a source of commercial cutin production. It grows as a green floating plant with no flower. Duckweed is globally endemic, growing in most parts of the world. It has a relatively fast rate of growth at 80–100 tonnes dry mass per hectare annually. Like many other aquatic plants, it grows at a much faster rate than terrestrial plants, up to 5 times as fast as corn (Lam et al. 2014). Although not yet commercially cultivated or harvested for cutin production on a large scale, this much faster rate of biomass generation indicates a potentially reliable availability of duckweed feedstock as a non-food source of cutin which does not require land space for cultivation.

FAO reports the global aquatic plant production rate to be 30.1 million tonnes as of 2016 (FAO 2018). Much of this present is seaweed since the commercial applications of the macroalgae have been well established, and therefore, seaweed has a well-established market and as such is more cultivated. As the industrial applications of other aquatic plants such as duckweed become better established, more aquatic farmers will be encouraged to cultivate and harvest them, thereby increasing their market value. Cuticle from wild and cultivated plants is globally estimated at 180–1500 kg per acre (Heredia 2003), and 8.9×10^9 acres of land are estimated to be available for plant cultivation globally (FAO, IFAD, UNICEF, WFP and WHO 2018). 40–80% of this cuticle by weight is cutin. Duckweed is one aquatic plant which serves as a source of cutin and the cutin content and composition in this aquatic plant has been studied (Borisjuk et al. 2018). 80–100 tonnes dry mass of duckweed is annually attainable per hectare of aquatic space. There is therefore potential abundant source of raw material for cutin production to be explored considering that aquatic plants grow at up to five times faster rate than land plants and they are produced in spaces which do not compete with land spaces occupied by humans, in addition to the fact that the earth is about 71% water, and there is more space for expansion of aquatic plant.

Genetic selection can be carried out to optimize the production of specific plant metabolites by altering the growth parameters such as temperature, ultraviolet light radiation, pH, water salinity among others. Such can be applied to improve the cutin production by plants, aquatic plants included . As duckweed has attracted increasing attention for its potential applications such as bioactives and biofuel production as a cutin producing aquatic plant, this should also raise interest in optimizing such plant for cutin production.

15.5 Extraction of Cutin

A few methods have been presented for extraction of cutin from plants such as tomatoes and fewer on aquatic plants. The extraction methods presented either make use of various solvents or acid or alkali for extraction of cutin from parts of the plants such as the leaves and peels (Borisjuk et al. 2018; Cifarelli et al. 2016; Cigognini et al. 2015). The cuticle is separated from the plant epidermal wall by either mild chemical treatment or enzyme treatment. Ammonium oxalate is often used in the chemical treatment, and pectinases are used in the enzyme method to degrade the pectinaceous layer which serves as a "glue" to adhere the cuticle to the epidermal wall.

Extraction of cutin from duckweed as presented by Borisjuk et al. (2018) is used to extract cutin from the cuticle of aquatic plant for the main purpose of studying the biochemistry of cutin from species of aquatic plants. This was therefore not presented as a commercial process for cutin extraction; rather, it was the mildest means to obtain cuticle components for analysis. In the said process, the first step was to remove the wax fraction of the cutin matrix. The duckweed fronds were lyophilized and immersed in 2 ml chloroform in two consecutive processes. The first process involves a 60 s chloroform treatment of the lyophilized fronds twice then followed by another chloroform treatment for 30 s. This is then followed by treatment with alkane at a concentration of 10 mg/50 ml. The wax residue is separated at this point. This wax residue can also be a source of other useful compounds such as squalene. The extract is then dried under a stream of nitrogen. This was then followed by further exhaustive extraction with a mixture of chloroform and methanol at equal volume ratios for 2 weeks with the solvent changed daily. The residue is then transesterified with 1 N methanolic HCL for 2 h at 80 °C. Following transesterification-saturated sodium chloride and dotriacontane were added followed by subsequent triple extraction with hexane (Borisjuk et al. 2018).

In a method patented in 2015 (Cigognini et al. 2015), used for extraction of cutin from tomato peels, a less laborious process using much less solvent is presented. However, no work is yet presented which uses this method for cutin extraction from aquatic plants. It is assumed that the same method is applicable to extract cutin from aquatic plant leaves. The extraction is achieved with sodium hydroxide, heating and centrifugation as follows. It can be assumed that since the cutin from both aquatic and terrestrial plants has the same chemical components, similar methods can be employed in their extraction since the general goal of the extraction is the isolation of cutin from the other components of the cuticle through dissolution and partial degradation followed by precipitation.

In the said method, tomato peels are treated in sodium hydroxide at a concentration of 0.75 M (patent claim states that anywhere between 0.5 and 6 M can be used) and a temperature of 100 °C or anywhere between 65 and 130 °C for a period of 2 h with an acceptable time period range between 15 min and 6 h. The liquid alkali solution is added at a solid to liquid alkali ratio of 1:100 tomato peels: liquid. The mass is then filtered with sieve, and the solid residue is discarded while the brown liquid fraction

Fig. 15.3 Flowchart showing process for extraction of cutin from plant biomass

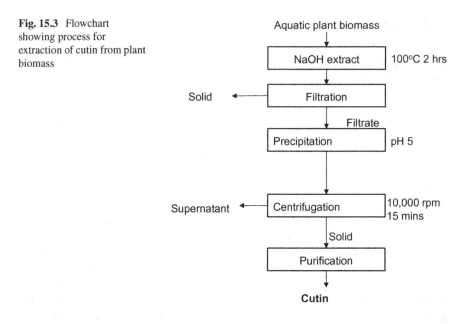

Cutin

is kept for further processing. The next stage involves addition of hydrochloric acid until the pH reaches between 5 and 6. This is achievable at 5% volume HCl of total volume at a concentration of about 6 M. The acidification is required in order to precipitate the dissolved cutaneous compounds. This is then centrifuged at an rpm of 10,000–14,000 for 15–20 min. The liquid part is discarded while the solid residue is repeatedly washed and centrifuged up to three times to finally obtain raw cutin. This method of extraction is summarized in Fig. 15.3.

Using this method, the obtained cutin is made up of mostly a mixture of cutin oligomers and monomers with a yield of 60–80%. The molecular weight of the cutin extracted according to GC-MS analysis is 650 g/mol thus indicating a presence of mostly dimers and trimers as the molecular weight of cutin monomer, 10,16-dihydroxyhexadecanoic acid and a C16 hydroxy fatty acid ($C_{16}H_{32}O_3$) is 284 g/mol. Advantage of this process is the ready availability of sodium hydroxide and hydrochloric acid, compared to the other extraction processes which require use of expensive solvents which also have a negative impact on health and environment.

15.6 Environmental Implication

The environmental impact of cutin stems from the harvesting of the cutin containing plants to the process of extraction and purification. Here, we take into consideration, the patented method by Cigognini et al. (2015) which requires use of acid and alkali. It is assumed that the same process is applicable to cutin from aquatic plants. Consumptions for the extraction of cutin are estimated in Table 15.1.

Table 15.1 Consumptions during cutin extraction

Consumptions	Amount
Sodium hydroxide	0.7 M, 100 ml/g cuticle
HCl	6 M, 5 ml/g cuticle
Water	>100 ml/g cuticle
Solid waste	1.5 g per gram cutin
Energy	
Heating	10,000 rpm for 15 min
Centrifugation	100 °C for 2 h

15.6.1 Use of Sodium Hydroxide

The process uses sodium hydroxide at a concentration of 0.7 M and for every gram of plant cuticle mass 100 ml of 0.7 M solution of sodium hydroxide is used. The sodium hydroxide does not take part in the reaction; rather, it acts as a catalyst providing the alkaline environment required for the process. Much of this is neutralized during the acidification of the liquid filtrate after the alkali treatment in the first stage of the extraction. However, the solid mass which is discarded remains alkali, and this will require neutralization before it is fed into the recycle stream to the feed or added to the next batch in a batch process. If the solid is discarded, more water will be required to neutralize it before it is safely released to the environment. Although this is not water directly used in the process of extraction of the cutin, it adds to the wastewater generated and water consumption during the process of producing cutin.

Following acidification, the liquid filtrate after centrifugation can also be sent back to the feed to extract any remaining cutin which might still be present. However, because of the alkali condition required for extraction, this higher pH allows the precipitation of the cutin components. This filtrate which has previously been acidified to a pH of ~5 now needs to be returned to a much higher pH of ~10, thus requiring additional sodium hydroxide to be used. The producer must then compare the cost of recycling to discarding. This will vary from place to place.

15.6.2 Use of Hydrochloric Acid

Hydrochloric acid is used in the acidification stage to precipitate the crude cutin extracted using the thermal alkali treatment in the first stage. The hydrochloric acid used makes up 5% of the liquid volume at a concentration of 6 M. Much of this goes into neutralization reaction, and the rest goes toward reducing the pH from 10 to pH 5–6. The solid residue following centrifugation needs to then be washed until neutral. This results in further water consumption. Part of this acidic wash water could be used to neutralize the water used in the washing of the discarded alkali solid residue from the first stage after alkali treatment. This mixing of high and low pH streams could minimize use of further acid/alkali for neutralization prior to further wastewater treatment and then released into the environment. The use of sodium hydroxide and hydrochloric acid replaces the use of chloroform and organic solvents such as hexane and methanol. Since no volatile solvents are used in the patented method, the acids and alkali used can be treated at the wastewater treatment.

15.6.3 Wastewater Generated

Most of the water used in the process eventually becomes wastewater since the final extract is in the solid form. The rest is evaporated during the drying process. The acids and bases used during the extraction are all dissolved in water. A liquid to solids ratio of 100:1 is used in the first extraction stage using high-temperature alkali treatment. For every gram of cuticle, 100 ml of water is required. Most of the water used here can be recycled back into the feed stream or fed in with the next batch. More water is used in the washing of the solid residue to obtain the final product. Additional water used in washing the waste solid residues is also directed to the wastewater stream. The water could either be treated on site or sent to the wastewater treatment plant. Most industrial laws require the acidic and alkali wastewater to be neutralized and not simply diluted. The level of treatment required will depend on the country of operations and the local laws which exist there.

15.6.4 Solid Waste Generated

The cuticle is made up of 40–80% cutin, and the rest of the cuticle comprises of the waxes and other components of the cuticular matrix such as cinnamic acid, but mainly waxes. The wax is made up of mostly aliphatic molecules such as ketones, long-chain fatty acids, primary alcohols, aldehydes and ketones. These are more soluble in acidic conditions and hence can be filtered off following acidification while the crude cutin precipitates as a gel-like solid form (Borisjuk et al. 2018; Cigognini et al. 2015). The plant biomass also contains other valuable biopolymers such as starch and protein. The cuticle can be first removed, and the rest of the plant further processed for extraction of the other biopolymers.

15.6.5 Aquatic Plant Harvesting

Cultivation of aquatic plants such as duckweed has benefits for the environment. They are utilized in the bioremediation of wastewater due to their ability to take up nutrients such as nitrogen and phosphate compounds from the water. Beyond acting as a feedstock for cutin, the process of extraction of cutin also yields waxes and essential fatty acids as by-products which have high commercial value for their bioactivities and other applications as discussed within the chapter. Having a mixed culture of aquatic plants could be beneficial in minimizing the risk of disease spread and allowing for biodiversity. The monomer composition of the cutin polymer varies for different species and plant parts; therefore, mixed cultivation of variety of species may cause some constraints. This is due to the variation in the monomer composition resulting in the variation in the mechanical properties of the cutin.

15.7 Applications

Much of the potential and existing application of cutin lies in its protective barrier formation. This makes it potentially applicable in packaging for food and pharmaceutical products. The cutin matrix also contains other compounds which are of commercial significance. This section therefore also includes the potential applications of these compounds in the cutin matrix.

15.7.1 Sunscreen

The cuticular matrix offers protection for the plant against UV radiation, and the phenolic compounds within the cutin matrix also provide UV attenuation. Cutin can be extracted from the plant and used in applications which mimic the UV protection. The waterproofing role of the cuticle could also serve in retaining skin moisture which also minimizes sun damage to skin. This has been explored by researchers (Zhang and Uyama 2016; Heredia-Guerrero et al. 2017a, b). The sunscreen market is a huge one, and as consumers increasingly demand more "natural" products, plant-based sunscreens are in demand and cutin-based sunscreens could offer an alternative.

15.7.2 Food Packaging and Additives

The food packaging industry is one where a lot of biopolymers find demand. There is an increasing demand for not just food component but also the packaging within which the food is enclosed, to be plant based and/or from an environmentally friendly source. The plant role serves a similar role as is required of food packaging, to protect the product encased within it from the environment. Nutrient retention property of the cutin matrix has potential for application in the food industry in for example retaining nutrients in animal or fish feed (Cheng and Stomp 2009). The application of cutin in the food industry is based around replicating the roles cutin/cuticle plays in nature.

Cutin contains some essential fatty acids which could form extracts for human feed supplements. Phytosterols, also found in the cutin matrix, have been FDA approved for use as anticholesterol. Phytosterols act by competing with cholesterol for active sites for absorption into the blood stream, thereby lowering the blood cholesterol level.

Cutin has also shown applicability in food as packaging films. Edible films with hydrophobic properties have been produced using cutin extracted from tomatoes (Manrich et al. 2017). At a ratio of 50/50 by eight cutin/pectin and high pH in the alkali range, well-dispersed films of cutin-pectin blend with good water repellency and physical properties that match those of tomato peels were achieved. The films remained stable at different pH and showed good thermal stability. This biomimicking cutin-based edible film has potentials in replacement of fossil-based food packaging films. The chemical composition of cutin allows it to play protective as well as bioactive role as edible food packaging. It is also important that such films could be produced using the conventional low cost, reproducible film casting methods used in industry.

Cutin can also be applied as a coating on metallic food packaging. Such protective coating separates the metal compound from the food. This could potentially replace bisphenol A diglycidyl ether (BADGE) resin which is presently being used for this application. BADGE has potentially harmful effects on human health (Montanari et al. 2014).

15.7.3 Cosmetics

The by-products of cutin extraction from the wax fraction contain squalene. These could be further explored for production of sunscreens. The transesterification of the fatty acids from the cutin matrix yields glycerol as a by-product. Glycerol is used in the cosmetics industry as a skin care product. Palmitic acid is also found in the cutin matrix in considerable amount, making up almost 95% of the fatty acids in the wax fraction. Palmitic acid is used in cosmetics as emulsifier and in soap production. The main source of palmitic acid at present is palm oil, a plant which only grows in a few countries in the world, the largest producer being Indonesia and Malaysia being second followed by countries in Western and Central Africa as well as Colombia and Thailand (Poku 2002). An alternative source of palmitic acid using a plant source that pretty much grows everywhere will have significant impacts on the cosmetic market.

15.7.4 Pharmaceutical

As it is applicable in food, cutin has potential for packaging in pharmaceutical products. The protective edible lipophilic coating property could be applied for capsules and tablets and other applications which require a biodegradable protective hydrophobic layer. The other components of the cutin matrix also have some potential pharmaceutical applications. The cutin matrix contains palmitic acid as well as a variety of other fatty acids, some of which have bioactive properties. One of such is linolenic acid. Linolenic acid, a monomer of cutin, has been shown to have antiinflammatory effect with potential to alleviate chronic migraine based on results on a 45-year-old human female patient (Santos and Weaver 2018). The study showed that when linolenic acid is applied to the base of the neck, it exhibits some prophylactic effect against chronic migraine. Topically applied treatments for chronic migraine are particularly preferred due to the convenience and the reduced side effects such as vomiting which is experienced in some people when using orally administered medications. This is one of the many potential applications of linolenic acid which is a monomer from cutin.

Linolenic acid has also been shown to have some anticancer effect, and it also has good potential for application in the treatment of osteoporosis, cardiovascular diseases, antioxidant effect and some neuroprotective properties. However, some of these are still being researched and are inconclusive.

15.7.5 Biomimicry of Cutin Matrix for Wastewater Treatment

The bioremediation properties of plant source of cutin matrix could be extracted or mimicked in wastewater treatment (Ziegler et al. 2016; Zhou et al. 2018). Although this is not a direct impact or application of the aquatic biopolymer, the process of cultivating these plants for extraction of cutin could in the process be applied to treat contaminated water prior to them being harvested for cutin production. Aquatic plants can either be grown wild or cultivated using aquaculture. Some aquaculture facilities are connected to seawater. Where the water comes in from the sea gets filtered and treated before being released to the plants. The water is then sent back out. This is mainly done to provide a controlled environment for cultivating aquatic plants. Such facilities could be extended to incorporate wastewater treatment by aquatic plants alongside cultivation of the plants for cutin extraction.

15.7.6 Biodiesel Production

Biodiesel production from triglycerides comes with one disadvantage of most triglycerides are used as foods. Furthermore, the transesterification reaction results in the formation of glycerol which needs to be separated from the main product. The fatty acids from the cutin matrix could serve as a source of biodiesel. Studies by Borisjuk et al. (2018) found that the wax fraction of the cutin matrix contained 95% fatty acids. These fatty acids can be transesterified with alcohols such as methanol and ethanol to produce biodiesel using appropriate catalysts. This will depend on obtaining a suitable method which is economical and obtains free fatty acids which are suitable for biofuel production. Experimental attempts have been made at esterification reactions of free fatty acids with methanol using 98% sulfuric acid as homogeneous catalyst (Javidialesaadi and Raeissi 2013). This is particularly significant as it acts as a source of non-edible oil for biodiesel production such that there is no competition with oils used in food. Other studies have investigated the use of other catalysts such as biological catalysts saccharomyces cerevisiae, ferric sulfate impregnated carbon from waste tyres, as for production of biodiesel from fatty acids (Yongjin et al. 2014; Hood et al. 2018). Cutin production could therefore yield fatty acids which could be used in production of biodiesel.

15.7.7 Biopolyester Production

Another approach to apply cutin is the esterification of its monomers, dimers and trimers into polyesters through the condensation reaction. The key to this is using the right reaction catalysts, conditions and solvents/disperse phase. This has been successfully carried out to the point of polyester oligomers (Gómez-Patiño et al. 2013). Using lipases as catalyst and carrying out the polymerization reaction at low temperature using organic solvents as dispersions, short-chain polyesters were obtained. Although these oligomers lack the desired properties to be applicable as polyesters, such process could serve as precursors to further polymerization reactions to obtain longer chain polyesters (Gómez-Patiño et al. 2015). Some polymerization reactions are carried out in two stages where short-chain oligomers are obtained in the first stage followed by a second stage to obtain a longer chain polymer. This could be for reasons such as controlling viscosity to avoid hot spots and allow proper mixing as viscosity increases with increasing molecular weight.

15.8 Commercial Production

Although there exists no known commercial production of cutin in the market to date. There is however rising research interest in the potentials for cutin from duckweed. As a fast-growing plant with bioremediation properties, duckweed has been a plant of interest and much of it is being grown and studied.

One of the better-explored plant polyesters is suberin found in terrestrial plants. Suberin is commercially exploited in the production of cork, most commonly in the production of wine stoppers. It is also used as stoppers in storage of some other products and chemicals. The monomers of suberin are also used in several applications. There is rarely traceable amount of suberin in aquatic plants, and therefore, it will not be considered as an aquatic biopolymer at this point. Nonetheless, the existing applications of suberin serve as a guide to the potential applications of cutin as research in this area advances.

15.9 Conclusion

The main polyester which can be sourced from the aquatic environment is cutin. It is obtainable from aquatic plants. Aquatic plants have the advantage of a higher productivity than terrestrial plants and therefore have potential as a more available source of cutin. This biopolyester is one of the lesser explore biopolymers available in the aquatic environment, and however, as discussed in the chapter, research studies have explored applications of cutin in a range of applications such as packaging films and UV sunscreens. Duckweed is one of the aquatic plants which have been explored for cutin production with reliable availability globally.

References

Angst G, Heinrich L, Kogel-Knabner I, Muelle CW (2016) The fate of cutin and suberin of decaying leaves, needles and roots—inferences from the initial decomposition of bound fatty acids. Org Geochem 95:81–92

Borisjuk N, Peterson AA, Lv J, Qu G, Luo Q, Shi L, Chen G, Kishchenko O, Zhuo Y, Shi J (2018) Structural and biochemical properties of duckweed surface cuticle. Front Chem. https://doi.org/10.3389/fchem.2018.00317

Cheng JJ, Stomp AM (2009) Growing duckweed to recover nutrients from wastewaters and for production of fuel ethanol and animal feed. CLEAN–Soil Air Water 37:17–26

Cifarelli A, Cigognini I, Bolzoni L, Montanari A (2016) Cutin isolated from tomato processing by-products extraction methods and characterization. In: Proceedings of CYPRUS 2016 4th international conference on sustainable solid waste management, pp 1–20

Cigognini IM, Montanari A, De la Torre Carreras R, Montserrat CB (2015) Extraction method of a polyester polymer or cutin from the wasted tomato peels and polyester polymer so extracted. WO2015/028299 A1

De Vries H, Bredemeijer G, Heinen W (1967) The decay of cutin and cuticular components by soil microorganisms in their natural environment. Acta Bot Neerl 16:102–110

FAO (2018) The state of world fisheries and aquaculture 2018—meeting the sustainable development goals. Rome. Licence: CC BY-NC-SA 3.0 IGO

FAO, IFAD, UNICEF, WFP and WHO (2018) The state of food security and nutrition in the World 2018. Building climate resilience for food security and nutrition. Rome, FAO. Licence: CC BY-NC-SA 3.0 IGO

Gómez-Patiño MB, Cassani J, Jaramillo-Flores ME, Zepeda-Vallejo LG, Sandoval G, Jimenez-Estrada M, Arrieta-Baez D (2013) Oligomerization of 10,16-dihydroxyhexadecanoic acid and methyl 10,16-dihydroxyhexadecanoate catalyzed by lipases. Molecules 18:9317–9333

Gómez-Patiño MB, Gutiérrez-Salgado DY, García-Hernández E, Mendez-Mendez JV, Andraca Adame JA, Campos-Terán J, Arrieta-Baez D (2015) Polymerization of 10,16-dihydroxyhexadecanoic acid, main monomer of tomato cuticle, using the lewis acidic ionic liquid choline chloride·2ZnCl$_2$. Front Mater 2:1–9

Graca J (2015) Suberin: the biopolyester at the frontier of plants. Front Chem 3(62):1–11. https://doi.org/10.3389/fchem.2015.00062

Graca J, Lamosa P (2010) Linear and branched poly(ω-hydroxyacid) esters in plant cutins. J Agric Food Chem 58:9666–9674

Heredia A (2003) Biophysical and biochemical characteristics of cutin, a plant barrier biopolymer. Biochim Biophys Acta (BBA) Gen Subj 1620:1–7

Heredia-Guerrero AJ, Heredia A, Dominguez E, Cingolani R, Bayer IS, Athanassiou A, Benitez JJ (2017a) Cutin from agro-waste as a raw material for the production of bioplastics. J Exp Bot 68(19):5401–5410

Heredia-Guerrero JA, Benítez JJ, Cataldi P, Paul UC, Contardi M, Cingolani R, Bayer IS, Heredia A, Athanassiou A (2017b) All-natural sustainable packaging materials inspired by plant cuticles. Adv Sustain Syst 1:1600024

Hood ZD, Adhikari SP, Evans SF, Wang H, Li Y, Naskar AK, Chi M, Lachgar A, Paranthaman MP (2018) Tyre-derived carbon for catalytic preparation of biofuels from feedstocks containing free fatty acids. Carbon Resour Convers 1:165–173

Javidialesaadi A, Raeissi S (2013) Biodiesel production from high free fatty acid-content oils: experimental investigation of the pretreatment step. Procedia APCBEE 5:474–478

Kolattukudy PE (2001a) Suberin from plants. In: Steinbtichel A, Doi Y (eds) Biopolymers. Wiley-VCH Verlag GmbH, Weinheim, pp 41–73

Kolattukudy PE (2001b) Polyesters in higher plants. In: Babel W, Steinbüchel A (eds) Biopolyesters. Springer Berlin Heidelberg, Berlin, Heidelberg, pp 1–49

Lam E, Appenroth KJ, Michael T, Mori K, Fakhoorian T (2014) Duckweed in bloom: the 2nd international conference on duckweed research and applications heralds the return of a plant model for plant biology. Plant Mol Biol 84:737–742. https://doi.org/10.1007/s11103-013-0162-9

Li Y, Beisson F, Koo AJ, Molina I, Pollard M, Ohlrogge J (2017) Identification of acyltransferases required for cutin biosynthesis and production of cutin with suberin-like monomers. Proc Natl Acad Sci 104:18339–18344

López-Casado G, Matas AJ, Domínguez E, Cuartero J, Heredia A (2007) Biomechanics of isolated tomato (*Solanum lycopersicum* L.) fruit cuticles: the role of the cutin matrix and polysaccharides. J Exp Bot 58:3875–3883

Manrich A, Moreira FKV, Otoni CG, Lorevice MV, Martins MA, Mattoso LHC (2017) Hydrophobic edible films made up of cutin and pectin. Carbohyd Polym 164:83–91

Montanari A, Bolzoni L, Cigognini IM, de la Torre Carreras R (2014) Tomato bio-based lacquer for sustainable metal packaging. Agro Food Ind Hi Tech 25:50–54

Poku K (2002) Small scale palm oil processing in Africa. In: FAO agricultural service bulletin, vol 148. Rome. ISSN 1010-1365

Santos C, Weaver DF (2018) Topically applied linoleic/linolenic acid for chronic migraine. J Clin Neurosci 58:200–201

Yeats TH, Buda GJ, Wang Z, Chehanovsky N, Moyle LC, Jetter R, Schaffer AA, Rose JKC (2012) The fruit cuticles of wild tomato species exhibit architectural and chemical diversity, providing a new model for studying the evolution of cuticle function. Plant J 69:655–666

Yongjin JZ, Bujis NA, Siewers V, Nielson J (2014) Fatty acid-derived biofuels and chemicals production in *Saccharomyces cerevisiae*. Front Bioeng Biotechnol 2:32

Zhang B, Uyama H (2016) Biomimic plant cuticle from hyperbranched poly(ricinoleic acid) and cellulose film. ACS Sustain Chem Eng 4:363–369

Zhou Y, Chen G, Peterson A, Zha X, Cheng J, Li S et al (2018) Biodiversity of duckweeds in Eastern China and their potential for bioremediation of industrial and municipal wastewater. J Geosci Environ Prot 6:108–116. https://doi.org/10.4236/gep.2018.63010

Ziegler P, Sree KS, Appenroth K-J (2016) Duckweeds for water remediation and toxicity testing. Toxicol Environ Chem 98:1127–1154. https://doi.org/10.1080/02772248.2015.1094701

Chapter 16
Others Aquatic Biopolymers

Abstract In this section, we discuss other polymers which are not covered in separate chapters for different reasons. Some of these biopolymers are less explored compared, some are more abundant in terrestrial plants and some are relatively newly discovered. Nonetheless, they are biopolymers present in the aquatic environment and have significance in different ways.

Keywords Bioluminescence · Biofluorescence · Phlorotannins · Lignin · Pectin · Mucin

16.1 Phlorotannins

These polymers are more commonly found in the Eckloma species of brown algae such as *Ecklonia stolonifera* and *Ecklonia cava*. They are made up of phloroglucinol repeating unit, and their molecular weight ranges between 126 Da and 650 kDa. Phlorotannins can also be produced synthetically through the polymerization of phloroglucinol (Seca and Pinto 2018). They are of much pharmaceutical interest for their antihypertensive properties. For example, studies carried out on laboratory rats administered with oral dosage at 20 mg per kg of body weight show that 6,6′-bieckol phlorotannins act as ACE I inhibitors which prevent vasoconstriction (Ko et al. 2017). This is important in the management of high blood pressure and other cardiovascular diseases. Different forms of this biopolymer exist, and they vary based on their chain length as well as the side groups and linkage between the phloroglucinol within the chain.

16.2 Bioluminescent Proteins

76% of aquatic organisms living below the abyssopelagic zone have the ability to produce light. The property referred to as bioluminescence, biological light, is used by these organisms for different reasons. Some use it for protection to hide from predators, some for luring food and some for attracting a mate. Organisms which

© Springer International Publishing 2020   349
O. Olatunji, *Aquatic Biopolymers*, Springer Series on Polymer and Composite Materials,
https://doi.org/10.1007/978-3-030-34709-3_16

possess bioluminescence include mollusks, copepods, shrimp, zooplankton and fish among others.

While some of the organisms produce this light through chemical reactions within their systems, others are able to create this light through symbiotic relationships with other organisms which they host within their bodies, mainly bacteria. They then have means through which they switch on and off the biological light. Some of this "living light" as it is often referred to, are continuous while some have a frequency which could range from a fraction of a second to every 10 s. There are fish known to produce light which remains luminous even hours after the organism itself has died.

Bioluminescence has almost exclusively been observed in marine aquatic organisms (also occurs in some terrestrial organisms such as the fireflies; Kricka et al. 2019), with the exception of one freshwater snail known as the *Latia neritoides* (Harvey 1952). The possibilities that there might be some more bioluminescent aquatic organisms that are yet to be discovered do exist.

The main proteins responsible for bioluminescence in aquatic organisms are luciferin and luciferase, the former being the substrate which takes part in the reaction, while the latter is the enzyme which catalyzes the reactions. Other essential components necessary for the production of biological light are oxygen, ATP and ions particularly Ca^{2+} and Mg^+ (Kricka et al. 2019; Brasier et al. 1989). The reactions are written out in Eq. (16.1).

$$\text{Luciferin} + \text{ATP} \xrightarrow{O_2, C_2^+} \text{Luciferin ATP} \xrightarrow{\text{Luciferase}} \text{Oxyluciferin} + \text{Light} \qquad (16.1)$$

There are five types of luciferin, and the most common, particularly in marine organisms, is the coelenterazine which is found in a wide range of organisms from copepods to fish (Altun et al. 2008).

Humans have always tried to find a way to either harvest, harness or mimic some features of some living organisms and find ways to use them to the benefit of the human species. This does not exclude bioluminescence. In the most basic way, some have been known to remove parts of a bioluminescent fish and use that as fishing bait to lure prey (Ganeri 1999; Altun et al. 2008).

16.3 Biofluorescent Proteins

Biofluorescent is the ability of an organism to absorb one form of light and use that to emit another form of light. This differs from bioluminescence in the sense that light is required for this process unlike bioluminescence which is the ability of an organism to make the light even in darkness.

Biofluorescent exists in a wide range of aquatic organisms such as fish, cnidarians, mantis shrimps, ctenophores, amphioxus and copepods (Sparks et al. 2014). In these aquatic organisms, their biofluorescence has mainly been attributed to the green fluorescent protein (GFP), fatty-acid-binding fluorescent proteins and GFP-like proteins and recently bromo-tryptophan kynurenines from catshark (Park et al. 2019). Cyanobacteria such as Spirulina produce phycobiliproteins which are used to capture light needed for photosynthesis and they also are biofluorescent (Pagels et al. 2019). The red algae Porphyridium also produces phycobiliproteins (Li et al. 2019). The main types of phycobiliproteins are phycocyanin, phycoerythrin, allo-phycocyanin and phycoerythrocyanin. These phycobiliproteins have also been found to have some bioactive properties such as anti-inflammatory, anticancer and antioxidant activities (Pagels et al. 2019). Such that understanding the mechanisms by which aquatic organisms fluorescence and the polymers which are used for it has lead to development in other areas.

16.4 Xyloglucan

In most plants, the cell wall comprises cellulose microfibrils in a matrix of hemicellulose and pectin. The hemicellulose is a complex mixture of different polymers. The most abundant of these polymers is xyloglucan. It is a polymer with a repeating unit of beta 1-4 D glucans with every sixth glucan replaced by an alpha-D-xylose. This xylose could further be attached to other units such as fucose, arabinose and galactose (Hayashi and Kaida 2011). The type of units attached to the xylose varies in different plant species and affects the property of the xyloglucan. Therefore, xyloglucan from different species can exhibit different properties such as solubility in water. Although predominantly in plants, xyloglucan has also been reported to be present in green algae cell wall (Domozych et al. 2010) although only lower amount compared to the other cell wall polymers.

Xyloglucans show some gelation properties which have been investigated for application in developing thermoresponsive drug delivery materials (Sakakibara et al. 2019). They play mainly load-bearing roles in the cell wall, and this is based on their interaction with cellulose. This interaction with cellulose could be explored for production of materials with desirable mechanical properties from xyloglucan and cellulose composites.

16.5 Pectin

Pectin is an anionic polysaccharide with a relatively complex structure. Its polymer structure comprises beta-1,4-D-galacturonic acid where some or all of the uronic acids present in the chain form an ester linkage with a methoxy group. This is referred to as the methoxyl content, and it affects the gel strength and gelling temperature of the pectin.

The most well-known application of pectin is for its use as a gelling agent in the production of jams. They also have biomedical applications such as drug delivery vehicles, anti-inflammatory, wound healing and anticoagulant. It is mentioned in the chapter on polyesters that the cuticle of the plant is linked to the epidermal wall by layer of pectin such that its abundance correlates with that of cutin in plants. Aquatic plants are the main source of pectin in the aquatic environment. The duckweed species *Spirodela polyrhiza*, for example, contains up to 57% polysaccharide most of which is pectin and xyloglucans (Sowinski et al. 2019) in their cell wall mass. Pectin is also present in the cell walls of certain algae belonging to the Charophyceae taxa (Domozych et al. 2012). This class of green algae is said to be the closest related to terrestrial plants, and this is evidenced by the presence of a cell wall with similar polymer composition as land plants.

16.6 Lignin

Lignin is a major evolutionary feature of plants. Terrestrial plants are believed to have developed lignified cell walls as a means of adapting to survive on land (Boyce 2004). It is therefore not expected to be needed in terrestrial plants. However, studies have found lignin in aquatic plants (Chen et al. 2012; Thomas et al. 2009) albeit at a lower amount compared to terrestrial plants. Lignins are complex aromatic heteropolymers used by the plant to bind the cellulose microfibrils which are present in the cell wall, and it also serves as a cross-linker for other cell wall components. This cross-linked structure provides the stiffness and fortification within the xylem tissues in the plant secondary cell walls that are needed for plants to have the erect structure.

Azolla has one of the highest lignin content at about 10%, while lignin content in corn, an example terrestrial plant, is about 16% (Konda et al. 2015). *A. pinnata*, a particular species of the *Azolla* plant, has up to 13.2% lignin (also contains 4.7% starch, 12.8% cellulose and 10.1% hemicellulose). Duckweed has a lignin content of 12.2% (Yadav et al. 2017). The discovery of the presence of lignin in primitive green algae and red algae strays from the accepted concept that lignin is an evolutionary feature which allowed aquatic plants to evolve and adapt to terrestrial environment (Martone et al. 2009).

16.7 Peptides from Frog Skin

Recent years have seen the increasing interest in the peptides within the skin of some frog for their bioactive properties. These peptides include magainin-AM2, tigerinin-1R and esculentin-2CHa. One of the most attractive bioactivities of these frog skin-derived peptides is the antidiabetic activities. Peptides from frog skin can stimulate insulin release, improve the sensitivity of the insulin, reduce the glucagon concentration in the pancreas and the blood and also promote the proliferation of the beta cells. It also has some lipid-lowering effects and promotes the expression of genes coding for insulin secretion. These have been confirmed in in vivo studies on mice (Conlon et al. 2018). Furthermore, recent studies have also shown that the bioactivity of peptides from frog skin also extends to antimicrobial activity. Peptides from frog skin have the ability to kill a range of human influenza virus strains (Crowe 2017). However, these were based on studies carried out at the cellular level. Peptides from the skin of frog also have potential to serve as antioxidant agents as recent studies point to antioxidant activity in particular peptide called the antioxidant peptide, antioxidant-I. These antioxidant proteins which were obtained from the skin of the *Physalaemus nattereri* frog have also been shown to possess antioxidant activity (Barbosa et al. 2018). This peptide also shows the ability to improve the protective response to hypoxia (low oxygen levels). Other bioactivities which have been recently detected in frog skin peptides include activity against cystic fibrosis and cystic fibrosis-related MRSA (methicillin-resistant *S. aureus*) (Yuan et al. 2019).

Frog skins have been designed to protect the organism from external factors such as temperature, moisture, UV light, pathogenic microbes and mechanical stress. Much of these properties can be attributed to proteins present in the skin secretions which serve this purpose. Although some live on land, frogs spend a substantial part of their life in or around water, particularly at the tadpole stage, and therefore, they are considered here as aquatic inhabitants.

Other polysaccharides present in aquatic plants include glucoronan present as minor polysaccharides in sea lettuce (Kim et al. 2011; Lahaye and Robic 2007), Rhamnus, arabinogalactans and mannans which are produced by green algae strains *Monostroma* and *Codium* (Tabarsa et al. 2012; Kim et al. 2011; Li et al. 2015). Cyanobacterium, the only known photosynthesizing eukaryotes, produces glycogen, a storage carbohydrate (Chao and Bowen 1971; Yoo et al. 2007). Laminarin and mannitol produced by brown algae have been explored for fermentation ethanol production, and the microbes for such have been identified by Horn et al. (2000a, b).

16.8 Conclusion

Most of the polymers discussed in this section are either present in less abundance to those that have been discussed in dedicated chapters in this book or they have generated less commercial or research interest. Nonetheless, these contribute to the diversity of biopolymers in the aquatic ecosystem. Even the polymers present in small traces are of importance as they aid in expanding knowledge on the origins of certain organisms.

References

Altun T, Celi F, Danabas D (2008) Bioluminescence in aquatic organisms. J Anim Vet Adv 7(7):841–846

Barbosa EA, Oliveira A, Placido A, Soccodato R, Leite JRSA (2018) Structure and function of a novel antioxidant peptide from the skin of tropical frogs. Free Radic Biol Med 115:68–79

Boyce CZ (2004) Evolution of xylem lignification. Proc Natl Acad Sci 101:17555–17558

Brasier AR, Tate JE, Habener JF (1989) Optimized use of the firefly luciferase assay as a reporter gene in mammalian cell lines. Biotechniques 7(11):1116–1122

Chao L, Bowen CC (1971) Purification and properties of glycogen isolated from a blue-green alga, *Nostoc muscorum*. J Bacteriol 105(1):331–338

Chen Q, Jin Y, Zhang G, Fang Y, Xiao Y, Zhao H (2012) Improving production of bioethanol from duckweed (*Landoltia punctata*) by pectinase pretreatment. Energies 5:3019–3032

Conlon JM, Mechkarska M, Abdel-Wahab YH, Flatt PR (2018) Peptides from frog skin with potential for development into agents for type 2 diabetes therapy. Peptides 100:275–281

Crowe JE (2017) Treating flu with skin of frogs. Immunity 46:517–518

Domozych DS, Sorensen I, Pettolino FA, Bacic A, Willats GTW (2010) The cell wall polymers of the Charophycean green alga *Chara corallina*: immunobinding and biochemical screening. Int J Plant Sci 171(4):345–361

Domozych DS, Ciancia M, Fangel JU, Mikkelsen MD, Ulvskov P, Willats GTW (2012) The cell walls of green algae: a journey through evolution and diversity. Front Plant Sci 3:1–7

Ganeri A (1999) Creatures that glow. McClanahan Book Co., p 30

Harvey EN (1952) Bioluminescence. Academic Press, New York, p 649

Hayashi T, Kaida R (2011) Functions of xyloglucan in plant cells. Mol Plant 4(1):17–24

Horn SJ, Aasen IM, Østgaard K (2000a) Ethanol production from seaweed extract. J Ind Microbiol Biotechnol 25(5):249–254

Horn SJ, Aasen IM, Østgaard K (2000b) Production of ethanol from mannitol by *Zymobacter palmae*. J Ind Microbiol Biotechnol 24(1):51–57

Kim JK, Cho MI, Karnjanaprakorn S, Shin IS, You SG (2011) In vitro and in vivo immunomodulatory activity of sulfated polysaccharides from Enteromorpha prolifera. Int J Biol Macromol. 49:1051–1058

Ko S, Kang MC, Kang N, Kim H, Lee SH, Ahn G, Jung W, Jeon Y (2017) Effect of angiotensin I-converting enzyme (ACE) inhibition and nitric oxide (NO) production of 6,6′-bieckol, a marine algal polyphenol and its anti-hypertensive effect in spontaneously hypertensive rats. Process Biochem 58:326–332

Konda M, Singh S, Simmons BA, Klein-Marcuschamer D (2015) An investigation on the economic feasibility of macroalgae as a potential feedstock for biorefineries. Bioenergy Res 8:1046–1056

Kricka LJ, Smith ZM, Adcock JL, Barnett NW (2019) Bioluminescence. In: Encyclopedia of analytical science, 3rd edn., pp 291–299

Lahaye M, Robic A (2007) Structure and functional properties of Ulvan, a polysaccharide from green seaweeds, Biomacromolecules 8:1765–1774

Li N, Mao W, Yan M, Liu X, Xia Z, Wang S (2015) Structural characterization and anticoagulant activity of sulfated polysaccharide from green alga Codium divaricatum. Carbohydrates Poly 121:175–182

Li T, Xu J, Wu H, Jiang P, Chen Z, Xiang W (2019) Growth and biochemical composition of *Porphyridium purpureum* SCS-02 under different nitrogen concentrations. Mar Drugs 17(124):1–16

Martone PT, Estevez JM, Lu F, Ruel K, Denny MW, Somerville C, Ralph J (2009) Discovery of lignin in seaweed reveals convergent evolution of cell-wall architecture. Curr Biol 19(2):169–175

Pagels F, Guedes CA, Amaro MH, Kijoa A, Vasconcelos V (2019) Phycobiliproteins from cyanobacteria: chemistry and biotechnological applications. Biotechnol Adv 37(3):422–443

Park HB, Lam CY, Gaffney JP, Weaver JC, Krivoshik SR, Hamchand R, Pieribone V, Gruber DF, Crawford JM (2019) Bright green biofluorescence in sharks derives from bromo-kynurenine metabolism. iScience (in press). https://doi.org/10.1016/j.isci.2019.07.019

Sakakibara NC, Sierakowski MR, Ramirez RR, Chassenieux C, Riegel-Vidotti I, de Freitas RA (2019) Salt-induced thermal gelation of xyloglucan in aqueous media. Carbohyd Polym 223:115083

Seca AML, Pinto DCGA (2018) Overview on the antihypertensive and anti-obesity effects of secondary metabolites from seaweeds. Mar Drugs 16(237):1–18

Sowinski EE, Gilbert S, Lam E, Carpita NC (2019) Linkage structure of cell-wall polysaccharides from three duckweed species. Carbohydrate Poly 223(115119): 284–298

Sparks JS, Schelly RC, Smith WL, Davis MP, Tchernov D, Pieribone VA, Gruber DF (2014) A covert world of fish biofluorescence: a phylogenetically widespread and phenotypically variable phenomenon. PLoS One 9(1):1–9 e83259

Tabarsa M, Han JH, Kim CY, You SG (2012) Molecular characteristics and immunomodulatory activities of water-soluble sulfated polysaccharides from Ulva pertusa. J Med Food 15:135–144

Thomas EK, Gao L, Huang Y (2009) A quantitative and qualitative comparison of aquatic and terrestrial plant lignin phenols: critical information for paleoecological reconstructions. In: American Geophysical Union fall meeting 2009

Yadav D, Barbora L, Bora D, Mitra S, Rangan L, Mahanta P (2017) An assessment of duckweed as a potential lignocellulosic feedstock for biogas production. Int Biodeterior Biodegradation 119:253–259

Yoo S, Keppel C, Spalding M, Jane JL (2007) Effects of growth condition on the structure of glycogen produced in cyanobacterium Synechocystis sp. PCC6803. Int J Biol Macromol 40(5):498–504

Yuan Y, Zai Y, Xi X, Ma C, Wang L, Zhou M, Shaw C, Chen T (2019) A novel membrane-disruptive antimicrobial peptide from frog skin secretion against cystic fibrosis isolates and evaluation of anti-MRSA effect using galleria mellonella model. Biochim Biophys Acta (BBA) Gen Subj 1863(5):849–856

Chapter 17
Future Perspectives

The aquatic ecosystem serves as host to a diverse range of organisms which produce a diverse range of polymers. Due to the ubiquity of polymers, in the process of exploring the different polymers in the aquatic ecosystem from raw materials, application to final products, we have inevitable touched on different industries. These include the polymers used in the medical field such as HIV microbicides to those used in the energy industry in the production of the third-generation biofuels. We have also covered the diverse chemistry of the different polymers and explored a range of extraction processes. In the process, some current issues of the aquatic environment are discussed, and we discuss possible solutions to addressing such. In addition to serving as a source of biopolymers, understanding the role of the polymers within the organisms from which they are sourced and the role in the aquatic environment has also enhanced our understanding of certain biochemical process. For example, the role of polymers allows organisms to survive in extreme environments and protection against diseases.

Although the production rate of some of these biopolymers goes into thousands of tonnes annually, this is still very low in comparison to the current production rate of synthetic non-biodegradable plastics which are produced in hundreds of millions of tonnes annually. Through disseminating knowledge and aiding the development of understanding of the aquatic biopolymers, further interest will be generated in commercial production of these biopolymers from aquatic resources, particularly those which can be produced from waste resources or underutilized excess resources.

Extraction of biopolymers from aquatic resources should focus more on sustainable utilization of waste generated from fisheries' industry with the goal of value addition to fish resource and waste minimization rather than farming of aquatic resources for the sole purpose of production of biopolymers. In other words, food production from aquatic waste should be synergized with biopolymer production. With valuable resources such as biopolymers being produced from aquatic resources, the risk of over exploitation should be avoided as this then defeats the whole idea of having a renewable source of polymers.

© Springer International Publishing 2020
O. Olatunji, *Aquatic Biopolymers*, Springer Series on Polymer and Composite Materials,
https://doi.org/10.1007/978-3-030-34709-3_17

From the review of the environmental impact of the process for producing these biopolymers, we see that many of these are not as benign as their end product. Therefore, future industries need to focus efforts in ensuring the materials and energy source used in the extraction, purification and packaging of these polymers which are as sustainable as possible. Otherwise, the use of more fossil fuels and non-renewable source of energy and materials in the production of renewable polymers is rather counterproductive.

The allure of aquatic biopolymers is the third-generation polymers which do not compete with humans and terrestrial life for land space, utilizes waste materials and contributes to lowering the CO_2 levels in the atmosphere. In addition to these, aquatic biopolymers also serve as a source of polymers with more diverse chemistry thereby opening doors of possibilities for a range of materials and bioactive compounds and at the same time providing solutions to existing global problems such as food shortage, plastic pollution and deterioration of the global aquatic ecosystem.

As our fossil source of polymers are continually depleting, the organisms of the ocean, seas, rivers and lakes promise a renewable source of polymers which either already do, or could potentially, play a significant role in human survival. The ideal aquatic polymer resource will remove carbon dioxide from the atmosphere, take up nutrients from the aquatic environment it grows in thereby cleaning wastewater, will be completely renewable, require minimal amount of energy and chemicals to extract and be as effective in use as the terrestrial- or fossil-derived polymers.

Table 17.1 gives a summary of the polymers which has been covered within this book; their sources, applications and repeating units form their polymeric structure. The applications listed include both the existing applications and the potential applications which are still in early stage research.

Table 17.1 Summary of aquatic polymers covered within this book

Polymer	Source	Applications (existing/potential)	Repeating unit
Cellulose	Aquatic plants and algae	Textiles, biofuel, paper, nanomaterials	Glucose
Chitin	Crustaceans, fish scales, fungi	Water treatment, antimicrobial, packaging films, biomedical sutures, antioxidant, chelation, paper, antiacne, antiwrinkle	Acetyl glucosamine
Alginate	Brown algae	Food additives, textile printing, tissue engineering	Mannuronic acid and guluronic acid
Fucoidan	Brown algae	Anticancer, tissue engineering, antipathogenic, anticoagulant, antiinflammatory	Fucose, xylose, ribose
Ulvan	Green algae	Immune modulation, antioxidant, anticancer, anticoagulant	Sulfated rhamnose and xylose disaccharide/glucuronic acid
Starch	Aquatic plants and algae	Food thickener, biofuel, bioplastics	Glucose
Laminarin	Brown algae	Anticancer, biofuel, immunomodulatory	Glucose
Polyester (cutin)	Aquatic plants	Bioplastics, sunscreen agent	C16 and C18 ester-linked fatty acids
Agar	Red algae	Tissue culture, gelling agent	Galactose and L-galactopyranose
Carrageenan	Red algae	Gelling agent, toothpastes	Galactose and D-galactopyranose
Collagen	Fish, jellyfish, squids	Topical agents, tissue engineering, drug delivery	Amino acids
Enzymes	Fish viscera, fungi, bacteria, algae, aquatic plants	Biofuel catalysts, food processing	Amino acids